D1574040

Advances
in Electrochemical
Science and Engineering

Volume 9
Diffraction and
Spectroscopic Methods
in Electrochemistry

Advances in Electrochemical Science and Engineering

Advisory Board

Prof. Elton Cairns, University of California, Berkeley, California, USA

Prof. Adam Heller, University of Texas, Austin, Texas, USA

Prof. Dieter Landolt, Ecole Polytechnique Fédérale, Lausanne, Switzerland

Prof. Roger Parsons, University of Southampton, Southampton, UK

Prof. Laurie Peter, University of Bath, Bath, UK

Prof. Walter Schultze, University of Düsseldorf, Düsseldorf, Germany

Prof. Sergio Trasatti, Università di Milano, Milano, Italy

Prof. Lubomyr Romankiw, IBM Watson Research Center, Yorktown Heights, USA

In collaboration with the International Society of Electrochemistry

Advances in Electrochemical Science and Engineering

Volume 9
Diffraction and Spectroscopic Methods
in Electrochemistry

Edited by
Richard C. Alkire, Dieter M. Kolb, Jacek Lipkowski,
and Philip N. Ross

WILEY-VCH Verlag GmbH & Co. KGaA

Editors

Prof. Richard C. Alkire
University of Illinois
600 South Mathews Avenue
Urbana, IL 61801
USA

Prof. Dieter M. Kolb
University of Ulm
Department of Electrochemistry
Albert-Einstein-Allee 47
89081 Ulm
Germany

Prof. Jacek Lipkowski
University of Guelph
Department of Chemistry
N1G 2W1 Guelph, Ontario
Canada

Prof. Philip N. Ross
Lawrence Berkeley National Laboratory
Materials Science Department
1 Cyclotron Road MS 2-100
Berkeley, CA 94720-0001
USA

■ All books published by Wiley-VCH are carefully produced. Nevertheless, authors, editors, and publisher do not warrant the information contained in these books, including this book, to be free of errors. Readers are advised to keep in mind that statements, data, illustrations, procedural details or other items may inadvertently be inaccurate.

Library of Congress Card No.: applied for

British Library Cataloguing-in-Publication Data
A catalogue record for this book is available from the British Library.

**Bibliographic information published by
Die Deutsche Bibliothek**
Die Deutsche Bibliothek lists this publication in the Deutsche Nationalbibliografie; detailed bibliographic data is available in the Internet at <http://dnb.ddb.de>.

© 2006 WILEY-VCH Verlag GmbH & Co. KGaA, Weinheim, Germany

All rights reserved (including those of translation into other languages). No part of this book may be reproduced in any form – by photoprinting, microfilm, or any other means – nor transmitted or translated into a machine language without written permission from the publishers. Registered names, trademarks, etc. used in this book, even when not specifically marked as such, are not to be considered unprotected by law.

Typesetting K+V Fotosatz GmbH, Beerfelden
Printing betz-druck GmbH, Darmstadt
Binding Schäffer GmbH, Grünstadt

Printed in the Federal Republic of Germany
Printed on acid-free paper

ISBN-13: 978-3-527-31317-4
ISBN-10: 3-527-31317-6

Series Preface

This ninth volume results from the merger of two well-established monograph series. One point of origin began in 1961 with the editorial collaboration of Paul Delahay and Charles Tobias, and which continued in 1976 with Heinz Gerischer and Charles Tobias. Their efforts led in 1987 to the series "Advances in Electrochemical Science and Engineering", which was continued in 1997 by Richard Alkire and Dieter Kolb. The second point of origin is the series "Frontiers in Electrochemistry" established in 1992 by Jacek Lipkowski and Philip N. Ross. With this volume the concept of topical volumes is introduced to the new series of Advances that resulted from the merger. The favourable reception of the Frontiers in Electrochemistry and the first eight volumes of Advances, and the steady increase of interest in electrochemical science and technology provide good reasons for the continuation of these editions with the same high standards. The purpose of the series is to provide high quality advanced reviews of topics of both fundamental and practical importance for the experienced reader.

Richard Alkire
Dieter Kolb
Jacek Lipkowski
Philip Ross

Advances in Electrochemical Science and Engineering Vol. 9.
Edited by Richard C. Alkire, Dieter M. Kolb, Jacek Lipkowski and Philip N. Ross
Copyright © 2006 WILEY-VCH Verlag GmbH & Co. KGaA, Weinheim
ISBN: 3-527-31317-6

Contents

Advances in Electrochemical Science and Engineering Vol. 9.
Edited by Richard C. Alkire, Dieter M. Kolb, Jacek Lipkowski and Philip N. Ross
Copyright © 2006 WILEY-VCH Verlag GmbH & Co. KGaA, Weinheim
ISBN: 3-527-31317-6

Volume Preface

Recent decades have witnessed spectacular developments in *in-situ* diffraction and spectroscopic methods in electrochemistry. The synchrotron-based X-ray diffraction technique unraveled the structure of the electrode surface and the structure of adsorbed layers with unprecedented precision. *In-situ* IR spectroscopy became a powerful tool to study the orientation and conformation of adsorbed ions and molecules, to identify products and intermediates of electrode processes, and to investigate the kinetics of fast electrode reactions. UV-visible reflectance spectroscopy and epifluorescence measurements have provided a mass of new molecular-level information about thin organic films at electrode surfaces. Finally, new non-linear spectroscopies such as second harmonics generation, sum frequency generation, and surface-enhanced Raman spectroscopy introduced unique surface specificity to electrochemical studies.

The aim of this edition is to provide an up-to-date account of these recent advances. The first chapter describes a fascinating application of the X-ray diffraction technique to the study of the structure-reactivity relationship in electrocatalysis. The next two chapters illustrate the power of UV-visible spectroscopy and epifluorescence microscopy to explore electric field-driven transformations of thin organic films. Two chapters are devoted to non-linear spectroscopies at the liquid-liquid and liquid-solid interfaces, demonstrating the uniqueness of these techniques for revealing the structural details of these buried interfaces. Four chapters give a comprehensive description of applications of infrared spectroscopy to *in-situ* studies of electrified semiconductor-solution and metal-solution interfaces. The volume is concluded by a chapter that describes the emerging new technique of STM tip-induced surface-enhanced Raman spectroscopy.

This volume is addressed to a wide audience of scientists interested in electrochemistry, materials, and interfacial science. Each chapter provides enough background material for it to be read by non-specialist and specialist alike, and each chapter concludes with an overall assessment of future directions.

Guelph, June 2006   J. Lipkowski

Advances in Electrochemical Science and Engineering Vol. 9.
Edited by Richard C. Alkire, Dieter M. Kolb, Jacek Lipkowski and Philip N. Ross
Copyright © 2006 WILEY-VCH Verlag GmbH & Co. KGaA, Weinheim
ISBN: 3-527-31317-6

List of Contributors

Steven Baldelli
University of Houston
Department of Chemistry
136 Fleming Building
Houston, TX 77204
USA

Dan Bizzotto
University of British Columbia
Department of Chemistry
Advanced Materials and Process
Engineering Laboratory (AMPEL)
Vancouver, BC
Canada, V6T 1Z1

Jean-Noël Chazalviel
CNRS-École Polytechnique
Laboratoire de Physique de la Matière
Condensée
Route de Saclay
91128 Palaiseau
France

David J. Fermín
Universität Bern
Departement für Chemie
und Biochemie
Freiestrasse 3
3012 Bern
Switzerland

Andrew A. Gewirth
University of Illinois at Urbana-
Champaign
Department of Chemistry
600 S. Mathews Avenue
Urbana, IL 61801
USA

Carol Korzeniewski
Texas Tech University
Department of Chemistry
and Biochemistry
Lubbock, TX 79409-1061
USA

Jacek Lipkowski
University of Guelph
Department of Chemistry
Guelph, Ontario, N1G 2W1
Canada

Christopher A. Lucas
University of Liverpool
Oliver Lodge Laboratory
Department of Physics
Liverpool, L69 7ZE
UK

Advances in Electrochemical Science and Engineering Vol. 9.
Edited by Richard C. Alkire, Dieter M. Kolb, Jacek Lipkowski and Philip N. Ross
Copyright © 2006 WILEY-VCH Verlag GmbH & Co. KGaA, Weinheim
ISBN: 3-527-31317-6

Nenad M. Marković
University of California
Materials Sciences Division
Lawrence Berkeley National
Laboratory
Berkeley, CA 94720
USA

Masatoshi Osawa
Hokkaido University
Catalysis Research Center
N21-W10, Kita-ku, Sapporo 001-0021
Japan

François Ozanam
CNRS-École Polytechnique
Laboratoire de Physique de la Matière
Condensée
Route de Saclay
91128 Palaiseau
France

Bruno Pettinger
Fritz Haber Institute
of the Max Planck Society
Department of Physical Chemistry
Faradayweg 4–6
14195 Berlin
Germany

Takamasa Sagara
Nagasaki University
Department of Applied Chemistry
Bunkyo 1–14
Nagasaki 852-8521
Japan

Jeff L. Shepherd
University of British Columbia
Department of Chemistry
Advanced Materials and Process
Engineering Laboratory (AMPEL)
Vancouver, BC
Canada, V6T 1Z1

Vlad Zamlynny
Acadia University
Chemistry Department
6 University Avenue
Wolfville
Nova Scotia, B4P 2R6
Canada

1

In-situ X-ray Diffraction Studies
of the Electrode/Solution Interface

Christopher A. Lucas and Nenad M. Marković

1.1
Introduction

Since the early days of modern surface science, the main goal in the electrochemical community has been to find correlations between the microscopic structures formed by surface atoms and adsorbates and the macroscopic kinetic rates of a particular electrochemical reaction. The establishment of such relationships, previously only developed for catalysts under ultrahigh vacuum (UHV) conditions, has been broadened to embrace electrochemical interfaces. In early work, determination of the surface structures in an electrochemical environment was derived from *ex-situ* UHV analysis of emersed surfaces. Although such *ex-situ* tactics remain important, the relationship between the structure of the interface in the electrolyte and that observed in UHV was always problematic and had to be carefully examined on a case-by-case basis. The application of *in-situ* surface-sensitive probes, most notably synchrotron-based surface X-ray scattering (SXS) [1–6] and scanning tunneling microscopy (STM) [7, 8], has overcome this "emersion gap" and provided information on potential-dependent surface structures at a level of sophistication that is on a par with (or even in advance of) that obtained for surfaces in UHV.

In this chapter, we review some applications of the SXS technique for exploring surface atomic structure in an electrochemical environment. While the choice of topics is extracted mainly from recent studies, the survey will provide a full perspective into the potential value of SXS for exploiting many fundamental issues pertaining to the development of surface electrochemistry as a science. SXS is an ideal technique for probing, in detail, the surface atomic structure, and such structural details can be important in understanding some electrochemical systems. In this article, however, we highlight the use of SXS in potentiodynamic measurements where the aim is to correlate directly the electrochemical reactivity with atomic-scale structural changes at the electrode surface. The review begins with a description of the potential/adsorbate-induced changes in a metal surface, focusing on those cases where detailed and rather complete surface coverages and/ or structures have been determined. Central to the adsorbate-induced changes in

Advances in Electrochemical Science and Engineering Vol. 9.
Edited by Richard C. Alkire, Dieter M. Kolb, Jacek Lipkowski and Philip N. Ross
Copyright © 2006 WILEY-VCH Verlag GmbH & Co. KGaA, Weinheim
ISBN: 3-527-31317-6

monometallic (Pt metals and IB group metals) and some of their bimetallic single-crystal surfaces are the concepts of relaxation and reconstruction of near-surface atoms. The data for adsorbate-induced restructuring effects on Pt and Au single-crystal surfaces far outnumber those for Ag, Cu, and bimetallic surfaces; the references in this chapter, however, show them side by side for comparison. The next section describes the atomic/molecular scale structures of adsorbed anions and cations on single-crystal surfaces and summarizes applications of the SXS technique for exploring the behavior of molecules in the temperature range which is important for the development of the polymer electrolyte fuel cell. Most of adsorbate surface structures are indicated in the text, so that they can be directly related to the nature of the substrate and the supporting electrolytes. The review concludes with some examples of how SXS can be utilized for elucidating the structure changes at the metal-oxide/electrolyte interface.

1.2
Experimental

Surface X-ray diffraction is now a well-established technique for probing the atomic structure at the electrochemical interface, and, since the first *in-situ* synchrotron X-ray study in 1988 [1], several groups have used the technique to probe a variety of electrochemical systems [1, 2, 9]. It is beyond the scope of this article to provide a comprehensive description of basic X-ray diffraction from surfaces. Readers are referred to the excellent reviews by Feidenhans'l [10], Fuoss and Brennan [11], and Robinson and Tweet [12] for explicit details. It should be noted that, throughout this review, the acronym SXS is used to describe the X-ray measurements, although all of the results are obtained by X-ray diffraction.

The majority of the results presented in this report are X-ray diffraction studies of the three low-index surfaces of metal crystals with the face-centered cubic (*fcc*) crystal structure. The close-packed (111) surface has a hexagonal unit cell that is defined such that the surface normal is along the $(0, 0, l)_{hex}$ direction and the $(h, 0, 0)_{hex}$ and $(0, k, 0)_{hex}$ vectors lie in the plane of the surface and subtend 60°. The units for h, k and l are $a^* = b^* = 4\pi/\sqrt{3}a_{NN}$ and $c^* = 2\pi/\sqrt{6}a_{NN}$ where a_{NN} is the nearest-neighbor distance in the crystal. Because of the stacking of the ABC along the surface normal direction, the unit cell contains three monolayers, and the Bragg reflections are spaced apart by multiples of three in l. The (001) surface is more open than the (111) surface and is indexed to a surface tetragonal unit cell. This is related to the conventional cubic unit cell by the transformations $(1, 0, 0)_t = 1/2(2, 2, 0)_c$, $(0, 1, 0)_t = 1/2(2, -2, 0)_c$ and $(0, 0, 1)_t = (0, 0, 1)_c$. The units for h, k and l are $a^* = b^* = 2\pi/a_{NN}$ and $c^* = 4\pi/\sqrt{2}a_{NN}$. Finally, the (110) reciprocal surface unit cell is rectangular and the reciprocal lattice notation is such that h is along [1 −1 0], k along [0 0 1] and l is along the [1 1 0] surface normal. The units for h, k and l are $a^* = c^* = 2\pi/a_{NN}$ and $b^* = 4\pi/\sqrt{2}a_{NN}$.

As in the UHV environment [13], the extraction of structural information, such as surface coverage, surface roughness and layer spacings (both adsorbate-substrate distances and the expansion/contraction of the substrate surface atoms themselves) at the electrified solid/liquid interface relies on measurement of the crystal truncation rods (CTRs). By combining specular CTR results (where the momentum transfer, Q, is entirely along the surface normal direction) with non-specular CTR results (where Q has an additional in-plane contribution) it is possible to build up a 3-dimensional picture of the atomic structure at the electrode surface. If the surface or adlayer adopts a different symmetry from that of the underlying bulk crystal lattice, then the scattering from the surface becomes separate from that of the bulk in reciprocal space and it is possible to measure the surface scattering independently. This independent structural information can be combined with CTR analysis to give the registry of the surface layers with respect to the bulk lattice [10, 11, 13].

Apart from the standard analysis of CTR data, perhaps the most interesting application of SXS is in potentiodynamic measurements, i.e. when the scattered X-ray intensity is measured at a particular reciprocal lattice position as the electrode potential is cycled over a given range. We have termed this technique X-ray voltammetry (XRV), although a few alternative terms have been used in the literature. By measuring the XRV at a number of different CTR positions, an insight into the nature of the structural changes at the surface can be obtained without recourse to the detailed measurement of the CTR profiles [14, 15]. If scattering due to an ordered surface layer with a different symmetry to that of the underlying bulk crystal is obtained, then measurement of the potential dependence of this scattering directly indicates the potential range of stability of the structure. This information is particularly useful for comparison with cyclic voltammetry (CV) results, as features in the CV can be directly correlated with structural changes at the electrode surface.

A key aspect in the study of single-crystal metal electrodes is the preparation of the surface prior to the experiment and the transfer of the crystal into the electrochemical X-ray cell. One approach is to prepare the surface in a UHV environment by cycles of ion sputtering and annealing. This methodology has the advantage that the surface quality can be checked during preparation by standard surface science techniques such as LEED and Auger Electron Spectroscopy. UHV preparation is important for bimetallic surfaces for which precise surface compositions are dependent on annealing temperatures. An alternative method for surface preparation for some crystals is to use the flame-annealing technique [16–18]. After pretreatment to produce a flat, well-oriented surface, the crystal is heated in a hydrogen or butane flame and then allowed to cool in air or hydrogen/argon before transfer to the electrochemical X-ray cell. This procedure has been used successfully for the preparation of Au(*hkl*) and Pt(*hkl*) surfaces. Other monometallic metal surfaces, such as Cu and Ni, cannot be prepared by flame-annealing methods. Other than UHV preparation, the most successful method for these surfaces, including Ag, has been to use a long pre-anneal of the surface in a forming gas, such as hydrogen, followed by a short electrochemical etch, rinsing with water, and then direct transfer into the electrochemical X-ray cell [19–21].

1.3
Adsorbate-induced Restructuring of Metal Substrates

Modern surface crystallographic studies have shown that on the atomic scale most clean metals tend to minimize their surface energy by two kinds of surface atom rearrangements, *relaxation* and *reconstruction*, which collectively may be called *restructuring*. Relaxation of metal surfaces is usually defined as small interlayer spacing changes relative to the ideal bulk lattice [22–26]. The displacements should be small compared to near-neighbor distances, such that no bond-breaking/bond-making events take place within the substrate. Adsorbate-induced relaxations occur in many varieties: as *interlayer spacing changes*, where the top layer of metal atoms undergo inward or outward relaxation; *lateral relaxation*, in which surface atoms are shifted parallel to the surface (for example, adsorbate-induced collective rotation of substrate atoms around an adsorbate site), and *layer buckling*, whereby a coplanar atomic layer loses its coplanarity because of preferred adsorption on a particular surface site. Reconstruction, on the other hand, involves large atomic displacements both perpendicular to and parallel to the surface plane leading to re-bonding and a change in the periodicity of the surface with respect to the underlying substrate. The periodicity of the surface can be defined by Woods' notation; for example an unreconstructed Pt(110) surface would be termed as (1×1) whereas if the surface unit cell size was doubled in one of the primary vector directions it would be termed as (2×1) etc.

The present knowledge of surface restructuring at electrified metal/solution interfaces would not have been possible without certain key advances, for example, the many excellent UHV [24] and theoretical studies [27], which will not be reviewed in this article. A further advance was the advent of *in-situ* surface-sensitive techniques and the development of efficient methods for surface preparation of monometallic and bimetallic single-crystal surfaces and their clean transfer from UHV into the electrochemical environment [28, 29]. Within the framework of this article, it would be impossible to provide a comprehensive review of these experiments. For these reasons, the surface reconstruction and relaxation of monometallic Pt(*hkl*) and Au(*hkl*) single-crystal surfaces, induced by the adsorption of hydrogen[1] ($H^+ + e^- = H_{upd}$ in acid electrolytes and $H_2O + e^- = H_{upd} + OH^-$ in alkaline solutions) and oxygenated species[2] ($H_2O = OH_{ad} + H^+ + e^-$ in acid solution and $OH^- = OH_{ad} + e^-$ in alkaline solution), are not discussed in great detail. In preference, the focus is on systems of catalytic interest, i.e., hydrogen- and oxygen-in-

---

1) In electrochemistry, adsorbed hydrogen is denoted as either the underpotentially deposited hydrogen, H_{upd}, that is the H adlayer formed under thermodynamic equilibrium conditions where the coverage is changed reversibly with the potential applied or the overpotentially deposited hydrogen, H_{opd}, as defined by Conway and co-workers [102] for the hydrogen evolution reaction, i.e. $2H^+ + 2e^- = H_2$.

2) Two different types of chemisorbed oxygen-containing species are proposed to form on *fcc* metals, a reversible form (denoted hereafter as OH_{ad}) and an irreversible form, which we shall call "oxide" hereafter.

duced restructuring of bimetallic surfaces as well as CO-induced changes in the surface structure of Pt(*hkl*), Au(*hkl*), and their bimetallic alloys.

1.3.1
Surface Relaxation

There have been numerous experimental and theoretical studies of the relaxation of metal surfaces in UHV, and a significant data base now exists based mainly on LEED studies. In UHV, on unreconstructed clean low-index single-crystal surfaces it is often found that the outermost layer of atoms is *contracted* toward the second atomic layer. The tendency for surface contraction has been explained by Finnis and Heine using the Hellman-Feynmen theorem, which states that the force on an ion is just the electrostatic force from the other ions and the self-consistent electron density. Adsorption onto clean metal surfaces in UHV often reverses the surface contraction such that the outermost layer of atoms is *expanded* away from the second atomic layer. These concepts can be extended to the metal/electrolyte interface, and, indeed, surface relaxation has been observed at coinage-metal surfaces (Cu, Ag, Au) as well as at Pt-group monometallic and bimetallic single-crystal surfaces in the electrochemical environment. Key results are presented below with the aim of providing a suitable framework for the discussion of adsorbate-induced interlayer spacing changes, lateral relaxation, and layer buckling.

1.3.1.1 **Pt Monometallic and Bimetallic Surfaces**
The restructuring of Pt single-crystal surfaces has previously been described in some detail [28, 29], and so in this section new results for adsorbate-induced restructuring of Pt bimetallic surfaces are presented. Two types of bimetallic surfaces will be used as model systems: the surface of a $Pt_3Sn(111)$ bulk alloy crystal and ultra thin metal films of Pd deposited onto Pt(111) and Pt(100), hereafter denoted as Pt(*hkl*)-Pd_x ($0 < x < 4$ ML) systems. Key to the success of these experiments is the fact that both the UHV-prepared p(2×2) structure of $Pt_3Sn(111)$ [30, 31] as well as electrochemically- or UHV-prepared thin films of Pd on Pt(*hkl*) [32–34] can be transferred into an electrochemical SXS cell without altering the surface structure and composition. The surfaces are also stable between the potential for the hydrogen evolution reaction and that for reversible oxide formation, indicating that the energy of adsorption of H_{upd}, OH_{ad}, and anions is not strong enough to induce surface reconstruction.

As noted in Section 1.2, accurate determination of adsorbate-induced changes in surface-normal structure, i.e. the Δd_{12} interplanar spacing between the first and the second atomic layers, can be achieved by measuring the CTRs [1–4, 10, 35]. Previous reviews summarized adsorbate-induced relaxation and reconstruction on well-defined Pt(*hkl*) and Pt–bimetallic surfaces in aqueous electrolytes at electrode potentials at which a maximum surface coverage of adsorbed species is established [28, 29]. The data revealed that either close to the hydrogen evolution potential (0.05 V) or close to the onset of irreversible oxide formation (ca. 1 V) the surface expansion increases in the sequence Pt(111) < Pt(100) << Pt(110) in both H_2SO_4 and KOH solutions. It has been proposed that the observed differences in relaxation of Pt(*hkl*) arise because of the interplay between the Pt-adsorbate bonding and the coordination of surface atoms (for details see [28]).

Direct confirmation of the relationship between surface relaxation and binding interaction is obtained in experiments in which relatively weakly bonded H_{upd} is displaced from the surface by strongly bonded CO. As shown previously [28], following the adsorption of CO onto Pt(111) at 0.05 V, the surface expansion increases from 2% on the H_{upd}-covered surface to 4% on the CO-covered surface, supporting the proposition that the difference in relaxation can be correlated with metal-adsorbate bonding, the Pt-CO interaction being much stronger than the Pt-H_{upd} interaction. The surface coverage by CO at 0.05 V is 0.75 ML and a p(2×2)-3CO structure is formed that can be measured by SXS [ML=amount of monolayer adsorption]. As will be discussed in Section 1.4, the potential stability of the p(2×2)-3CO structure is strongly affected by the oxidation of a small fraction of the CO adlayer, so that at 0.65 V a ($\sqrt{19}\times\sqrt{19}$)-13 CO structure (hereafter dubbed the "$\sqrt{19}$" structure) forms with $\Theta_{CO} = 0.68$ ML. Very recently, the structural relaxation of the Pt(111)-CO surface has been studied in detail in the presence of the "$\sqrt{19}$" structure [36]. The structural model of the Pt(111) surface relaxation induced by the "$\sqrt{19}$" phase in Fig. 1.1 shows that different layer expansions and in-plane rotations occur for the Pt atoms un-

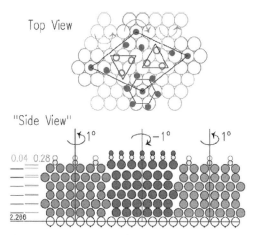

Fig. 1.1 Structural model of the Pt(111) surface relaxation in 0.1 M $HClO_4$ induced by the ($\sqrt{19}\times\sqrt{19}$)-13CO phase. Small solid and open circles represent the CO molecules in nearly top and bridge sites, respectively. Large circles in different shades of gray represent the Pt atoms in three groups with different rotation centers. The "side view" is not a projection, but a simplified picture to illustrate schematically the grouping and layer expansion. The top layer expansions of 0.28 and 0.04 from the lattice positions are amplified 4-fold for simplicity (taken from Ref. [36]).

der the near-top sites and the near-bridge site CO molecules. The top view (Fig. 1.1) shows how the 19 Pt atoms in the top-layer of the "$\sqrt{19}$" unit cell are grouped into a motif that contains 7 Pt atoms in a centered-hexagon region and 12 Pt atoms that are split between two triangular regions. Within each group, the Pt atoms are allowed to expand/contract laterally and to rotate around their center. For the underlying layers, the Pt atoms are assigned to either a centered-hexagon or a triangular region based on the ABC stacking sequence. This feature is illustrated schematically by the simplified "side view" shown in Fig. 1.1. These results demonstrate that, if adsorption is relatively strong, then subsurface atoms can move collectively to form a pattern on the nanometer scale that is determined by the unit cell of an adlayer. The ability to probe the relaxation in sub-surface layers at the metal/electrolyte interface is unique to the SXS technique, and this kind of surface restructuring cannot be studied by any other experimental methods.

The study of relaxation phenomena at monometallic/electrolyte interfaces has recently been extended to bimetallic surfaces. The $H_{upd}(OH_{ad})$-induced [3] changes in the surface relaxation of a bulk $Pt_3Sn(111)$ bimetallic alloy have been studied by SXS and CV measurements [31]. From fits to CTR data, using a structural model in which the vertical displacement (Δd_{Pt}, Δd_{Sn}), surface coverage (Θ_{Pt}, Θ_{Sn}), and roughness (σ_{Pt}, σ_{Sn}) of the Pt and Sn in the topmost two atomic layers were allowed to vary, it was found that expansion of the surface Pt atoms induced by the adsorption of hydrogen was very similar to that observed on Pt(111) [35, 37] (at 0.05 V $\Delta d_{Pt}^1 = +2\%$). As shown in Fig. 1.2, XRV at (1, 0, 3.6) and (1, 0, 4.3) indicate that the desorption of hydrogen, as well as the adsorption of bisulfate anions, leads to surface contraction, i.e. the interplanar spacing, shown as Δd in the insert of Fig. 1.2b, decreases monotonically by scanning the potential positively from 0.05 V. At 0.55 V the Pt surface atoms are unrelaxed whereas the Sn atoms in the topmost layer expand up to $\Delta d_{Sn}^1 = 8.5\%$ of the lattice spacing, i.e. the surface becomes increasingly *buckled* as the Sn atoms expand outwards. The onset potential of Sn dissolution is determined by the potential at which the bulk termination of the $Pt_3Sn(111)$ lattice begins to disorder. It is interesting that, in the presence of CO [4], the potential range of surface stability is wider than in CO-free solution, an effect which strongly resembles the CO-controlled "rec" $\leftrightarrow$ (1×1) transition of Au(hkl) in alkaline solution (the latter is discussed in Section 1.3.2). It has been proposed that the reason for the increased potential range of stability of the $Pt_3Sn(111)$ surface in the presence of CO arises because of the continuous consumption of OH in the Langmuir-Hinshelwood (L-H) reaction ($CO_{ad}+OH_{ad}=CO_2+H^++e^-$). In this reaction, OH is adsorbed on Sn surface atoms whereas CO is exclusively adsorbed on Pt atoms. Note that, in the H_{upd} region, the adsorption of CO on the $Pt_3Sn(111)$ surface does not lead to the large surface relaxation (4%) observed

---

3) The nature of the oxygenated species adsorbed onto Sn is unknown but, for the sake of discussion, this species will be collectively designated OH_{ad}.

4) Details about the vibrational properties of CO on platinum bimetallic surfaces including the $Pt_3Sn(111)$ system are presented by Korzeniewski in Chapter 7.

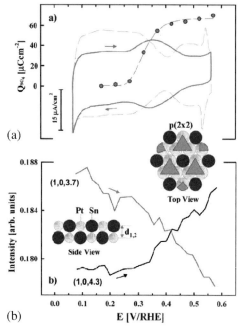

Fig. 1.2 (a) Cyclic voltammograms of Pt(111) (dashed line) and Pt₃Sn(111) (solid line) in 0.5 M H₂SO₄, scan rate 50 mV s⁻¹. Potential-dependent integrated charges for the adsorption of (bi)sulfate anions on the Pt₃Sn(111) surface are represented by circles. (b) The measured X-ray intensities at (1, 0, 3.7) and (1, 0, 4.3) as a function of the electrode potential. Top and side views represent the proposed p(2×2) structure. The gray circles are Pt atoms, the black circles are Sn atoms, and triangles are (bi)sulfate anions which are adsorbed on Pt sites. The side view indicates the surface normal spacing that is derived from CTR measurements (taken from Ref. [30]).

on the Pt monometallic surface. This agrees with DFT calculations in which the binding energy of CO on $Pt_3Sn(111)$ was calculated to be weaker than on Pt(111) [38]. As discussed below, however, on other bimetallic surfaces, in this case Pt/Pd, CO can produce large relaxation effects in the presence of H_{opd}.

The deposition of Pd onto Pt(*hkl*) and Au(*hkl*) surfaces has been studied by SXS, and a precise description of the Pd surface structure can be obtained by measuring and modeling the CTRs [32–34, 39]. This is particularly true for Pd/Pt(111) as, for coverages of up to 1 ML, Pd forms a pseudomorphic overlayer, i.e. it is fully commensurate with the Pt(111) lattice [28, 33]. The specular, (0, 0, *l*), and first-order, non-specular, (1, 0, *l*) and (0, 1, *l*), CTRs measured at 0.05 V for 1 ML of Pd on Pt(111) are shown in Fig. 1.3. The calculated CTRs for an ideally terminated Pt(111) surface are shown by the dashed curves, and this is close to the results obtained for the clean Pt(111) surface in electrolyte [37]. The data in Fig. 1.3 differ significantly from the model calculations at the "anti-Bragg" positions, midway between the Bragg reflections. At these positions, the scattering from the bulk Pt lattice is effectively canceled and the scattered inten-

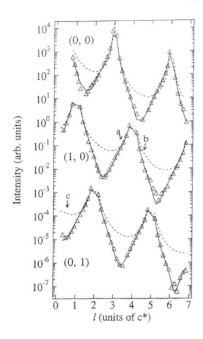

Fig. 1.3 The specular and first-order non-specular CTRs for the Pd/Pt(111) surface measured at an electrode potential of 0.05 V in CO-saturated solution. The dashed lines are calculated for an ideally terminated Pt(111) surface. The solid lines are a fit according to the structural model described in the text. The indicated positions (a, b, c) are those at which the data shown in Fig. 1.4 were measured (taken from Ref. [34]).

sity is due to the topmost Pt atomic layer and the Pd overlayer. Because of the difference in atomic form factor for Pt and Pd, this results in a significant decrease in intensity compared to the ideal Pt surface calculation [28]. The solid lines in Fig. 1.3 are a calculated best fit to the data using a structural model in which the Pd coverage (Θ_{Pd}), Pt-Pd distance (d_{Pt-Pd}), relaxation of the Pt surface (ε_{Pt}), and surface roughness parameters (σ_{Pd}, σ_{Pt}) were varied. The Pd atoms were located at Pt threefold hollow sites to continue the ABC stacking of the bulk lattice. The structural parameters obtained are $\Theta_{Pd} = 1.0 \pm 0.1$ monolayers (ML), $d_{Pt-Pd} = 2.32 \pm 0.02$ Å, expansion of the Pt lattice $\varepsilon_{Pt} = 0.02 \pm 0.02$ Å and static Debye-Waller-type roughness, $\sigma_{Pd} = 0.12 \pm 0.04$ Å and $\sigma_{Pt} = 0.09 \pm 0.04$ Å. The Pt-Pd layer spacing corresponds to an expansion of $\sim 2.5\%$ relative to the Pt-Pt layer spacing, which is similar to the expansion observed on the monometallic Pt surface in the presence of H_{upd}.

The potential dependence of the surface relaxation and the stability of the Pd film was probed by XRV measurements [32]. Desorption of H_{upd} caused a slight Pt-Pd contraction, but the surface became expanded again ($\sim 2.5\%$) after the adsorption of bisulfate anions. At positive potentials, i.e. in the potential region of OH adsorption, there was a significant (but *reversible*) change in the Pd structure associated with oxide formation. Importantly, if the Pt(111) monometallic surface had been cycled up to this potential, then a Pt oxide would be formed that would lead to irreversible roughening (which is the precursor to Pt dissolution) of the surface upon reduction.

Adsorption of CO onto the Pt/Pd surface at 0.05 V also led to no significant changes in the Pt-Pd layer spacing (compared to a 4% expansion of the Pt monometallic surface). This may indicate that the Pd-CO$_{ad}$ interaction is weaker than the Pt-CO$_{ad}$ interaction (an electronic effect) and/or is compensated by an even stronger Pt-Pd interaction, thus limiting the surface expansion. This conclusion is supported by the XRV measurements measured at three reciprocal lattice positions as the electrode potential was cycled over the range 0.05–1.15 V in CO-saturated solution (shown in Fig. 1.4). The (1, 0, 3.6) and (1, 0, 4.4) positions, on either side of the Pt (1, 0, 4) Bragg reflection, are principally sensitive to changes in the Pt-Pd spacing, provided that the Pd atoms remain commensurate with the Pt lattice. Thus, an increase in intensity at (1, 0, 3.6) and a corresponding decrease at (1, 0, 4.4) implies an expansion of the Pd-Pt surface and vice-versa. The (0, 1, 0.5) position is an "anti-Bragg" point, midway between the (0, 1, –1) and (0, 1, 2) Pt Bragg reflections and is primarily sensitive to the Pd coverage (as can be seen from the model calculations in Fig. 1.3). The data in Fig. 1.4 essentially show two structural effects which are separated in the figure by the vertical marker line at ∼0.85 V. Considering only the anodic sweep, it is clear from the (0, 1, 0.5) data that no changes in the Pd surface coverage occurs over the range 0–0.85 V, the upper limit being the potential at which CO oxidation is occurring. The intensities at (1, 0, 3.6) and (1, 0, 4.4), however, show dramatic changes over this potential region consistent with a significant contraction in the Pd-Pt layer spacing. Such changes are not observed in CO-free solution [32]. As the intensities shown are normalized to the scattered intensity measured at 0.05 V, the change in the Pt-Pd spacing necessary to reproduce the intensity ratios, R (R=0.8 and R=1.5 for l=3.6 and l=4.4 respectively), can be calculated. For a Pd-Pt spacing of 2.25 Å, the calculated ratios are R=0.80 and R=1.42, which are in reasonable agreement with the experimental results. It is apparent, therefore, that the onset of the CO

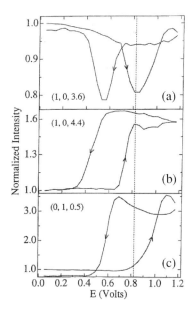

Fig. 1.4 The X-ray intensity measured at three reciprocal lattice positions (indicated in Fig. 1.3) as the potential was cycled over the range 0.05–1.15 V vs RHE (sweep rate=2 mV s^{-1}). At each position the intensities are normalized to the intensity measured at 0.05 V. The vertical line marker denotes the two potential regions of interest (see text; taken from Ref. [34]).

oxidation reaction causes a large contraction (–1% compared to the Pt-Pt bulk spacing) of the Pd-Pt surface. Given that the adsorbed CO molecules are reacting with OH species [presumably in an L-H reaction as observed for Pt(111)], it may be that the lattice contraction reflects the expected Pd-Pt bond length in the absence of strong chemisorption on the surface. In fact this bond length is in good agreement with a calculation of the Pt-Pd distance according to a hard sphere model using the atomic radii of Pt and Pd.

Pd films were also examined at successively higher levels of thickness, ranging from 1 ML to n ML (n > 2) regime. It was found that on the top of a pseudomorphic monolayer film, Pd deposited on Pt(111) forms three-dimensional is-

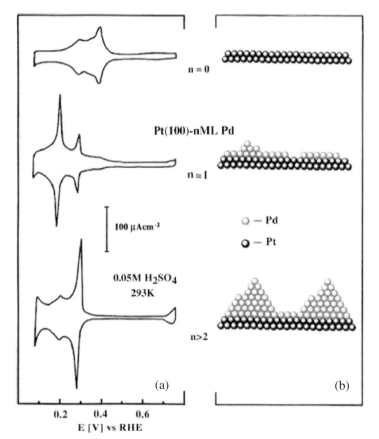

Fig. 1.5 (a) Cyclic voltammetry from the Pt(001) and Pt(001)-Pd surface in 0.05 M H₂SO₄ at a sweep rate of 20 mV s⁻¹. Subsequent to Pd deposition, sharp peaks arise at ∼0.2 and 0.29 V which correspond to hydrogen adsorption/desorption accompanied by (bi)sulfate desorption/ adsorption. The 0.29 V peak is due to adsorption/desorption at Pd step-terrace sites, which occur after the completion of the monolayer. (b) A schematic of the Pt(001)-Pd surfaces derived from CTR measurements at each level of Pd coverage (taken from Ref. [34]).

lands, implying that above 1 ML Pd deposition proceeds via a pseudomorphic Stranski-Krastanov growth mode [33]. The growth of Pd onto Pt(100) follows a similar mechanism to the Pt(111)-Pd system, with growth of the second Pd layer beginning after the first Pd layer reaches a coverage of 0.8 ML. As summarized in Fig. 1.5, further Pd deposition sees the formation of larger islands built onto the first Pd layer, characteristic of pseudomorphic S-K growth. For both Pt(111) and Pt(100) the SXS measurements provided the structural characterization that enabled interpretation of the voltammetric features (shown in Fig. 1.5). In the thick Pd films the lattice spacing in the Pd films was close to the bulk Pd lattice constant. Because of the complexity of the Pd film structure, XRV measurements showed no significant changes upon potential cycling, presumably as adsorption occurs at terrace and step sites on the defective Pd film and there is no concerted movement of surface atoms. Similarly, the effect of CO on the $Pd_{nML}/Pt(100)$ structure in the H_{upd} region was negligible, i.e. measurements of the X-ray scattering signal at the surface expansion-sensitive CTR positions showed no change after saturation of the electrolyte with CO, although H_{upd} is displaced from the Pd surface by CO. In contrast, however, a dramatic change in the X-ray intensity was observed at –0.04 V (the hydrogen evolution region and the formation of H_{opd}). For full details, see [34]. Given that the same effect is not observed in CO-free electrolyte, it was proposed that the observed relaxation (determined as expansion of the bulk lattice constant of the Pd film by detailed CTR measurements) is caused by CO-induced hydrogen *absorption* into the Pd lattice. A promoting effect was explained through the change in local surface electronic structure, which gives rise to electron transfer between adsorbed CO and H_{upd} through the underlying metal, an explanation previously used to rationalize the promotion of hydrogen absorption on Pt modified by poisoning species such as P, S, As, Se, Sb [34, 40, 41]. Hydrogen absorption is important for energy storage systems, and this may be a research field in which SXS can make a significant contribution to the design of new materials.

1.3.1.2 **Group IB Metals**

In this section we briefly discuss the surface expansion of the group 1B metals, Cu, Ag and Au, focusing on the close-packed (111) surfaces as they have been studied in the most detail (in fact there have been no published SXS studies of the Cu(100) and Cu(110) surfaces in the electrochemical environment). In terms of surface expansion effects, the 1B metals are more difficult to study than Pt as no H_{upd} is formed, and so it is difficult to correlate structural changes with well-defined adsorption processes. Furthermore, the Au(*hkl*) surfaces reconstruct at negative potential, which limits the potential window where the surfaces are in the unreconstructed state. Despite these difficulties, relaxation at the Au(111) surface was recently studied by a combination of SXS and surface stress measurements. For potentials on the positive side of the potential of zero charge (pzc), where the surface is unreconstructed, increasing positive surface charge causes a decrease in the tensile stress, a result that can be understood on the basis of a simple jellium model [42]. Correspondingly, one might expect that positive surface charge would lead to an increase in the surface expansion due to the apparent weakening of the surface bonds, but it was found that the opposite was true. Determination of the surface expansion of Au(111) in 0.5 M H₂SO₄ solution was obtained by XRV measurements at "expansion-sensitive" reciprocal lattice positions coupled with measurement and modeling of CTR data to calibrate the XRV results (in this case, the only parameter allowed to vary for data taken at different potentials was the expansion of the topmost Au atomic layer) [43]. The results show that close to the pzc the surface is expanded by ∼1.5% of the bulk layer spacing and that this expansion decreases with increasing surface charge to ∼0.2% at the positive potential limit. In order to rationalize the apparent conflict between the surface stress and surface expansion results, first principle density functional theory (DFT) calculations of Au(111) surface relaxation were made. For the uncharged surface the calculations gave an outward expansion of 1.3% of the bulk layer spacing, in remarkable agreement with the X-ray measurements. Based on these results, it was proposed that surface restructuring is a trade-off between cohesion between adjacent atoms and the band structure of the crystal. The density of states (DOS) for the relaxed surface is depleted near the Fermi level but enriched in the interval –4 eV to –1 eV below the Fermi level. The authors concluded that the shifting of the crystal bands to lower energies is the driving force for the outward expansion.

Surface restructuring at the Ag(111) electrode was examined as part of the study of the water structure at the metal/electrolyte interface by Toney and co-workers [44]. They measured the non-specular CTRs for the Ag(111) electrode in 0.1 M NaF to determine the Ag surface expansion independently of the incommensurate water layer and found that the surface was unrelaxed at a potential –0.23 V negative of the pzc and contracted by 0.03 Å (1.2% of the Ag(111) layer spacing ($d_{(111)} = 2.36$ Å) at a potential +0.52 V of the pzc). Although the importance of the electrostrictive effect was noted, it was not discussed in any detail; instead, the report focused on the interpretation of the specular CTR, which indicated the presence of a dense, "ice-like", water layer at the interface. The presence of such a water structure at the interface has yet to be confirmed by any other experimental technique. Indeed, similar CTR measurements were obtained on a Cu(111) crystal in 0.1 M HClO₄ [45], and the data were interpreted in terms of an adsorbed monolayer of oxygen or OH (with no long-range order). It was postulated that such an adsorbed monolayer would explain the high density of the water layer proposed by Toney et al. [44].

Recently, Ag(111) in 0.1 M KOH was also studied by SXS [46], and, as in the previous study, the surface was characterized by measurement of the non-specular CTRs and specular CTR. In alkaline solution there is no competition between OH and anions for adsorption sites, and the SXS data can be interpreted purely on the basis of the surface coverage by OH$_{ad}$. At –0.95 V (vs SCE) the best fit parameters to the CTR data indicated that the surface layer undergoes a small *inward* relaxation (contraction) of ∼0.8% of the Ag(111) layer spacing

$(d_{(111)} = 2.36$ Å$)$. At the positive potential limit (–0.11 V), where the coverage by OH is calculated to be ~ 0.45 ML from the CV, the surface contraction was increased to 1.2% of the Ag(111) layer spacing (identical to the measurement in 0.1 M NaF solution), and the modeling of the specular CTR indicated an OH coverage of ~ 0.45 ML. Key insight into the potential dependence of the surface relaxation is given by the XRV measurements that are shown in Fig. 1.6, together with the cyclic voltammetry in Fig. 1.6a. The solid black (anodic) and dashed black (cathodic) lines in Fig. 1.6b represent the background subtracted intensity as a function of potential during cycling measured at the (0, 0, 1.6) position on the specular CTR. The reversibility of the scattered intensity indicates that between these potential limits the adsorption of oxygenated species is a fully reversible process. Also shown is the OH coverage calculated from the charge under the CV (dash-dot line), showing a gradual increase in Θ_{OH}, implying that change in the specular CTR is associated with OH adsorption. The behavior of the non-specular CTRs (used to measure the surface relaxation), however, reveals a new structural effect at –0.75 V (indicated in the figure by a vertical dashed line), a potential that is very close to the pzc [47]. Fig. 1.6c shows the X-ray intensity at (0, 1, 0.5), a position on the (0, 1, l) CTR sensitive to the surface termination of the Ag(111) crystal. In the positive-going sweep there is a significant dip in the intensity beginning at 0.9 V, going to a minimum at ca. –0.75 V, and followed by a partial recovery by –0.55 V. Following this there is a small continuous decrease in intensity over the range –0.55 V to 0 V. The onset of the decrease in intensity at –0.9 V is coincident with the start of OH⁻ adsorption on the surface, indicating that significant changes to the Ag surface structure are occurring with a relatively small OH coverage of less than 0.05 ML. Note also that this potential is very close to the pzc. The negative-going sweep shows the XRV to be fully reversible over the entire potential range.

It is important to note that the changes in intensity at the (0, 1, 0.5) CTR position cannot be rationalized in terms of surface relaxation as this has a negligible effect on the intensity at an anti-Bragg position. The dip in intensity, therefore, must be attributed to a surface roughening type effect. *In-situ* UHV-STM measurements of low surface coverage $(\Theta_O < 0.05$ ML$)$ of OH on Ag(111) indicated that oxygen randomly adsorbs with a large near-neighbor exclusion zone of approx. 12 Ag atoms for each O atom [48], whereas at coverage greater than ~ 0.05 ML a p(4×4)-O phase is observed [49]. In the electrochemical environment, STM results indicate that, for the Ag(111) surface, strong adsorption of OH causes 2D (in-plane) stretching of the surface layer [50], and this may account for the changes observed in Fig. 1.6c.

For the group 1B metals there appears to be a correlation between the pzc (surface charge) and surface relaxation. In contrast to Pt(hkl) surfaces, the Ag(hkl) surfaces (Ag(100) and Ag(110) also exhibit inward relaxation [46]) undergo surface contraction that increases as the coverage by OH$_{ad}$ increases. In the case of Au(hkl), the trend toward surface relaxation is difficult to extract because of the reconstruction of the surfaces at negative potential. In the unreconstructed state, however, it appears that the surfaces also contract as the coverage

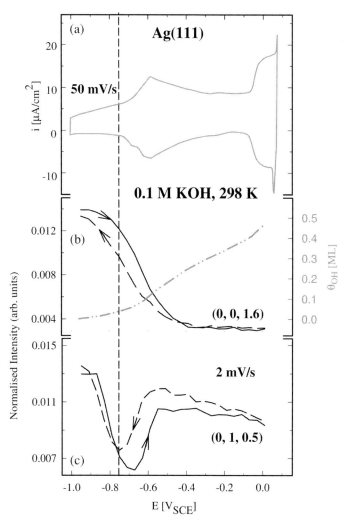

Fig. 1.6 (a) Cyclic voltammetry of Ag(111) in 0.1 M KOH recorded at a sweep rate of 50 mV s^{-1}, (b) X-ray voltammetry (XRV) measured at an "anti-Bragg" position on the specular CTR (solid and dashed lines), a position sensitive to the effects of adsorbates on the surface; the potential-dependent coverage by OH$_{ad}$, assessed from the CV, is shown as a dash-dot line, (c) XRV measured at (0, 1, 0.5) showing reversible structural changes triggered by adsorption of OH at low surface coverages ($\sim$0.05 ML) (to be published).

by oxygenated species increases. At present we do not have a definitive explanation for these effects; however, the differences in electronic structure between Pt group metals and the 1B group metals may be the key to understanding the relaxation behavior.

1.3.2
Surface Reconstruction

It is now well established that in UHV the clean low-index faces of some *fcc* metals tend to exhibit reconstruction under certain conditions of sample temperature and surface preparation. This is certainly true of Pt and Au, whereas most other *fcc* metals, such as Cu, Ni, and Ag, do not reconstruct. The tendency of the clean Pt and Au surfaces to reconstruct in UHV has been explained by Ho and Bohnen using first-principles calculations [51] and later by Norskov using a combination of local density functional calculations and modeling within the effective medium theory [52]. Both studies predicted that the participation of d electrons in bonding in the solid and the decrease of the kinetic energy of delocalized electrons at the surface play a decisive role in the stabilization of the experimentally observed missing-row reconstructions of Au(110) and Pt(110) surfaces. Several experiments also illustrated that, in the UHV environment, the adsorption of atoms and molecules on clean metal surfaces has a dramatic effect on the metal surface relaxation and can cause clean metal surfaces to reconstruct or to deconstruct ("lift") back to the (1×1) phase [28]. The thermodynamic driving force for adsorbate-induced restructuring is the formation of strong adsorbate-substrate bonds that are comparable to, or stronger than, the bonds between the substrate atoms in the clean surface. Although the same thermodynamic effects may be valid in electrochemistry, significant differences in adsorbate-induced restructuring in UHV and in electrochemical systems have been observed. As has been demonstrated [28, 29], these differences can be accounted for in terms of (a) the presence of co-adsorbed solvent and adsorbed species, and (b) differing surface potentials (φ) and (continuously adjustable) variations in electrode potential. The former effect results in the formation of an adsorbed layer of water, the adsorption of hydrogen, reversible/irreversible formation of oxygenated species, and the adsorption of anions from supporting electrolyte. The latter effect results in potential-induced changes of the surface electron density. As demonstrated below, although in an electrochemical environment thermodynamic and/or electronic effects are difficult to decouple, taken together these two effects play a major role in determining surface restructuring in electrochemical systems.

Given that adsorbate-induced lifting of reconstruction is observed only at Au(*hkl*)/solution interfaces, we focus on the transition between the reconstructed ("rec") and (1×1) phases of Au(*hkl*) surfaces in the electrochemical environment. Simplified real space models for the bulk-terminated and reconstructed low-index surfaces of gold are depicted in Fig. 1.7. The reconstructions of the Au(*hkl*) surfaces are observed at negative potentials where the surfaces are (almost) free of strongly adsorbing anions. For Au(111), the reconstruction is rather complex and involves a small increase in the surface density, which leads to a large unit cell, with a ($23 \times \sqrt{3}$) periodicity [53, 54]. The Au(110) surface exhibits either (1×2) or (1×3) periodicity [54, 55]. The (1×2) phase is called the "missing row" structure, since every other row is lost in going from the

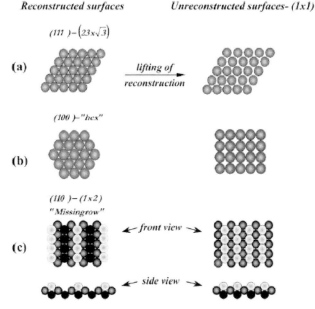

Reconstructed surfaces   **Unreconstructed surfaces- (1x1)**

Fig. 1.7 Simplified real space model of reconstructed
(left side) and bulk terminated (right side) Au(*hkl*) surfaces:
(a) Au(111)-(23×3) ↔ (1×1); (b) Au(111)-(5×20) ↔ (1×1);
(c) Au(110)(1×2) ↔ (1×1) (taken from Ref. [59]).

(1×1) phase to the (1×2) phase, and, as a result, the surface atomic density changes very significantly, by 50%! The Au(100) surface exhibits a hexagonal reconstruction which consists of a single, buckled, slightly distorted overlayer, which is aligned close to the [110] bulk direction and is often referred to as a "5×20" or "hex" reconstruction [2, 9]. The results presented in this section are for the Au(100) surface, and Fig. 1.8 shows representative in-plane X-ray diffraction results in the form of rocking scans through reciprocal lattice positions where scattering from the reconstruction is observed. In 0.1 M HClO$_4$ the scan always displays two peaks, which are rotated by ∼0.75° from the [110] direction. In 0.1 M KOH, however, the scan shows either a single broad peak aligned with the Au[110] direction [9] or two peaks rotated by 0.75° from the [110] direction. Based on the fact that the peaks are much broader, it was suggested that the structure in alkaline solution is more pinned to defects which may hinder the rotation [56].

The lifting of the reconstruction is found to be potential-dependent, and each surface has its own dynamics in the "rec" ↔ (1×1) conversion, which is associated with significant surface mass change. Of particular interest in this regard has been the behavior of Au(100) in alkaline as well as in acidic electrolytes, and this section therefore focuses on the adsorbate-induced "hex" ↔ (1×1) transition on this surface. Although only results for the adsorbate-induced recon-

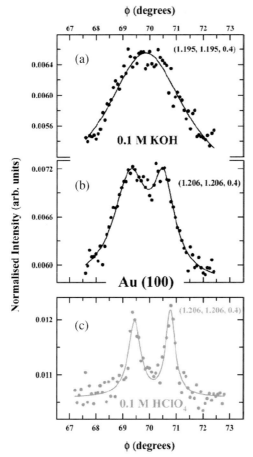

Fig. 1.8 Rocking scans through: (a) (1.195, 1.195, 0.4) and (b) (1.206, 1.206, 0.4) in 0.1 M KOH and (c) (1.206, 1.206, 0.4) in 0.1 M HClO₄ at 0.05 V vs RHE. At these reciprocal space positions, scattering from the reconstructed "hex" phase is observed (taken from Ref. [56]).

struction of the Au(100) surface are shown, the conclusions are rather general and equally applicable to the Au(111) and Au(110) systems (see [29]).

The potential range of stability as well as the dynamics of formation/lifting of the "hex" phase in alkaline solution have been analyzed by a combination of SXS and cyclic voltammetry [56]. A significant feature of the voltammetry in 0.1 M KOH, depicted in Fig. 1.9a, is the pseudocapacitance peak appearing at ∼1.0 V. The sharp peak at 1.0 V was initially suggested to be associated with the potential/(OH$_{ad}$)-induced lifting of the "hex" reconstruction, by analogy with the related peaks observed in acid solution containing strongly adsorbing anions [7]. It is interesting that XRV measurements at the reconstruction-sensitive reci-

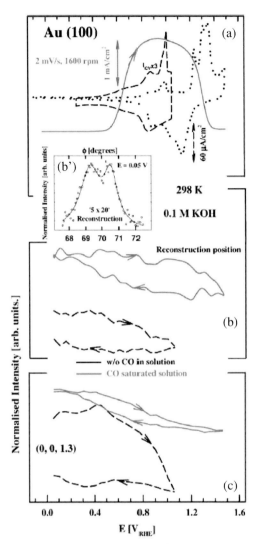

Fig. 1.9 (a) The cyclic voltammetry for Au(100) in 0.1 M KOH along with the polarization curve for CO oxidation in CO-saturated solution (continuous curve); (b) XRV measured at the reconstruction position (1.206, 1.206, 0.4) in CO-free (dashed curve) and CO-saturated (solid line) solution; (b') rocking scan through (1.206, 1.206, 0.4); (c) XRV measured at a position sensitive to the mass transfer involved in the surface reconstruction, (0, 0, 1.3), with and without CO in solution (taken from Ref. [59]).

procal space position in Fig. 1.9b unambiguously show that the onset of lifting of the "hex" phase takes place below 1.0 V. (Fig. 1.9b shows in-plane X-ray diffraction results in the form of rocking scans through reciprocal lattice positions where scattering from the reconstruction is observed.) As a result, the kinetics of "hex" formation is rather slow, as demonstrated in Fig. 1.9c at the specular CTR position sensitive to the mass transfer involved in the surface reconstruction. In fact the lifting of the reconstruction follows the adsorption isotherm of OH_{ad}. This implies that the lifting of the reconstruction is adsorption-induced rather than an electric field effect.

Further insight into the "electric field vs adsorption" issue has been obtained by monitoring the surface reconstruction under conditions in which strongly adsorbed OH_{ad} is removed from the surface by a relatively weakly adsorbed reactant, namely CO_{ad}. (For details of the vibrational properties of CO on Pt and Au see Korzeniewski's Chapter 7.) Surprisingly, it has been found that CO acts to catalyze the formation of uniform reconstructed domains. This phenomenon was established initially by STM studies of the Au(100) surface in alkaline solution [58] and was reexamined using SXS. The key results are shown in Fig. 1.9, which, in addition to the CV/XRV measurement in nitrogen-purged 0.1 M KOH solution, also summarizes the results in CO-saturated 0.1 M KOH. The new information provided by the SXS results was that the slow kinetics of the formation of the reconstruction observed in N_2-purged alkaline solution is much faster in CO-saturated solution (Fig. 1.9c), and, even more remarkably, the reconstruction is present over the entire potential range, i.e. up to 1.5 V in Fig. 1.9b. Gallagher et al. [59] showed that the CO-induced changes in the "rec" $\leftrightarrow$ (1×1) phase transition are not unique to the Au(100) surface but rather constitute a general phenomenon, which, on all three Au single-crystal surfaces, is induced by continuous removal of strongly adsorbed OH by weakly adsorbed CO in the L-H reaction. The fact that the potential range of stability of the reconstructed surfaces is so extended in the presence of CO (up to 1.7 V!) indicates that lifting of the reconstruction is determined by energy of adsorption rather than by the electric field.

Further evidence for this conclusion was obtained in acid solution, where anions and OH_{ad} are in strong competition for the adsorption sites. To minimize the effects of OH_{ad} and to emphasise the effect of anions, results were obtained for Au(100) in 0.1 M $HClO_4$ before and after Br^- anions were added to solution. An overview of the electrochemical and SXS results is given in Fig. 1.10. As has been discussed recently [60], similarly to the effect of Cl_{ad} on the thermodynamics/dynamics of the "hex" $\leftrightarrow$ (1×1) transition, the specific adsorption of Br^- shifts the equilibrium potential of the "hex" $\rightarrow$ (1×1) transition to less positive values (from Fig. 1.10b ca. 0.3 V) relative to Br^--free solution. From analysis of SXS data, Blizanac et al. concluded that, although the specific adsorption of Br acts to catalyze the mobility of gold surface atoms, the formation of uniform "hex" domains is not enhanced by this movement [60]. Further inspection of Fig. 10a and b reveals that the "hex" $\rightarrow$ (1×1) transition is completed at ca. 0.1 V, or, based on the X-ray intensity measured at (0.5, 1, 0.15) in Fig. 1.10c, just before the formation of the $c(\sqrt{2}\times2\sqrt{2})R45°$ Br_{ad} superstructure (the latter will be discussed in Section 1.4.1). SXS results obtained in solution containing CO are also plotted in Fig. 1.10. Interestingly, in contrast to alkaline media (pH = 13), Fig. 1.10b shows that CO has a negligible effect on the potential-de-

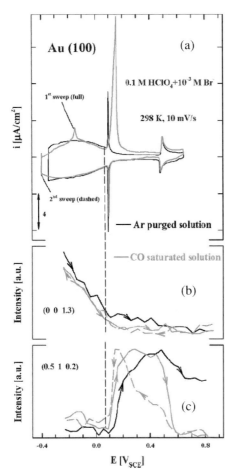

Fig. 1.10 (a) Current-potential curves (first scan) for Au(100) in 0.1 M HClO₄ + 10⁻³ M Br⁻ purged with either argon (black curve) or CO (grey curve) at 10 mV s⁻¹; (b) Corresponding XRV measurements (2 mV s⁻¹) at (0, 0, 1.3), a position on the specular CTR; (c) XRV measurements (2 mV⁻¹ s) at (0.5, 1, 0.2), where scattering from the c($\sqrt{2} \times 2\sqrt{2}$)R45° structure of Br$_{ad}$ is observed (taken from Ref. [60]).

pendent XRV measured at the (0, 0, 1.3) position. Thus, at pH = 1 in the presence of CO, the "hex" ↔ (1×1) transition is shifted *negatively* by ca. 40 mV. This apparently opposite effect from that observed in Fig. 1.9 for alkaline solution has been explained on the basis of a CO-induced increase in the equilibrium Br$_{ad}$ surface coverage relative to that at the same potential in CO-free solution. Given that the same effect is also observed for the other two low-index Au single crystals it is important to rationalize the opposite behavior observed for the "rec" ↔ (1×1) transition in acid versus alkaline electrolyte. Blizanac et al. proposed that the pH-dependent CO effect is controlled by a delicate balance between the nature of the interaction of adsorbates with the Au(*hkl*) surface (the energetic part) and the potential-dependent surface coverage by anionic species [60]. In particular, the energetic part is determined by the strength of the Au(hkl)-adsorbate interaction, the Au(hkl)-CO interaction being much weaker than the Au(hkl)-OH$_{ad}$ interaction and, in particular, than the Au(hkl)-Br$_{ad}$ interaction. As a result, OH$_{ad}$ and Br$_{ad}$ cannot be displaced from the Au(hkl) surface by CO, but, in contrast to Br$_{ad}$, OH$_{ad}$ can be oxidatively removed from the surface in the L-H reaction. On this basis it was suggested that the equilibrium surface coverage by Br$_{ad}$ and thus the "equilibrium" potential for the "rec" $\leftrightarrow$ (1×1) transition is a consequence of the CO-correction to the adsorption isotherm of Br$_{ad}$. As shown in Fig. 1.10c and discussed in Section 1.4, the fact that CO and OH$_{ad}$ can react on the Au(100)-Br$_{ad}$ surface determines the potential stability of the formation of the Br$_{ad}$ c($\sqrt{2} \times 2\sqrt{2}$)R45° structure.

1.4
Adlayer Structures

The chemisorption of atoms or molecules onto monocrystalline metal surfaces characteristically yields ordered adlayer structures. This ordering can be understood, in general terms, from a balance between the preference for adsorbate bonding in specific coordination sites and the dominant adsorbate-adsorbate interaction at higher coverages. This leads to a rich variety of structural behavior that is ideally suited to study by SXS because of the high resolution of the technique, and a comprehensive list of these studies is given in the References. In this section, a brief review of these experiments is given. The first part describes the adsorption of anions, which is, in addition to the substrate-anion interaction, controlled by the competitive adsorption between anions and OH$_{ad}$. The second part describes the adsorption of CO onto Pt(111) and the effects of solution temperature on the CO structure and kinetics. Section 1.4.3 describes the driving force for the ordering of metal adlayers, such as reversible formation of metal UPD adlayers including co-adsorption of UPD cations and anions. The elucidation of this structural behavior, from a standpoint of both ordering of adlayers on *fcc* metal electrodes as well as electrochemical reactivity, is as significant in model systems as it is in systems of practical importance. Some important limitations in terms of understanding electrocatalytic activity should, however, be pointed out:

- In many systems it is difficult to find a correlation between the adsorbate structures that are formed and the catalytic behavior, as well-defined adsorbate structures of spectator species tend to block active surface sites.
- Most reaction intermediates, particularly those of importance to fuel cell reactions, tend not to form ordered structures and cannot be studied by SXS.

1.4.1
Anion Structures

The adsorption of anions on metal electrodes has been one of the major topics in surface electrochemistry. Specific adsorption of anions occurs when the anion loses all or part of its solvation shell and forms a direct chemical bond with the substrate. In this situation the surface coverage by anions can be high and the adlayer tends to form a close-packed structure that depends critically on the surface atomic geometry of the underlying substrate and the balance between the anion-metal and anion-anion interaction energies. The structures of halide anions adsorbed onto Au(*hkl*), Ag(*hkl*), and Pt(*hkl*) low-index surfaces have been the most widely studied systems by SXS, and a comprehensive review of ordered anion adlayers on metal electrodes is given by Magnussen [57].

On the close-packed (111) surfaces, halide anions form hexagonal overlayers that undergo electrocompression, and a variety of structures are observed: commensurate, incommensurate, uniaxially incommensurate, and rotated phases. On the more open (100) and (110) metal surfaces the symmetry of the substrate differs from the energetically-favorable hexagonal symmetry of the halide adlayers, and, in addition, the corrugation potential is higher, and this favors the formation of low-order commensurate phases. In fact, on (100) surfaces the most commonly observed structure is the c(2×2) phase, a structure with a simple square symmetry and a surface coverage of 0.5 adatoms per metal surface atom. This structure has been observed on Ag(100), Cu(100), Pt(100), and Pd(100) for Cl^-, Br^- and I^- halide anions despite the fact that the atomic size mismatch can lead to large anion-anion bond lengths compared to the bond lengths found in the corresponding hexagonal phases [61–64].

It is only on the Au(100) surface that the anion-anion interaction is sufficient to perturb the influence of the substrate corrugation potential. For example, the bromide adlayer on Au(100) undergoes a commensurate-incommensurate transition where a commensurate c($\sqrt{2}$×2$\sqrt{2}$)R45° structure transforms continuously to an incommensurate c($\sqrt{2}$×p)R45° structure [65]. The CV for Au(001) in 0.05 M NaBr is shown in Fig. 1.11. Three sharp peaks (labeled by P1, P2, P3) are observed as the potential is swept between –0.25 V and 0.6 V; P1 corresponds to the lifting of the Au reconstruction to leave the surface in the (1×1) state, P2 corresponds to the formation of the c($\sqrt{2}$×2$\sqrt{2}$)R45° structure (as shown in Fig. 1.11 by the intensity changes at the (0, 1, 0.1) reciprocal space position, which is where scattering from such a structure would arise), and P3 corresponds to the commensurate-incommensurate phase transition. The c($\sqrt{2}$×2$\sqrt{2}$)R45° structure is shown schematically in Fig. 1.12a, which indicates that the surface coverage by bromide is $\Theta = 1/2$ and that the structure is close to a hexagonal arrangement, despite the square symmetry of the underlying substrate. This implies that the elastic interactions between the relatively large Br adatoms (which would favor hexagonal packing) dominate over the adsorbate-substrate interaction. The observed in-plane diffraction pattern is shown in Fig. 1.12c, where the squares correspond to substrate reflections and the circles

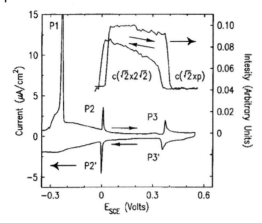

Fig. 1.11 The CV for an Au(100) electrode in 0.05 M NaBr solution and the corresponding X-ray intensity at the position (0, 1, 0.1) where scattering from a bromide adlayer was observed (taken from Ref. [65]).

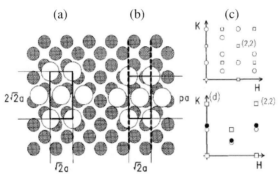

Fig. 1.12 Real space model (a, b) of the commensurate $c(\sqrt{2}\times2\sqrt{2})R45°$ and incommensurate $c(\sqrt{2}\times p)R45°$ bromide structures observed on Au(001). The open circles correspond to bromide atoms and closed circles correspond to gold atoms;

(c) and (d) show the corresponding in-plane reciprocal space patterns where the squares are scattering from the Au substrate and the circles are Br reflections. In (d) the peaks move outward along K with increasing potential (taken from Ref. [65]).

to Br reflections. At potentials positive of P3 (0.38 V vs SCE), the bromide adlayer undergoes a commensurate-incommensurate transition, which was signified by the movement of the low-order diffraction features continuously and uniaxially outward with increasing potential. As shown in the diffraction pattern (Fig. 1.12d), reflections were only observed at $(1, 0.5+\varepsilon/2)$, $(0, 1+\varepsilon)$ and $(2, 1+\varepsilon)$, along with symmetry equivalents (ε denotes the incommensurability). The real-space model of the structure is shown in Fig. 1.12b. For the domain shown, the bromide lattice is commensurate along H (the $\sqrt{2}$ direction) and incommensurate along K. As the potential increases above 0.38 V, the value of ε

increases continuously, and this corresponds to movement of the bromide atoms along the directions shown by the dashed lines in Fig. 1.12 b. In the in-commensurate phase the Br nearest-neighbor spacing is fixed at 4.078 Å, whereas the next-nearest-neighbor spacing decreases from 4.56 Å ($\varepsilon=0$, com-mensurate) to 4.14 Å ($\varepsilon=0.13$). This latter spacing is close to the minimum value observed on the Au(111) surface (4.02 Å).

Very recently, in an attempt to understand the competitive adsorption between Br_{ad} and OH_{ad} species and probe the role of OH_{ad} on the formation of the or-dered bromide adlayer on Au, the Au(100)-Br_{ad} system was reexamined in both acidic and basic media. In line with the discussion in Section 1.3.2, the role of OH_{ad} was established by monitoring the microscopic interface structures of Br_{ad} during the consumption of OH_{ad} in an electrochemical reaction in which strongly adsorbed OH_{ad} is removed from the surface by relatively weakly ad-sorbed CO_{ad}. This follows the same methodology used to probe the effect of Cl_{ad} and Br_{ad} on the Au(100) surface reconstruction, as described in Section 1.3.2. We may recall that, because of oxidative removal of a small amount of OH_{ad} in the L-H reaction and an equivalent increase in the equilibrium Br_{ad} surface coverage, in the presence of CO the "hex" $\leftrightarrow$ (1×1) transition is shifted negatively by ca. 40 mV (Fig. 1.10 c). Although the increase in $\Theta_{Br_{ad}}$ and the shift of the "hex" $\leftrightarrow$ (1×1) transition in solution saturated with CO is rather small, adsorption of CO has a significant effect on the ordering of the $c(\sqrt{2}\times2\sqrt{2})R45°$ structure, as shown in Fig. 1.10 c. Analysis of rocking curves through the (0.5, 1, 0.2) reflection measured at 0.4 V, with and without CO in solution (Fig. 1.13), revealed that the domain size of the structure increases from 200 Å in CO-free solution to 230 Å in solution saturated with CO. As shown in Fig. 1.10 c, in the presence of CO the onset of formation of the com-mensurate $c(\sqrt{2}\times2\sqrt{2})R45°$ structure is shifted negatively by ca. 40 mV, equal to the value of the thermodynamic potential shift of the "hex" $\leftrightarrow$ (1×1) transi-tion in Fig. 1.10 b. This reflects the fact that the complete transition of the "hex" phase into the (1×1) phase is required in order to form the $c(\sqrt{2}\times2\sqrt{2})R45°$ structure. Further inspection of Fig. 1.10 c reveals that, in the presence of CO, the $c(\sqrt{2}\times2\sqrt{2})R45°$ structure develops/disappears more rapidly than in solu-tion free of CO, i.e., comparison of the XRV data shows that both the disorder-order transition and the commensurate-incommensurate transition are much sharper in the presence of CO. The observed CO effects on the formation of the $c(\sqrt{2}\times2\sqrt{2})R45°$ structure can be rationalized with the same argument as for the reconstruction phase transition in Fig. 1.9 c, i.e., at the same electrode po-tential the Br_{ad} coverage may increase in CO-saturated solution because the competing OH_{ad} species are removed from the surface in the L-H reaction. As a consequence, the surface coverage by Br_{ad} required to form the $c(\sqrt{2}\times2\sqrt{2})R45°$ order adlayer ($\Theta_{Br_{ad}}=0.5$ ML) may be established at lower po-tentials in the presence of CO. Such a mechanism would correlate nicely with the weak adsorption of CO on Au surfaces [23, 66], i.e., the Au(100)-Br_{ad} and Au(100)-OH_{ad} interaction is much stronger than the Au(100)-CO interaction and only Br_{ad} and OH_{ad} are competing for the surface sites.

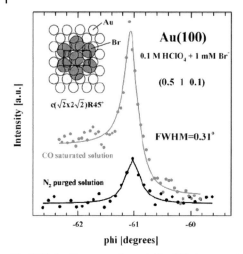

Fig. 1.13 Rocking scans through the (0.5, 1, 0.15) reciprocal lattice position measured at 0.4 V in the absence of CO (upper curve) and in the presence of CO (lower curve), where scattering from the commensurate $c(\sqrt{2}\times2\sqrt{2})R45°$ structure of the Br_{ad} adlayer in 0.1 M $HClO_4 + 10^{-3}$ M Br^- is observed. The solid lines are fits of a Lorentzian line shape to the data which gives a coherent domain size for the $c(\sqrt{2}\times2\sqrt{2})R45°$ structure of $\sim$150 Å. Also shown is a schematic picture of the $c(\sqrt{2}\times2\sqrt{2})R45°$ unit cell (taken from Ref. [60]).

The proposed effect of CO on competitive adsorption between Br^- and OH^- was further examined in alkaline solution. In particular, one may expect that the removal of OH_{ad} from the Au(100) surface by weakly adsorbed CO may allow the Br_{ad} adlayer to order, as it does in acid solution. To probe the CO effect on the ordering of the Br_{ad} adlayer, an "*h*" scan through the (1/2, 1, 0.15) peak due to the $c(\sqrt{2}\times2\sqrt{2})R45°$ structure was measured at 0.2 V, before and after saturating the solution with CO (Fig. 1.14). From Fig. 1.14, whereas in the absence of CO no ordered structure with the $c(\sqrt{2}\times p)R45°$ symmetry is observed, in the presence of CO a sharp peak is found at the commensurate position ($h=0.5$). The insert of Fig. 1.14 shows a rocking scan at a Bragg reflection due to the $c(\sqrt{2}\times2\sqrt{2})R45°$ adlayer of Br_{ad} that is formed in the presence of CO. From the Lorentzian line shape fit (solid line) to this data, a domain size of ca. 200 Å is calculated, which is close to the domain size measured in acid solution. The potential dependence of the X-ray intensity at the (1/2, 1, 0.15) position (not shown, see Fig. 7 in [67]) revealed that a commensurate $c(\sqrt{2}\times2\sqrt{2})R45°$ structure is stable within ca. 0.3 V. At rather positive potentials, however, the commensurate-incommensurate transition is observed to be as reversible as in acid solution. In CO-free solution, neither a commensurate nor an incommensurate structure is found, presumably because of the strong competition between Br_{ad} and OH_{ad} in alkaline solution. Based on these observations, it has been proposed that the competitive adsorption between halide anions and oxygenated species may also play a significant role in the continuous transition

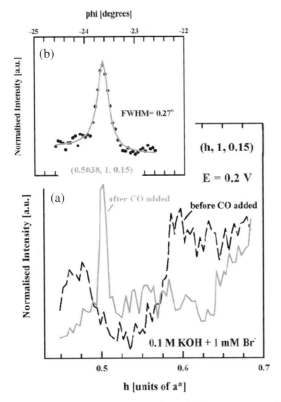

Fig. 1.14 (a) *h*-scans at E=0.2 V through the (1/2, 1, 0.15) position before (dotted curve) and after (continuous curve) saturating 0.1 KOH+10^{-3} M Br$^-$ solution with CO. (b) Rocking scans through the (1/2, 1, 0.15) reciprocal lattice position, where scattering from the commensurate c($\sqrt{2}\times2\sqrt{2}$)R45° structure of the Br$_{ad}$ adlayer is observed. The peak is only observed in CO-saturated solution. The solid lines are fits of a Lorentzian line shape to the data, which gives a coherent domain size for the c($\sqrt{2}\times2\sqrt{2}$)R45° structure of ca. 175 Å (taken from Ref. [60]).

from the commensurate c($\sqrt{2}\times2\sqrt{2}$)R45° structure to the incommensurate c($\sqrt{2}\times$p)R45° structure [68]. While a detailed understanding of the role of OH$_{ad}$ on the phase transition in the Br$_{ad}$ adlayer is not yet possible, the present discussion suggests that Br$_{ad}$ compressibility may be determined not only by the adsorbate-adsorbate interaction [65, 68] but, in addition, by strong competition for Au adsorption sites between Br$_{ad}$ and OH$_{ad}$. Competitive adsorption between anions and OH also plays a key role in the potential stability and ordering of many adsorbates, the prominent example being the interaction between CO and UPD metal adlayer structures on Pt(*hkl*). This is described in more detail in Section 1.4.3.

1.4.2
CO Ordering on the Pt(111) Surface

The surface electrochemistry of CO adsorbed on transition metal surfaces has been the subject of intense theoretical and experimental work (for an overview, see [28, 69, 70]). In these studies there has been an emphasis on linking the microscopic structure of the CO adlayer to the thermodynamics and other macroscopic electrochemical responses at electrified interfaces. Whereas information regarding macroscopic properties has come from classical electrochemical techniques [71–73], the chemical, physical, and structural properties on the atomic scale have been obtained from either a combination of *in-situ* STM and vibrational spectroscopy [74, 75] or SXS and vibrational spectroscopy [76]. The adsorptive and catalytic properties of CO adsorbed on platinum single crystals have been most widely studied. This is because the system offers an opportunity to gain basic understanding, which could ultimately lead to the design of new catalysts and also to an understanding of the activity pattern of Pt metal nanoparticles employed in fuel cells in the size range of a few nanometers. The upper part of Fig. 1.15 displays polarization curves for the oxidation of dissolved CO on Pt(111) in acid solution. According to Fig. 1.15, two potential regions can be distinguished, i.e., the so-called pre-ignition potential region is followed by the ignition potential region. The term "ignition potential" is analogous to the term "ignition temperature" in gas-phase oxidation. It is the potential at which the rate becomes entirely mass transfer limited. Although the rate of CO oxidation changes with electrode potential, it has been suggested that in both the pre-ignition potential region and at or above the ignition potential, the mechanism for CO oxidation obeys an L-H type reaction in which CO_{ad} reacts with OH=ad [77]. It was also proposed that the active sites for OH adsorption are defects in the Pt(111) surface [77]. This was supported by the fact that the activity in the pre-oxidation region is strongly dependent on the pre-history of the electrode, i.e., on the "CO-annealed" surface (second sweep) the activity is significantly reduced because of the removal of surface defects in the presence of CO. As shown below, defect sites also play a role in CO ordering on platinum single-crystal surfaces.

The first *in-situ* determination of CO_{ad} structure was reported for a CO adlayer on Pt(111) in acidic electrolytes. Using *in-situ* STM, Villegas and Weaver [74, 78] observed a hexagonal close-packed (2×2)-3CO adlayer structure at potentials below 0.25 V (vs SCE), with a CO coverage of $\Theta_{CO}=0.75$ ML. The z-corrugated pattern evident in STM images indicated the presence of adsorbed CO at two threefold hollow Pt sites and one atop Pt site per (2×2) unit cell. At potentials above 0 V (up to the onset of CO oxidation), a markedly different adlayer arrangement was formed, having a $(\sqrt{19}\times\sqrt{19})R23.4°-13CO$ unit cell with $\Theta_{CO}=13/19$. Following these earlier STM studies, direct information regarding the CO_{ad} structure was obtained in SXS measurements. While holding the potential at 0.05 V and with a continuous supply of CO to the X-ray cell, a diffraction pattern consistent with p(2×2) symmetry was observed, first by us [37, 71] and later by the Argonne [79] and Brookhaven [36] groups. Once formed, the

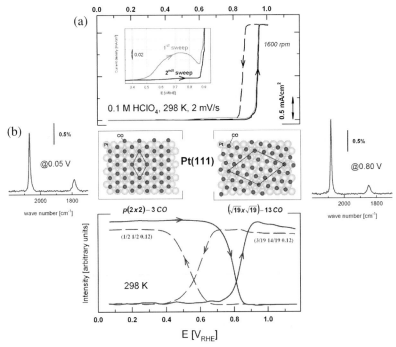

Fig. 1.15 (a) The polarization curve for CO oxidation on Pt(111) in CO-saturated 0.1 M HClO$_4$ solution at 298 K (sweep rate 2 mV/ s). (b) A close-up of the pre-oxidation potential region showing the first and second anodic sweeps. (c) X-ray voltam-metry measured at the (½, ½, 0.12) and

(3/19, 14/19, 0.12) positions where X-ray scattering arises due to the p(2×2)-3CO and (√19×√19)-13CO structures respectively. Schematics of the CO structures, indicating the unit cells, are shown in between the two panels (unpublished data).

structure was stable only when the CO was continuously supplied to the X-ray cell, i.e., when CO was replaced by nitrogen the p(2×2) structure slowly vanished. It is also worth mentioning that the potential range of stability of the p(2×2)-3CO phase was strongly affected by the oxidation of a *small* fraction (~15%) of CO in the pre-ignition potential region. The potential window of stability as well as the domain size of this structure can be controlled by the nature of the anions present in supporting electrolytes. For example, in one study the CO coherent domain size increased from KOH (ca. 30 Å), to HClO$_4$ (ca. 140 Å), to HClO$_4^+$Br$^-$ (ca. 350 Å) electrolytes. Given that the potential range of stability increased in the same order, it was suggested [77] that CO ordering at low overpotentials is determined by the competition between OH$_{ad}$ and specta-tor anions for the defect/step sites, i.e., the less active the surface is toward CO oxidation the larger the ordered domains and the greater the potential range of stability of the p(2×2) structure.

Fig. 1.15a shows that the disappearance of the p(2×2)-3CO structure at 0.7 V is accompanied by formation of the "√19" structure, which is stable even in the ignition potential region. This is a unique example of an adsorbate structure being present at a potential where maximum catalytic activity for an electrochemical reaction is observed. The observation of the "$\sqrt{19}$" structure by SXS was reported first by Tolmachev et al. [79] in solution containing strongly adsorbing Br anions, and then by Wang et al. in perchloric acid solution [36]. The derived structural models for both the p(2×2) and "$\sqrt{19}$" structures reveal that, while the p(2×2) structure consists of three CO molecules per unit cell, the unit cell of the "$\sqrt{19}$" structure contains 13 CO molecules, thereby having a coverage, Θ_{CO}, of 13/19 ($\Theta_{CO}=0.685$ per surface Pt atom). The *potential-induced* p(2×2) $\leftrightarrow \sqrt{19}$ phase transition, discovered by Villegas and Weaver using STM experiments, has thus been confirmed by SXS results. Considering that under practical conditions in fuel cells the electrooxidation of impure hydrogen (i.e. containing a trace of CO) may take place over the temperature range 273–363 K, an intriguing question to answer is how the solution temperature affects the phase transitions in the CO structure. Temperature effects have long been studied in traditional (UHV-based) surface science, but, until recently, there had been no such attempts in interfacial electrochemistry. New results describing the effect of temperature on the p(2×2) $\leftrightarrow \sqrt{19}$ phase transition in the CO adlayer on Pt(111) are presented below.

As in earlier studies at room temperature, the *temperature-controlled* p(2×2) $\leftrightarrow$ "$\sqrt{19}$" phase transition (depicted in Fig. 1.16) was monitored by measuring the scattered X-ray intensities at (1/2, 1/2, 0.12) and (3/19, 14/19, 0.12) at 280 K, 293 K and 319 K for the same conditions as in Fig. 1.15. These results yield two significant new pieces of information that give new insight into the interaction of CO with OH on the Pt(111) surface.

1. The ordering of the p(2×2)-3CO structure is frustrated under both "cold" (280 K) as well as "hot" (319 K) conditions. The fact that the integrated intensities and the widths (inversely related to coherent domain size) of the p(2×2) peaks show a "volcano" relationship with the temperature of electrolyte may indicate that the balance between the rate of CO ordering and the surface coverage by OH (rate of CO oxidation) reaches a maximum at room temperature (293 K). Note also that the potential window of stability of the p(2×2) structure, shown in the middle part of Fig. 1.16, decreases linearly on increasing the temperature, reflecting the negative shift in the onset of CO oxidation on increasing temperature (observed in a top part of Fig. 1.16).

2. The ordering of the $\sqrt{19}$ structure (i.e. coherent domain size) increases linearly on increasing the temperature, a consequence of enhanced OH adsorption at high temperatures. It is also interesting that at 319 K the $\sqrt{19}$ phase exists even at the onset of the hydrogen evolution reaction (as shown in the bottom part of Fig. 1.16 at ca. 0 V). It is important to note that in the potential range where CO oxidation depends entirely on the rate of CO diffusion from the bulk of the solution to the Pt surface, the $\sqrt{19}$ structure is rather stable. This, in turn, suggests that the surface coverage by CO, in the potential region where diffusion limiting currents for CO oxidation are observed, is rather high (0.68 ML), a rather surprising result (!). The results presented in the fol-

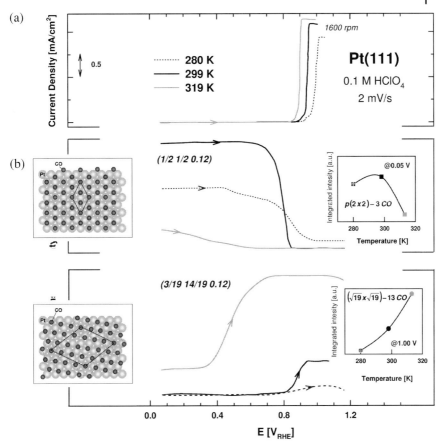

Fig. 1.16 (a) Polarization curves for CO oxidation for Pt(111) in CO-saturated 0.1 M HClO$_4$ solution measured at temperatures of 280 K (dotted line), 299 K (black line) and 319 K (grey line) (sweep rate 2 mV s^{-1}). Only the anodic sweeps are shown. (b) The corresponding XRVs measured at (½, ½, 0.12) and (3/19, 14/19, 0.12). The same color convention is used. The insets show the integrated intensities at the two positions as a function of the temperature, measured at 0.05 V and 1.0 V for the p(2×2)-3CO and ($\sqrt{19}\times\sqrt{19}$)-13CO structures respectively (unpublished results).

lowing section further demonstrate the richness of CO surface electrochemistry in a variety of metal-adsorbate systems.

1.4.3
Underpotential Deposition (UPD)

Metal UPD corresponds to the electrochemical adsorption, often of one monolayer, that occurs at electrode potentials positive with respect to the Nernst potential below which bulk metal adsorption occurs [80]. Numerous experiments have shown that the UPD layer can dramatically alter the chemical and electronic properties of the interface. The UPD layer is also the first stage of bulk metal deposition, and its structure can therefore strongly influence the structure of the bulk deposit. UPD was the first process to be studied using synchrotron X-ray diffraction in the late 1980s [1]. Quite a few systems have now been studied using this technique, and this has led to a greater understanding of the physics determining the structure of the UPD layer, in particular with regard to the role of the electrode potential and of various other adsorbing species that can be present in solution. In fact, considerable work was devoted to the co-adsorbate structures formed during UPD in the presence of halide anions, and results for Pt(hkl) and Au(hkl) surfaces have recently been reviewed [29].

To illustrate the complexity of structural behavior that can be observed in such systems, Fig. 1.17 summarizes X-ray diffraction results obtained during the UPD of Tl onto Au(111) in the presence of bromide anions [81, 82]. The top panel shows the cyclic voltammetry for Au(111) in 0.1 M HClO$_4$ containing 1 mM TlBr along with schematic models of the structures that were observed in

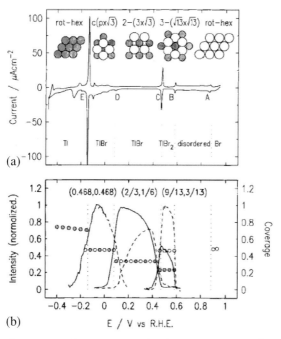

Fig. 1.17 (a) Voltammetry curve for the UPD of Tl on Au(111) in 0.1 M HClO$_4$ containing 1 mM TlBr. Sweep rate 20 mV s^{-1}. The in-plane and surface normal structural models are deduced from the SXS measurements. (b) Potential-dependent diffraction intensities at the indicated positions for the three coadsorbed phases. Scan rate 0.5 mV s^{-1}, except for the c(2.14×3) phase at (0.468, 0.468), where it is 0.2 mV s^{-1}, started after holding the potential around −0.1 V for several hours. Coverages, in units of monolayers of the Au substrate, shown by the open and filled circles for Br and Tl, respectively, are calculated from the adlayer lattice constants (taken from Refs. [81, 82]).

each potential range. The voltammetry shows many peaks that are not observed in the absence of bromide, and these peaks can be attributed to the coadsorption of bromide during Tl UPD. In the potential region on the positive side of peak A, bromide forms the rotated-hexagonal, close-packed adlayer, as observed in solution free of Tl. Similarly in the potential region on the negative side of peak E, thallium forms a rotated-hexagonal, close-packed monolayer that is also observed in solution free of bromide. At intermediate potentials, however, three superlattice structures consisting of mixed TlBr adlayers were observed. The lower part of Fig. 1.17 shows the measured X-ray intensity at the superlattice reciprocal lattice positions corresponding to each structure as a function of the applied potential and also the bromide and thallium coverage. The fact that the $(\sqrt{13} \times \sqrt{13})$ and $(3 \times \sqrt{3})$ structures are commensurate with the Au lattice highlights the importance of the substrate-adlayer bond, although the composition of the adlayers is best understood in terms of surface compound formation. Similar results were also observed for Tl-I, T-Cl and Pb-Br compounds formed on Au(111) [83].

In contrast to the Tl-halide systems, the structures formed during the UPD of Cu onto Au(hkl) and Pt(hkl) in the presence of halide (or sulfate) anions are dictated by the anion structures. For example, on Pt(111), Cu UPD in the presence of bromide is a two-stage process in which the hexagonal anion adlayer remains on the surface, causing an incommensurate CuBr bilayer to be formed at intermediate potentials (a structure that can accommodate a range of Cu coverage) until the Cu monolayer is formed at lower potentials prior to bulk Cu deposition [84]. Similarly, on Pt(100), formation of the Cu UPD monolayer leads to ordering of the bromide adlayer into a c(2×2) structure because of the strength of the Cu-Br bond [85]. The Cu-Br interaction can be considered to be cooperative in nature compared to the Pb-Br interaction, in which the Pb and Br adatoms compete for the Pt surface sites [86]. For example, on the Pt(100) surface this competition causes a negative shift in the electrode potential for the formation of the Pb c(2×2) structure and prevents the completion of the Pb monolayer as Br remains adsorbed on the Pt surface [87].

In determining the electrochemical reactivity of a surface toward a particular chemical reaction, it is clear that the electrode potential is key in determining the relative coverage by anion and/or cation species. For example, the coverage by anion species is an important factor in determining the rate of the oxygen reduction reaction on transition metal surfaces [29]. It is also important, however, to remember that the electrochemical interface is in chemical equilibrium and that adsorption/desorption processes are determined by the energies of adsorption unless kinetic barriers are present. This latter effect was illustrated in recent studies of the adsorption and oxidation of CO on Pt(hkl) surfaces modified by UPD metals. In these studies, RRDE and SXS measurements showed that UPD Cu and Pb are almost completely displaced from the Pt(100) and Pt(111) surfaces in perchloric acid (free of halide anions) by CO [14, 15]. Although these results are somewhat surprising and would not be observed in UHV studies, metal displacement can be understood from a simple thermodynamic analysis by calculating the Gibbs energy change (ΔG) for the component steps of the process [88]. These calculations predict the spontaneous displacement of Cu and Pb by CO from the Pt electrode due to the corresponding exothermicity of the overall ΔG.

For the Pb/Pt(111) system, some insight into the displacement mechanism was obtained by studying the temporal evolution of the Pb-$(3 \times \sqrt{3})$ structure as CO was introduced to the solution. The inset to Fig. 1.18a shows a rocking scan through the (4/6, 1/6, 0.2) position, where scattering from the $(3 \times \sqrt{3})$ structure occurs. The solid line is a Lorentzian fit to the line shape which enables a coherent domain size of ca. 160 Å to be calculated for this structure. The main part of Fig. 1.18a shows the time dependence of the peak intensity after CO was introduced to the solution at $\tau \approx 200$ s. The presence of CO in solution initially caused a large increase in the intensity due to the $(3 \times \sqrt{3})$ phase. This could be due to displacement of Pb that is adsorbed on the Pt surface in defect

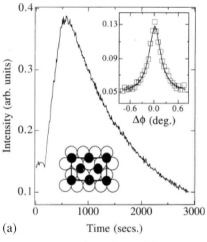 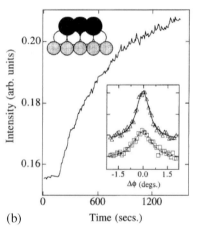

(a) Time (secs.)

(b) Time (secs.)

Fig. 1.18 (a) Time dependence of the X-ray scattering intensity at (4/6, 1/6, 0.1), a Bragg reflection due to the $(3 \times \sqrt{3})$-Pb structure on Pt(111). CO was introduced to the solution at $\tau \approx 200$ sec. Inset is a rocking curve measured at (4/6, 1/6, 0.1) at the beginning of the experiment. The $(3 \times \sqrt{3})$ unit cell is illustrated schematically in the figure, where the open circles are surface Pt atoms and the filled circles are Pb adatoms. The Pb adlayer is uniaxially compressed relative to a hexagonal phase. (b) Time dependence of the X-ray scattering intensity at (1/2, 1/2, 0.1), a Bragg reflection due to a c(2×2) Br adlayer on a p(1×1) Cu monolayer on Pt(100). CO was introduced to the solution

at $\tau \approx 200$ s. Inset are two rocking scans through the (1/2, 1/2, 0.1) position (the data are displaced for clarity); squares represent data measured at the beginning of the experiment and triangles after the intensity level had reached an equilibrium value after introduction of CO. The fitted Lorentzian line shape indicate an increase in domain size from ~ 50 Å to ~ 70 Å and a 25% increase in the integrated intensity. A side view of the structure is shown schematically, where the Cu forms a pseudomorphic monolayer on the Pt surface and the Br atoms are in the c(2×2) symmetry (taken from Ref. [29]).

sites, i.e., sites that are not commensurate with the $(3 \times \sqrt{3})$ phase. The displaced Pb is incorporated into the $(3 \times \sqrt{3})$ structure because of its increased mobility on the Pt surface, and thus the surface is populated by segregated domains of CO and the $(3 \times \sqrt{3})$ Pb structure. This interpretation is supported by two additional X-ray measurements, performed when the intensity in Fig. 1.18a had reached the maximum value. Firstly, a rocking curve through the (4/6, 1.6, 0.2) reflection indicated an increase in integrated intensity by a factor of three and an increase in domain size to ca. 200 Å. Secondly, a rocking curve at (0, 1, 0.5), a position which is sensitive to the Pb coverage, indicated that the surface coverage by Pb had been reduced by ca. $10 \pm 5\%$, i.e., only a small fraction of the Pb monolayer had been displaced by CO. Initially, displacement of Pb by CO causes increased ordering in the $(3 \times \sqrt{3})$ structure. Presumably the Pb atoms initially displaced are "extra" atoms not in the $(3 \times \sqrt{3})$ domains, probably at the domain boundaries. Replacement of some of the Pb atoms by CO (a smaller molecule) appears to allow additional ordering of the remaining atoms. The nature of the Pb-CO interaction is, however, repulsive. Given that there is also a repulsive Pb-Pb interaction in the $(3 \times \sqrt{3})$ adlayer due to the uniaxial compression which decreases the Pb-Pb interatomic spacing [89], it is apparent that the repulsive forces weaken the Pt-Pb bond, which creates a kinetic pathway for the displacement [90]. Pb in the $(3 \times \sqrt{3})$ phase is then displaced from the surface by CO, and the intensity of the $(3 \times \sqrt{3})$ peak decreases until, at $\tau \approx 3000$ s, it has almost disappeared.

The effect of halide anions on the CO-metal displacement process can be dramatic. For the Cu-Br/Pt(100) system, the inset of Fig. 1.18b shows rocking curves through (1/2, 1/2, 0.1), i.e. the strongest c(2×2) reflection, in the case of the open squares for the structure formed in solution free of CO. The Lorentzian line shape (dashed line) indicates a domain size in the region of ca. 50 Å. The intensity was then monitored at this position as CO was introduced to the X-ray cell (similarly to Fig. 1.18a). Surprisingly, CO appears to enhance the c(2×2) structure, and the rocking curve shown by triangular symbols in the inset to Fig. 1.18b was measured when the intensity at (1/2, 1/2, 0.1) had reached a stable level. The fit to this peak gives a domain size of ca. 70 Å and also indicates an approximately 25% increase in integrated intensity. It appears, therefore, that bromide in solution prevents the complete displacement of UPD Cu that was observed in Br-free solution. In addition, coadsorption of CO appears to enhance the ordering in the c(2×2) structure. At this point it is likely that the surface contains segregated domains of CO and the Cu-Br c(2×2) structure. Cycling the potential anodically lifted the Cu-Br structure, and the surface became fully populated by CO. Once this had occurred, the UPD process was completely blocked, and the surface remained covered by CO over the whole potential range during cycling (see [15] for more details). For the Pb-Br system, the Pb c(2×2) structure was formed at 0.0 V before CO was introduced to the X-ray cell. In this case the scattering at (1/2, 1/2, 0.1) disappeared immediately and XRV measurements indicated that CO had completely displaced Pb and Br from the surface as potential cycling caused no change in the measured X-ray intensity. It appears, therefore, that because Pb and Br are competing for the Pt surface adsorption sites, which presumably weakens the bonds between the adsorbing species and the substrate, CO displaces the Pb-Br adlayer, i.e., there is no energetic barrier to displacement, as suggested for the Cu-Br system. Almost identical effects were observed for the Pt(111) surface, i.e., CuBr blocked CO adsorption prior to potential cycling whereas PbBr was completely displaced and the ordered p(2×2)-3CO structure was formed. These results show that the nature of the metal-anion interaction, and hence the strength of the adlayer-Pt bond, is crucial in determining the interaction with solution CO.

The goal of linking surface atomic structure to reactivity is shared by the UHV heterogeneous catalysis community. The above results, however, highlight some clear differences between these two research areas that must be considered if common themes are to emerge. In particular, modification of an electrode surface by a UPD metal monolayer is not always equivalent to the UHV deposition counterpart, as the energetics of adatoms can be very different at the solid/liquid interface. This is clearly illustrated by the phenomenon of surface displacement, which cannot be observed under non-equilibrium UHV conditions.

1.5
Reactive Metals and Oxides

So far in this article we have focused on the surface structures observed on Pt, Pt-based alloys, and Au single-crystal electrodes. A key quality that has emerged from these studies is that the surfaces are stable, provided that the applied potential is restricted to a certain range remote from irreversible reactions such as oxide formation [91]. Experimentally, this is advantageous, in that detailed SXS measurements from a particular system can be obtained in a time scale over which the surface undergoes no structural modification. More reactive metal surfaces present a greater challenge because of the problems associated with the transfer of a prepared crystal into the X-ray electrochemical cell and the stability of the surface in the electrochemical environment. Of course, the structure and growth of passive films on metal electrodes has been an extensively studied subject because of the influence on corrosion properties of materials. In this section, we briefly review SXS measurements of passive film formation on reactive metal surfaces and then finally describe some recent measurements of a material with widespread technological application, namely ruthenium oxide.

The structure of the passive oxide film formed on iron has been the subject of much controversy dating back to the discovery of the phenomenon in the 1700s because of the difficulty in characterizing the thin film in the aqueous environment, and it is only recently that SXS has been able to resolve some of the issues. M. F. Toney and co-workers [92] used SXS to study the passive oxide films formed on single-crystal Fe(001) and Fe(110) substrates at a high anodic formation potential. Fig. 19a and b show the measured diffraction patterns from the passive film on Fe(001) and Fe(110) respectively. For growth on Fe(001), the oxide (001) planes

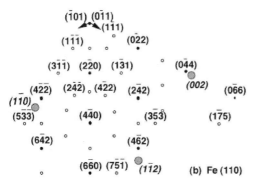

Fig. 1.19 One quadrant of the diffraction pattern for the passive film on (a) Fe(001) and (b) Fe(110). For clarity only some of the out-of-plane peaks are shown. In (a) these are collectively denoted by L. The diffraction peaks from the oxide are represented by small filled and open circles (which are purely in-plane peaks and peaks that are nearly in plane, respectively). Substrate peaks are represented by the large, lightly shaded circles (taken from Ref. [92]).

lie parallel to the Fe surface, while for Fe(110) the oxide (111) planes are parallel to the surface. For both surfaces the oxide [–1 1 0] direction is parallel to Fe[100]. Qualitatively, the diffraction patterns are consistent with those of spinel oxides (γFe$_2$O$_3$, Fe$_3$O$_4$, or related structures). The detailed atomic structure, however, was deduced by modeling of the structure factors of 68 symmetry-inequivalent diffraction peaks, and a comparison of the measured and best-fit structure factors for a subset of these peaks is shown in Fig. 1.20. The calculated structure factors for the Fe_3O_4 and γFe_2O_3 structures are shown by open square symbols and open circle symbols respectively, and the clear disagreement with some of the measured structure factors establishes that they are not present in the passive film. The filled diamonds give the best agreement with the measured data, the so-called LAMM structure that is based on a spinel oxide structure with random cation vacancies and interstitials (see [92] and [93]). Determination of the LAMM passive film structure resolves many of the discrepancies in interpretations of experimental data based on other structural models and highlights the applicability and strength of the *in-situ* SXS technique in the study of oxide films.

The formation of passive oxide films on the (111) surfaces of Cu and Ni has also been studied in detail by SXS [94, 95]. Measurements of Cu(111) in 0.1 M $NaClO_4$ (at pH 4.5) showed that the oxide exhibited a crystalline cuprite structure (Cu_2O) that was epitaxially aligned with the underlying Cu substrate [94]. Although a similar oxide structure was observed for oxidation in air, there were some key differences in the structure of the aqueous oxide. In particular it was found that a preferred "reversed" orientation of the oxide film was formed, and this indicated that oxide growth occurs at the interface between the oxide and the Cu(111) surface

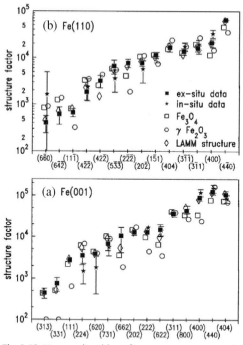

Fig. 1.20 Measured and best-fit structure factors for a subset of the diffraction peaks for the passive film grown on (a) Fe(001) and (b) Fe(110). Measured data are shown by black-filled squares. The best fits for the model structures are open squares for Fe_3O_4, open circles for γFe_2O_3, and filled diamonds for the LAMM phase (taken from Ref. [93]).

rather than at the oxide/electrolyte interface. The passive oxide film formed on an Ni(111) electrode in sulfuric acid electrolyte consisted of a duplex structure with a crystalline inner NiO(111) layer and a porous, amorphous hydroxide phase at the interface between the NiO surface and the electrolyte [96]. Similarly to Cu(111), there were differences between the air-formed and aqueous oxides, although in the case of Ni the air-formed oxide underwent a slow conversion to the aqueous oxide following immersion in electrolyte. The existence of the duplex oxide structure is key in understanding the behavior of the Ni electrode in alkaline solution. In this case a combined SXS/AFM study revealed that cycling the electrode potential over the $Ni(OH)_2/NiOOH$ redox peaks led to an increase in thickness of both the compact NiO layer and the amorphous hydroxide phase [97]. The thickening of the dense NiO phase limits further growth of the hydroxide phase, because of either limited Ni diffusion from the Ni substrate or the reduction in potential gradient across the NiO film.

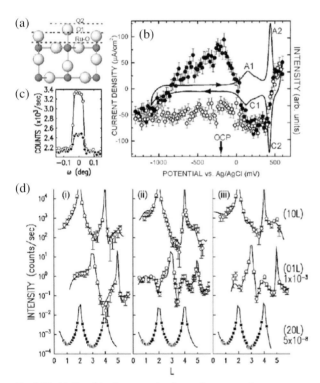

Fig. 1.21 (a) Possible three terminations of RuO_2 structure discussed in Ref. [99]. (b) Cyclic voltammogram (solid line) for the $RuO_2(110)$ surface in 0.1 M NaOH. The arrows indicate the sweep directions. Solid and open circles are the X-ray intensity measured at (0, 1, 4) during cathodic and anodic sweeps respectively. (c) ω scans at (014) (d) (1, 0, L) and (0, 1, L) "oxygen rods" and the (2, 0, L) CTR measured at (i) 330 mV, (ii) 500 mV and (iii) −200 mV. The solid lines are fits to the data according to the structural models shown in Fig. 1.22 (taken from Ref. [99]).

Conducting metal oxides are of widespread technological importance in electrochemistry, and their study by synchrotron SXS has recently been described in a review of the mineral/water interface [98]. Ruthenium dioxide is of particular interest from the catalytic point of view, partly because Ru is an important material in Pt-Ru electrocatalysts for fuel cell applications, but also because RuO_2 is routinely used in industrial electrolysis for chlorine, oxygen, and hydrogen production. H. You and co-workers have carried out a series of SXS experiments on well-defined RuO_2 single-crystal surfaces [99–101], and some of the key results are summarized in Fig. 1.21 for the RuO_2(110) electrode in 0.1 M NaOH electrolyte. Fig. 1.21 b shows the cyclic voltammetry which displays two distinct reversible reactions at electrode potentials of ca. 150 mV and 400 mV. Also shown is the X-ray scattering signal measured at (1, 0, 4) for the cathodic (solid circles) and anodic (open circles) sweeps. These data indicate that structural changes are associated with both of the peaks in the CV, the reaction at ca. 150 mV showing considerable hysteresis. The XRV measurements define three electrode potentials for potentiostatic CTR measurements, and these data are shown in Fig. 1.21 b. A particularly advantageous feature of the CTR data is that, for the RuO_2(110) bulk structure, the scattering contribution from the two Ru sublattices vanishes when H + K is odd, and so for these CTRs the scattering is due only to the oxygen atoms. These so-called "oxygen rods" are thus extremely sensitive to commensurate surface oxygen, i.e. commensurate water,

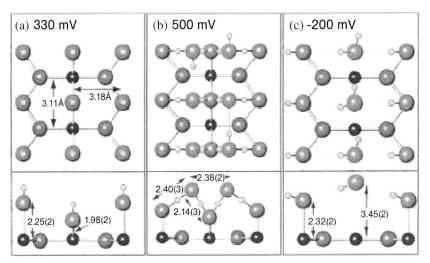

Fig. 1.22 Derived structural models of the RuO_2(110) surface at the potentials corresponding to the data shown in Fig. 1.21 b. The black balls represent the Ru atoms. The grey balls represent oxygen atoms in the bulk, on the surface bonded to Ru, and in the water molecules, respectively. The small light grey balls represent the hydrogen atoms, conjectured to show conceivable hydrogen bonds. Horizontal direction: (010) [(1–10)$_{bulk}$]. Vertical direction: (100) [(001)$_{bulk}$] for upper panels (top views); (001) [(110)$_{bulk}$] for lower panels (side views) (taken from Ref. [99]).

and this enabled a detailed modeling of the water structure at the interface [99]. It should be noted that previous attempts at identifying the water structure at the charged electrochemical interface have been somewhat controversial (as mentioned in Section 1.3.1.2) because of the relative weakness of the scattering from the water layers compared to the underlying metal substrate [44]. Although the (2, 0, L) CTR is relatively unchanged at the three electrode potentials, the "oxygen rods" show dramatic changes. Analysis of the CTR data led to the structural models for the surface terminations shown in Fig. 1.22. These models indicate that, at potentials close to oxygen evolution (500 mV), a layer of water molecules is vertically compressed to the surface oxygen layer by the strong electric field and that the formation of these layers may be a precursor to oxygen evolution. As the potential decreases, the bilayer converts to a simple OH layer bonded to Ru atoms (330 mV) and then to a layer of commensurate water stabilized by an O...H–O hydrogen network anchored to atop OH molecules (–200 mV).

1.6
Conclusions and Future Directions

It is apparent that during approximately the last 15 years a great deal has been learned from the application of SXS techniques to the study of the electrode/solution interface. Particular (but not exclusive) examples include (a) surface reconstruction and relaxation phenomena, (b) understanding the potential dependence of surface structures, (c) correlations that have been established between surface expansion and the energetics of adsorption, the pzc and its relation to structural properties, etc., (d) the extension of the potential range from H_{upd} to H_{opd} to learn about hydrogen absorption effects, (e) the fact that CO adsorption can lead to some surprising effects in electrochemistry, such as the enhancement of surface reconstruction and the displacement of adsorbed metal layers, (f) the *in-situ* monitoring of active surface sites during electrochemical reactions, and (g) studies of bulk metal deposition, corrosion, and conducting metal oxides.

As the emphasis of surface science studies using synchrotron radiation continues to shift away from the UHV characterization of surface structures to explore the relationship between structure and functionality, the electrochemical interface will continue to be at the forefront of such research. We envisage a number of areas that are ripe for investigation.

1. The study of temperature effects (see Section 1.4.2), which has already begun, should continue to explore both the possible existence of temperature-induced phase transitions and surface stability issues in materials of technological importance.

2. There should be increased study of bimetallic surfaces to establish structural trends in surface behavior across the periodic table and the correlation between these trends and electrocatalytic reactivity.

3. Further studies of semiconductor materials relevant to the development of photo-voltaic devices.
4. A better understanding of the electronic properties of selected surface atoms probed via resonant X-ray diffraction techniques.

Finally, we hope that the success of the collaboration between electrochemists and the SXS community can be extended to other interdisciplinary research areas, such as bioelectrochemistry.

Acknowledgments

This work was supported by the Director, Office of Science, Office of Basic Energy Sciences, Division of Materials Sciences, U.S. Department of Energy under Contract No. DE-AC03-76SF00098. CAL would like to acknowledge the support of an EPSRC Advanced Research Fellowship. Original data was measured at SSRL, which is funded by the Division of Chemical Sciences (DCS), US DOE, and the EPSRC-funded XMaS CRG beamline (BM28) at the ESRF, Grenoble. Other data is referenced in the figure captions. We would like to acknowledge the following people for their invaluable contributions to this work; Mathias Arenz, Matthew Ball, Berislav Blizanac, Ben Fowler, Mark Gallagher, Philip Ross, Vojislav Stamenkovic, and Paul Thompson.

References

1 M. G. Samant, M. F. Toney, G. L. Borges, K. F. Blurton, and L. M. O. R. Blum, J. Phys. Chem., 92 (1988) 220.

2 B. M. Ocko, J. Wang, A. Davenport, and H. Isaacs, Phys. Rev. Lett., 65 (1990) 1466–1469.

3 M. F. Toney and B. M. Ocko, Synchrotron Radiation News, 6 (1993) 28–33.

4 I. M. Tidswell, N. M. Markovic, and P. N. Ross, Phys. Rev. Lett., 71 (1993) 1601–1604.

5 C. Lucas, N. M. Markovic, and P. N. Ross, Surf. Sci., 340 (1996) L949–L954.

6 C. Lucas, N. M. Markovic, and P. N. Ross, Phys. Rev. Lett., 77 (1996) 4922–4925.

7 D. M. Kolb, Prog. Surf. Sci., 51 (1996) 109–173.

8 K. Itaya, Prog. Surf. Sci., 58 (1998) 121–247.

9 I. M. Tidswell, N. M. Markovic, C. Lucas, and P. N. Ross, Phys. Rev. B, 47 (1993) 16542.

10 R. Feidenhans'l, Surf. Sci. Rep., 10 (1989) 105–188.

11 P. H. Fuoss and S. Brennan, Annu. Rev. Mater. Sci., 20 (1990) 360.

12 I. K. Robinson and D. J. Tweet, Rep. Prog. Phys., 55 (1992) 599.

13 I. K. Robinson, Phys. Rev. B, 33 (1986) 3830.

14 C. A. Lucas, N. M. Markovic, B. N. Grgur, and P. N. Ross Jr., Surf. Sci., 448 (2000) 65–76.

15 C. A. Lucas, N. M. Markovic, and P. N. Ross Jr., Surf. Sci., 448 (2000) 77–86.

16 J. Clavilier, J. Electroanal. Chem., 107 (1980) 211–216.

17 N. M. Markovic, M. Hanson, G. McDougal, and E. Yeager, J. Electroanal. Chem., 241 (1986) 309.

18 L. A. Kibler, M. Cuesta, M. Kleinert, and D. M. Kolb, J. Electroanal. Chem., 484 (2000) 73–82.

19 M. G. Samant, M. F. Toney, G. L. Borges, L. Blum, and O. R. Merloy, Surf. Sci., 193 (1988) L29.

20 Y. S. Chu, I. K. Robinson, and A. A. Gewirth, Phys. Rev. B, 55 (1997) 7945.

21 O. M. Magnussen, J. Scherer, B. M. Ocko, and R. J. Behm, J. Phys. Chem., 104 (2000) 1222.

22 G. A. Somorjai and M. A. Van Hove, Prog. Surf. Sci., 30 (1989) 201.

23 G. A. Somorjai, Introduction to Surface Chemistry and Catalysis, John Wiley & Sons, New York, 1993.

24 P. A. Thiel and P. J. Estrup, in A. T. Hubbard (Ed.), The Handbook of Surface Imaging and Visualization, CRC Press, Boca Raton, 1995.

25 R. I. Masel, Principles of Adsorption and Reaction on Solid Surfaces, John Wiley & Sons, Inc., 1996.

26 M. A. Van Hove, in Landolt-Börnstein (Ed.), Physics of Covered Solid Surfaces, 1999.

27 J. Greeley, J. K. Norskov, and M. Mavrikakis, Annu. Rev. Phys. Chem., 53 (2002) 319–348.

28 N. M. Markovic and P. S. Ross, Surf. Sci. Rep., 45 (2002) 117–230.

29 C. A. Lucas and N. M. Markovic, in Ernesto J. Calvo (Ed.), The Encyclopedia of Electrochemistry, Vol. 2, Section 4.1.2.1.2, Wiley-VCH, 2004.

30 V. Stamenkovic, M. Arenz, C. Lucas, M. Gallagher, P. N. Ross, and N. M. Markovic, J. Am. Chem. Soc., 125 (2003) 2736–2745.

31 M. E. Gallagher, C. A. Lucas, V. Stamenkovic, N. M. Markovic, and P. N. Ross, Surf. Sci., 544 (2003) L729–L734.

32 N. M. Markovic, C. Lucas, V. Climent, V. Stamenkovic, and P. N. Ross, Surf. Sci., 465 (2000) 103–114.

33 M. Ball, C. Lucas, V. Stamenkovic, P. N. Ross, and N. M. Markovic, Surf. Sci., 518 (2002) 201–209.

34 M. Ball, C. A. Lucas, N. M. Markovic, V. Stamenkovic and P. N. Ross, Surf. Sci., 540 (2003) 295–302.

35 I. M. Tidswell, N. M. Markovic, and P. N. Ross, J. Electroanal. Chem., 376 (1994) 119–126.

36 J. Wang, I. K. Robinson, B. M. Ocko, and R. R. Adzic, J. Phys. Chem., 109 (2005) 24–26.

37 C. A. Lucas, N. M. Markovic, and P. N. Ross, Surf. Sci., 425 (1999) L381–L386.

38 T. E. Shubina and M. T. M. Koper, Electrochim. Acta, 47 (2002) 3621–3628.

39 H. Naohara, S. Ye, and K. Uosaki, J. Phys. Chem. B, 102 (1998) 4366–4373.

40 G. Jerkiewicz, Prog. Surf. Sci., 57 (1998) 137–186.

41 Y. Sung, W. Chrzanowski, A. Wieckowski, A. Zolfaghari, S. Blais, and G. Jerkiewicz, Electrochim. Acta, 44 (1998) 1019.

42 W. Haiss, R. J. Nichols, J. Sass, and K. Charle, J. Electroanal. Chem., 452 (1998) 199.

43 R. J. Nichols, T. Nouar, C. A. Lucas, W. Haiss, and W. A. Hofer, Surf. Sci., 263 (2002) 513.

44 M. F. Toney, J. N. Howard, J. Richer, G. L. Borges, J. G. Gordon, O. R. Merloy, D. Wiesler, D. Yee, and L. B. Sorensen, Nature, 368 (1994) 444.

45 S. Y. Chu (1997) PhD thesis, Department of Physics, University of Illinois.

46 M. E. Gallagher, C. A. Lucas, B. B. Blizanac, P. N. Ross, and N. M. Markovic, Surf. Sci., to be submitted.

47 S. L. Horswell, A. L. Pinheiro, E. Savinova, B. Danckwers, M. S. Zei, and G. Ertl, Langmuir, 20 (2004) 10970.

48 C. I. Carlisle, D. A. King, M.-L. Bockuet, J. Cerda, and P. Sautet, Phys. Rev. Lett., 470 (2000) 15.

49 C. I. Carlisle, D. A. King, M.-L. Bockuet, J. Cerda, and P. Sautet, Phys. Rev. Lett., 84 (2000) 3899.

50 J. Kunze, H. H. Strehblow, and G. Staikov, Electrochem. Comm., 6 (2004) 132.

51 K. M. Ho and K. P. Bohnen, Phys. Rev. Lett., 59 (1987) 1833.

52 J. K. Norskov, Surf. Sci., 299 (1994) 690.

53 J. Wang, J. Devenport, H. Isaacs, and B. M. Ocko, Science, 255 (1992) 1416–1418.

54 I. M. Tidswell, N. M. Markovic, and P. N. Ross, Surf. Sci., 317 (1994) 241–252.

55 B. M. Ocko, B. C. Schardt, and J. Wang, Phys. Rev. Lett., 69 (1992) 3350–3353.

56 B. B. Blizanac, C. Lucas, M. Gallagher, M. Arenz, P. N. Ross, and N. M. Markovic, J. Phys. Chem., 108 (2003) 625–634.

57 O.M. Magnussen, Chem. Rev., 102 (2002) 679–725.

58 G.J. Edens, X. Gao, M.J. Weaver, N.M. Markovic, and P.N. Ross, Surf. Sci. Lett., 302 (1994) L275–L282.

59 M. Gallagher, B.B. Blizanac, C.A. Lucas, P.N. Ross, and N.M. Markovic, Surf. Sci., 582 (2005) 215–226.

60 B.B. Blizanac, C.A. Lucas, M. Gallagher, M. Arenz, P.N. Ross, and N.M. Markovic, J. Phys. Chem., 108 (2004) 625–634.

61 B.M. Ocko, J. Wang, and T. Wandlowski, Phys. Rev. Lett., 79 (1997) 1511.

62 D.W. Suggs and A.J. Bard, J. Phys. Chem., 99 (1995) 8349.

63 C.M. Vitus, S.C. Chang, B.C. Schardt, and M.J. Weaver, J. Phys. Chem., 95 (1991) 7559.

64 K. Sashikata, H. Matsui, K. Itaya, and M.P. Soriaga, J. Chem. Phys., 100 (1996) 20027.

65 B.M. Ocko, O.M. Magnussen, J.X. Wang, and T. Wandlowski, Phys. Rev. B, 53 (2003) 7654–7657.

66 J.M. Gottfried, K.J. Schmidt, S.L.M. Schroeder, and K. Christmann, Surf. Sci., 536 (2003) 206–224.

67 B.B. Blizanac, M. Arenz, P.N. Ross, and N.M. Markovic, J. Am. Chem. Soc., 126 (2004) 10130–10141.

68 T. Wandlowski, J.X. Wang, O.M. Magnussen, and B.M. Ocko, J. Phys. Chem., 100 (1996) 10277–10287.

69 R.J. Nichols, in J. Lipkowski and P.N. Ross (Eds.), Adsorption of Molecules at Metal Electrodes, VCH Inc., New York, 1992, Ch. 7.

70 C. Korzeniewski, in A.Wieckowski (Ed.), Interfacial Electrochemistry – Theory, Experiment and Applications, Marcel Dekker, Inc, New York, Basel, 1999, Ch. 20.

71 N.M. Markovic, B.N. Grgur, C.A. Lucas, and P.N. Ross, J. Phys. Chem. B, 103 (1999) 487–495.

72 N.P. Lebedeva, M. Koper, E. Herrero, J.M. Feliu, and R.A. van Santen, J. Electroanal. Chem., 487 (2000) 37–44.

73 N.P. Lebedeva, A. Rodes, J.M. Feliu, M.T.M. Koper, and R.A. van Santen, J. Phys. Chem. B, 106 (2002) 9863–9872.

74 I. Villegas and M.J. Weaver, J. Chem. Phys., 101 (1994) 1648.

75 K. Yoshima, M. Song, and M. Ito, Surf. Sci., 368 (1996) 389.

76 N.M. Markovic, in W. Vielstich, A. Lamm, and H.A. Gasteiger (Eds.), Handbook of Fuel Cells; Fundamentals, Technology and Application, pp. 368–393, John Wiley & Sons Inc., 2003.

77 N.M. Markovic, C.A. Lucas, A. Rodes, V. Stamenkovic, and P.N. Ross, Surf. Sci., 499 (2002) L149–L158.

78 I. Villegas, X. Gao, and M.J. Weaver, Electrochim. Acta, 40 (1995) 1267–1275.

79 Y.V. Tolmachev, A. Menzel, A. Tkachuk, S.Y. Chu, and H. You, Electrochem. Solid-State Lett., 7 (2004) E23.

80 A.J. Bard and L.R. Faulkner, Electrochemical Methods, Wiley & Sons, New York, 1980.

81 R.R. Adzic and J. Wang, J. Phys. Chem., 102 (1998) 6305.

82 J. Wang, I.K. Robinson, and R.R. Adzic, Surf. Sci., 412/413 (1998) 374.

83 J. Wang, I.K. Robinson, J.E. DeVilbiss, and R.R. Adzic, J. Phys. Chem. B, 104 (2000) 7951.

84 N.M. Markovic, C. Lucas, H.A. Gasteiger, and P.N. Ross Jr., Surf. Sci., 372 (1997) 239–254.

85 C.A. Lucas, N.M. Markovic, B.N. Grgur, and P.N. Ross, Surf. Sci., 448 (1998) 65–76.

86 N.M. Markovic, B.N. Grgur, C.A. Lucas, and P.N. Ross, J. Electroanal. Chem., 448 (1998) 183.

87 N.M. Markovic, B.N. Grgur, C. Lucas, and P.N. Ross, J. Chem. Soc. Faraday Trans., 94 (1998) 3373–3379.

88 N.M. Markovic, B.N. Grgur, C.A. Lucas, and P.N. Ross Jr., Langmuir, 16 (2000) 1998–2005.

89 R.R. Adzic, J. Wang, C.M. Vitus, and B.M. Ocko, Surf. Sci. Lett., 293 (1993) L876–L883.

90 J.M. White and S. Akhter, Crit. Rev. Solid State Mater. Sci., CRC, 1988.

91 Z. Nagy and H. You, Electrochim. Acta, 47 (2005) 3037.

92 M.F. Toney, A. Devenport, M.P. Oblonsky, P.M. Ryan, and C.M. Vitus, Phys. Rev. Lett., 79 (1997) 4282.

93 A. Davenport, M.P. Oblonsky, P.M. Ryan, and M.F. Toney, J. Electrochem. Soc., 147 (2000) 2162.

94 S. Y. Chu, I. K. Robinson, and A. A. Gewirth, J. Phys. Chem., 110 (1999) 5952.

95 J. Scherer, B. M. Ocko, and O. M. Magnussen, Electrochim. Acta, 48 (2003) 1169.

96 O. M. Magnussen, J. Scherer, B. M. Ocko, and R. J. Behm, J. Phys. Chem., 104 (2000) 1222.

97 S. L. Medway, C. A. Lucas, A. Kowal, R. J. Nichols, and D. Johnson, J. Electroanal. Chem., submitted for publication (2005).

98 P. Fenter and N. C. Sturchio, Prog. Surf. Sci., 77 (2004) 171.

99 S. Y. Chu, T. E. Lister, W. G. Cullen, H. You, and Z. Nagy, Phys. Rev. Lett., 86 (2001) 3364.

100 T. E. Lister, S. Y. Chu, W. G. Cullen, H. You, R. M. Yonco, J. F. Michell, and Z. Nagy, J. Electroanal. Chem., 524/525 (2002) 201.

101 T. E. Lister, Y. V. Tolmachev, S. Y. Chu, W. G. Cullen, H. You, R. M. Yonco, and Z. Nagy, J. Electroanal. Chem., 554/555 (2003) 71.

102 B. E. Conway, in S. Davison (Ed.), Prog. Surf. Sci., Vol. 16, Pergamon Press, Fairview Park, NY, 1984.

2
UV-visible Reflectance Spectroscopy of Thin Organic Films at Electrode Surfaces

Takamasa Sagara

2.1
Introduction

In the burgeoning field of nanoscience, functionalization of electrified solid/liquid interfaces is becoming more and more important. Nano-regulated immobilization of organic molecules on well-defined solid electrode surfaces, leading to the emergence of macroscopic functions and signals, is an area of intensive research. Both the technology of the modification of the solid electrode surfaces with thin organic films and their characterization at molecular to macroscopic levels are interdependent and of paramount importance. The characterization should be conducted *in situ*, because we should know the structure of the electrified interface and the behavior of the molecules during their actual functioning. In the spectroelectrochemical methods, spectroscopic techniques are used at the same time as traditional electrochemical measurements in electrolyte solution. Spectroelectrochemistry is a powerful tool for the *in-situ* characterization of the electrified interface.

In 1954, Dr. Shikata, who had been a coworker of Heyrovsky, pointed out that "The polarographic method and the spectral method should be in future combined with each other" [1]. In modern electrochemical laboratories, electrochemists are daily characterizing the electrode/solution interfaces by a number of spectroscopic and scanning probe microscopic methods under potentiostatic and galvanostatic control. Among the *in-situ* spectroelectrochemical methods, reflection of ultraviolet (UV)-visible wavelength region is historically one of the first methods applied to electrochemical interfaces [2]. The electroreflectance effect was first reported for metals in 1967 by Hansen and Prostak [3, 4]. The application of *in-situ* UV-visible reflectance spectroscopy to see the thin organic films at electrode surfaces has been developed for over 30 years together with *in-situ* ellipsometry [5, 6].

In this chapter, we are concerned with UV-visible reflectance spectroscopy for an electrode covered with a thin organic film. The UV-visible reflectance spectroscopy is a simple optical measurement. What one needs is a sensitive UV-

Advances in Electrochemical Science and Engineering Vol. 9.
Edited by Richard C. Alkire, Dieter M. Kolb, Jacek Lipkowski and Philip N. Ross
Copyright © 2006 WILEY-VCH Verlag GmbH & Co. KGaA, Weinheim
ISBN: 3-527-31317-6

visible spectrophotometer and the appropriate optics to observe the reflection from the electrode surface. With the help of a modulation technique, we can achieve very sensitive detection of the change of reflectance of ca. ten parts per million due to the change of the state of the thin organic film. The data acquisition is rather easy. Using the potential sweep and modulation techniques, for example, one can obtain, respectively, a voltammogram of the reflection signal (i.e. a plot of the reflection signal as a function of the electrode potential) and a complex plane plot of the signal, just as in ac impedance spectroscopy. This means that a UV-visible reflection measurement is well suited to follow the dynamics and to clarify the kinetics of the electrode surface processes. On the other hand, the interpretation of the spectrum is sometimes a difficult task. Compared to IR reflection spectroscopy, UV-visible reflection spectroscopy lacks easily obtainable information which gives access to the molecular structural details.

This chapter is devoted to describing the basic aspects of the measurement, instrumentation, measurement techniques, and practical applications of potential-modulated UV-visible spectroscopy as a representative spectroelectrochemical tool to characterize thin organic films on electrode surfaces and to track the kinetics of the electrode surface processes. At the same time, miscellaneous features of the measurement, which may be important for those who intend to apply for the first time the potential-modulated UV-visible spectroscopic method in their experiments, will also be included. However, because of the limit to the chapter length as well as the existence of superior review articles on UV-visible reflectance spectroscopy at electrode/solution interfaces [2, 6–9], detailed comprehensive description is minimized. With the intention of overviewing the UV-visible spectroscopic method for the benefit of experimental electrochemists, optical issues, especially optical reflection theory, are not detailed.

The remainder of this chapter is broken down into Sections 2.2 through 2.16. The description of the general features of the UV-visible reflection spectral measurements (Section 2.2) is followed by the introduction to modulation techniques (Sections 2.3–2.5). Then, with regard to potential-modulated UV-visible reflectance spectroscopy, the instrumentation (Section 2.6), applications to the characterization of the redox processes at the electrodes covered with thin organic films (Sections 2.7 and 2.8) are described. After a comment on the reflection measurement at special electrode configurations (Section 2.9) and the technique to probe the molecular orientation (Section 2.10), the application of potential-modulated UV-visible reflectance spectroscopy to the measurement of electron transfer rate is detailed (Section 2.11). Furthermore, non-faradaic reflectance signal (Section 2.12), non-linear potential-modulated response (Section 2.13), the method to distinguish two simultaneously occurring processes (Section 2.14), and some recent application examples (Section 2.15) are briefly reviewed. To conclude this chapter, future scope together with some examples of very new applications are described in Section 2.16.

2.2
The Basis of UV-visible Reflection Measurement at an Electrode Surface

In UV-visible reflection spectroscopy at a solid electrode/electrolyte solution interface, a light beam is impinged onto the electrode surface at a certain incident angle, and the light reflected back from the electrode surface is detected by a detector positioned at the specular reflection angle (Fig. 2.1 a). The light input to the detector from the electrode surface may involve not only an optical reflection signal but also emitted light, non-linear optical signal, and scattered light including both elastic Rayleigh scattering and non-elastic Raman scattering. However, except for the elastic scattering, the intensity of the other contaminant lights from the irradiated electrode surface is usually far weaker than the reflected light or can be eliminated by the use of a suitable optical filter or grating. The elastic scattering sometimes gives rise to very useful information of the surface state of the electrode substrate or molecular aggregation state on the electrode surface [10], while it is out of our scope in this chapter. It is desirable in the reflection measurement to exclude or minimize the scattering components. To prepare an electrode sample which gives solely the reflection signal, flatness of the electrode surface is an important prerequisite. In order for this condition to be fulfilled, it may be useful to keep the so-called "Rayleigh criterion" in mind [11]. When the level of the surface roughness of the electrode surface is randomly distributed, let us assume that the average height difference between peak and bottom levels is h (Fig. 2.1 b). When the wavelength of the incident light is λ and the incident angle is θ with respect to the surface normal, the maximal phase difference of the light is described as

$$\Delta\varphi = 4\pi h \cos\theta/\lambda.\qquad(1)$$

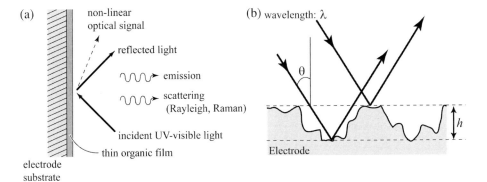

Fig. 2.1 (a) Reflection and other optical processes upon the illumination by UV-visible light of an electrode surface covered with a thin organic film in an electrolyte solution.
(b) Reflection of the light at a rough surface.

The strict Rayleigh criterion tells us that h should be smaller than $20 \cos\theta/\lambda$ to enable us to ignore the surface roughness in reflection measurement [11]. When our target of the reflection measurement is a thin organic film on an electrode surface, we need to use not only a mirror-like flat electrode substrate but also need to prepare a film with two-dimensionally homogeneous thickness (of the order of less than λ) and composition. It is also worthwhile to note that the effect of the surface roughness is a maximum when one uses perpendicular incidence of the light.

The basic optical theory of reflection for a solid surface covered with a thin film has already been well established. We do not intend to review it here in full detail. Instead, readers can refer to the textbook of optics [12] or review articles on UV-visible reflection spectroscopy at electrode surfaces [2, 6–8]. It must be noted that for some electrode/solution interfaces it is still an extremely difficult task to establish the modeling of the interface through the use of classical light reflection theory, as described in the later sections.

2.3
Absolute Reflection Spectrum versus Modulated Reflection Spectrum

It would be ideal in reflection spectral measurement if one could obtain experimentally the absolute reflection spectrum at a given condition within a sufficiently short time period. The absolute reflectance spectral measurement at a constant electrode potential, however, is hampered by several technical problems in many cases. First, the difference between the reflected light intensity of a thin-film-covered electrode and a film-free (bare) electrode is usually so small that we often only detect a difference of a few parts per million of the total reflected light intensity. To gain access to the optical properties of thin organic films, a very sensitive and stable spectrophotometer is necessary. However, the drift of the spectrophotometer itself, such as a time-dependent change in the source light intensity, for example, greatly affects the result. Secondly, a nearly perfect reference sample (viz. a film-free electrode) whose optical properties are already well known is almost unobtainable. Reproducibility of the polishing pretreatment of the electrode surface as well as the positioning of the sample electrode with respect to the optical coordinate can hardly be perfect.

Nevertheless, dc reflectance measurement can be accomplished if one is interested in the relative change of the reflectance but not in the absolute reflectance. Usually, the measured relative reflectance change is defined in a general form:

$$(\Delta R/R)_{\mathrm{dc}} = (R_{\mathrm{m}} - R_0)/R_0 . \tag{2}$$

In one case, R_{m} is the reflectance at a sample potential and R_0 is the reflectance at a reference potential. In another case, R_{m} is the reflectance of a film-covered electrode and R_0 is the reflectance of a bare electrode substrate. The spectral measurement of $(\Delta R/R)_{\mathrm{dc}}$ has been applied for a number of systems including

bare metal electrodes as well as oxide and thin organic films on metal electrodes [2]. In these measurements, one often needs to improve S/N ratio by
using the accumulation average of a number of repeated wavelength scans. An
example of the measurement of $(\Delta R/R)_{dc}$ as a response to a potential step will
be given in Section 2.15.

The modulation technique resolves the main technical difficulties of the constant potential spectral measurement at the electrode surface. Basically, in the
modulation spectroscopic measurement, only the change associated with the
change of the modulated parameter is detected. Thanks to lock-in amplification,
we can significantly increase the sensitivity of the optical signal detection. It
must be noted that what we can obtain is the change of the spectrum with respect to the modulated parameter but not the absolute reflection spectrum. In
other words, modulation methods give the difference or differential spectrum.
From the modulation spectrum, we cannot obtain explicitly the absolute spectrum at a unique condition unless a perfect reference absolute reflection spectrum is already in our hands. One should be careful in the interpretation of the
spectral curves at this point.

The strict prerequisite in the use of the modulation measurement is that the
change of the spectrum of interest with the change of the modulated parameter
should be chemically reversible. It should be also pointed out that the time
scale or frequency of the modulation cannot be set far beyond the time scale or
frequency of the intrinsic cell response. The time scale is usually determined by
the uncompensated solution resistance and the double-layer capacitance of the
electrochemical cell system.

The most frequently used modulation parameter is the electrode potential
(Fig. 2.2). Modulation of wavelength (Fig. 2.3 a) [13–15], polarization state of the
incident light, and photo-excitation state [16] have been also used.

Among modulation measurements, the ac potential modulation method
appears to be one of the superior ones. When using the potential modulation
method for a faradaic process, we can keep the redox process near to equilibrium. Dynamic spectroelectrochemical methods using potential modulation

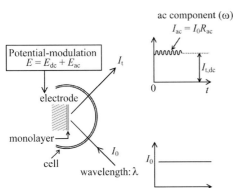

Fig. 2.2 Schematic representation of
potential-modulated reflectance
spectroscopy at an electrode surface
covered with an organic monolayer.

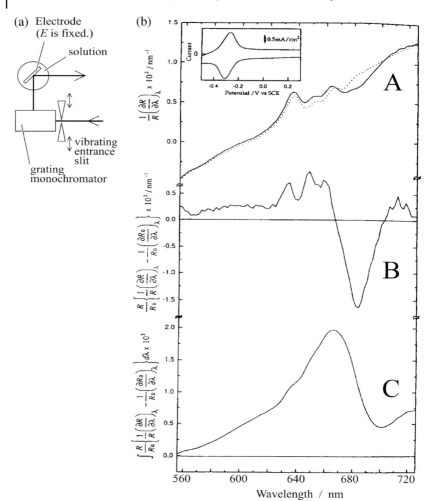

Fig. 2.3 (a) An example of a set-up for wavelength-modulated reflectance spectroscopy using a vibrating entrance slit. (b) Result of the wavelength-modulated UV-visible reflectance spectroscopic measurement at a graphite electrode with adsorbed methylene blue (MB) on its surface. Inset, Cyclic voltammogram of MB monolayer; (A) wavelength modulated spectrum obtained with unpolarized light before adsorption of MB (dotted line) and after adsorption of MB. The vertical axis title reads:

$\dfrac{1}{R}\left(\dfrac{\partial R}{\partial \lambda}\right)_{\lambda} \times 10^2$ nm^{-1}; (B) normalized difference wavelength modulated spectrum obtained from (A). The vertical axis title reads:

$\dfrac{R}{R_0}\left[\dfrac{1}{R}\left(\dfrac{\partial R}{\partial \lambda}\right)_{\lambda} - \dfrac{1}{R_0}\left(\dfrac{\partial R_0}{\partial \lambda}\right)_{\lambda}\right] \times 10^3$ nm^{-1}; (C) integrated spectrum of (B) with respect to λ.

Reproduced with permission from Ref. [15]. Copyright 1992 American Chemical Society.

have been applied from the early years of the development of *in-situ* spectroelec-trochemical techniques [2]. Potential modulation can be used not only in the UV-visible wavelength region but also in IR, microwave, and others. The potential modulation method can find many applications in optical processes such as absorption, scattering, emission, non-linear optical processes, and others.

In polarization modulation, *p*- and *s*-polarized light beams are alternately irradiated onto the electrode surface and the difference spectrum between the two polarizations is measured. This technique is widely used in IR reflection spectroscopy because of the existence of the surface selection rule on a metal substrate [17, 18]. In the UV-visible wavelength region, *s*-polarization light has a non-zero value of the standing wave electric field at the metal electrode surface [2, 19, 20], in contrast to the IR wavelength region. The polarization modulation method in the UV-visible wavelength region may be useful for the analysis of the adsorption orientation of a chromophore on the electrode surface. The method for estimating the chromophore orientation using potential modulation UV-visible measurement with polarized light will be described in Section 2.10.

Excitation modulation is useful for a semiconductor electrode or for the observation of the photo-induced change in the reflectance [16].

After the wavelength modulation technique is briefly described in the next section, the potential modulation method is described in full detail.

2.4
Wavelength-modulated UV-visible Reflectance Spectroscopy

If one vibrates a mirror or the entrance slit position (Fig. 2.3 a) of a grating mono-chromator, the wavelength of the light exiting from the monochromator can be modulated. The lock-in detection of such a modulated reflection signal represents the differential spectrum with respect to λ. Therefore, integration of the obtained wavelength modulation spectrum with respect to λ gives an apparent reflection spectrum. Scherson and co-workers applied this technique for the measurement of the apparent reflectance spectrum of a graphite electrode covered with a methylene blue (MB) adsorption layer (Fig. 2.3 b) [15]. They used a mechanical perturbation to the galvanometer-driven grating/mirror assembly of a rapid-scanning monochromator. The wavelength-modulated spectrum gave the first derivative of the relative reflection spectrum at a constant potential both in the absence and presence of the adsorbed MB molecule. The difference between two states was integrated with respect to λ to obtain the plot of $(\Delta R/R)_{dc}$ as a function of λ. Here, the difference in $(\Delta R/R)_{dc}$ corresponds to the difference between film-covered and film-free electrode surfaces. Fig. 2.3 b shows the representative spectral data. From upper through lower spectra in Fig. 2.3, we can see the wavelength-modulated spectrum for a film-free graphite electrode and the electrode covered with an oxidized form of MB, resulting in the difference spectrum and the difference-integrated spectrum. The last spectrum represents the change of reflection spectrum by the formation of the adsorption film of MB.

2.5
Potential-modulated UV-visible Reflectance Spectroscopy

Potential-modulated UV-visible reflectance spectroscopy is often called electro-reflectance (ER) spectroscopy. The electroreflectance effect originates from the change of reflection in response to the change of surface electronic state of a metal or semiconductor caused by an external potential control. In a narrower sense, the electroreflectance effect means the reflection change associated with the change of surface electron density. But even when the reflection signal originates from a thin organic film on an electrode surface, many electrochemists use the term "electroreflectance", as we do in this chapter.

The waveform of potential modulation is a sine wave in most of the cases. Using the sine wave, direct correspondence of the results of the potential-modulated spectral measurements to those of the ac impedance measurements can be made straightforwardly, as described in later sections.

In the case of sine wave modulation, the electrode potential E is described using the complex representation as

$$E = E_{dc} + E_{ac} = E_{dc} + \Delta E_{ac}^{0-P} \exp(j\omega t) \qquad (3)$$

where E_{dc} is the dc potential, E_{ac} is the ac potential, ΔE_{ac}^{0-P} is the zero-to-peak ac amplitude, $j = \sqrt{-1}$, $\omega = 2\pi f$, which is the angular frequency (f is the frequency of the potential modulation), and t is the time. Under steady-state illumination of monochromatic light onto the electrode surface with an intensity of I_0 (Fig. 2.2), the total reflected light intensity I_t under potential modulation with an angular frequency of ω is written in a general form as

$$I_t = I_{t,dc} + \Delta I_1 \exp[j(\omega t + \phi_1)] + \Delta I_2 \exp[j(2\omega t + \phi_2)] + \Delta I_3 \exp[j(3\omega t + \phi_3)]$$
$$+ \text{higher harmonic terms} \qquad (4)$$

where $I_{t,dc}$ is the dc intensity, ΔI_i ($i = 1, 2, 3, \ldots.$) is the amplitude of the i-th harmonic signal, and ϕ_i is the phase of the i-th harmonic signal with respect to the phase of E_{ac}. In the linear response system, only the first two terms appear. Even though a nonlinear response with the oscillation of 2ω and/or higher frequencies is involved, the lock-in amplification of the fundamental frequency gives rise solely to the second term of the right hand side of Eq. (4). The ac part of the reflectance R_{ac} of the angular frequency of ω (Fig. 2.2) can be then written as

$$R_{ac} = \Delta I_1 \exp[j(\omega t + \phi_1)]/I_0 . \qquad (5)$$

Division of R_{ac} by $E_{ac}/|E_{ac}|$ and further by the time-averaged reflectance, which is equal to $R_{dc} = I_{t,dc}/I_0$, results in an expression:

$$(\Delta R/R)_{ER} = (\Delta I_1/I_0) \exp(j\phi_1)/(I_{t,dc}/I_0) = (\Delta I_1/I_{t,dc}) \exp(j\phi_1) \qquad (6)$$

which is the electroreflectance (ER) signal. Note that we can experimentally obtain both $I_1\exp(j\phi_1)$ and $I_{t,dc}$ directly. The former is exactly the signal from a phase-sensitive lock-in amplifier and the latter is the time-averaged intensity of the reflected right. Normalization of the signal eliminates I_0. Unless either I_1/I_0 or $I_{t,dc}/I_0$ depends on the light intensity, $(\Delta R/R)_{ER}$ is a quantity independent of I_0. It is clear that $(\Delta R/R)_{ER}$ is a complex quantity and thus has both a real part (in-phase component with respect to E_{ac}) and an imaginary part ($90°$ out-of-phase component). They can be written as

$$\mathrm{Re}[(\Delta R/R)_{ER}] = (\Delta I_1/I_{t,dc})\cos\phi_1 \tag{7}$$

$$\mathrm{Im}[(\Delta R/R)_{ER}] = (\Delta I_1/I_{t,dc})\sin\phi_1 \tag{8}$$

When the reflection signal has a phase delay in reference to E_{ac} but the delay is smaller than $\pi/2$, the value of ϕ_1 is in the range $0°>\phi_1>-\pi/2$. In this case, $\mathrm{Im}[(\Delta R/R)_{ER}]$ has a different sign than that of $\mathrm{Re}[(\Delta R/R)_{ER}]$ at a given wavelength.

Ac impedance measurements of an electrochemical interface enable us to follow the kinetics of electrode surface processes in the frequency domain, because any electrochemical event occurring at a non-infinite rate causes a phase shift in the measured ac current signal with respect to the ac potential modulation. This situation is the same when using an ac optical signal instead of an ac current signal, since the electrochemical event producing reflection signal does not necessarily follow the modulation without any delay. In analogy to ac impedance measurement, phase-sensitive detection of the ac reflectance signal enables us to obtain kinetic information of the electrode processes of interest.

The spectrum of an ER signal, i.e. the plot of $(\Delta R/R)_{ER}$ as a function of λ, is called the electroreflectance (ER) spectrum. Electroreflectance measurement can also be carried out during linear potential scan or potential step of E_{dc}. Under linear potential scan, one obtains the voltammogram of the ER signal, the so-called ER voltammogram.

2.6
Instrumentation of the Potential-modulated UV-visible Reflection Measurement

Figure 2.4 depicts the instrumentation and optics for ER measurement in the author's laboratory. The light from a highly stabilized 300 W xenon-halogen lamp is monochromated through a grating monochromator and focused on the electrode surface. The deviation angle of the incident light is less than ca. $2°$. The irradiated area on the electrode surface is approximately 2×2 mm^2. A polarizer may be inserted if necessary. The reflected light from the electrode surface is selectively directed to a photomultiplier (PM) through an iris to exclude the reflected light from the cell wall. The light intensity signal is amplified and converted to a voltage signal. It is then sent simultaneously to a lock-in amplifier for the phase-sensitive detection of $I_{ac}=I_0R_{ac}$ and to an A/D converter equipped on a personal computer

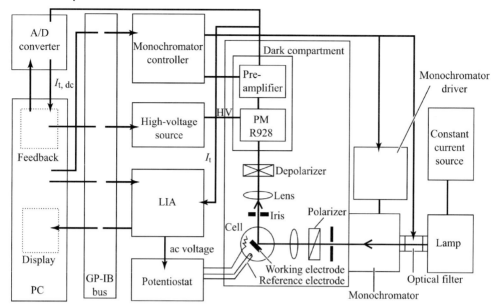

Fig. 2.4 Schematic diagram of the instrumentation for the ER measurements. Notations: LIA, lock-in amplifier; PM, photo-multiplier; PC, personal computer.

for the recording of time-averaged reflectance $I_{t,dc}$. The lock-in amplifier should be able to cut off the commercial frequency noise and white noise. The cell, made of quartz, is put on a cell holder with which one can adjust its height, position, and tilt angle. The position of the PM should be variable to achieve the measurement in the incident angle range from 20 to 70°. To improve the dynamic range of the signal detection, the source high voltage (HV voltage) supplied to the PM is controlled by negative feedback so that the value of the PM response level can be kept almost constant. When the resulting change in the value of $I_{t,dc}$ with λ is large, the linearity of the PM response to the input light intensity should be calibrated. Because of the specific characteristic of the PM response, it is desirable to minimize the HV voltage in order to attain a higher S/N ratio. For this purpose, the use of stronger incident light intensity is always recommended as far as the sample is not photo-damaged.

Other considerations for attaining a higher S/N ratio include the minimization of the physical vibration of the whole instrument, elimination of the stray light coming into the dark compartment, optimization of the connections of the electric ground lines, switching-off of the cooling fan of the light source during the measurement, and minimization of the density of the dust particles hanging in the air in the optical path. In the use of a lock-in amplifier, correct phase adjustment for the phase-sensitive detection is highly important. Usually, the grating in the monochromator may be controlled by a stepper motor. One must be aware that the stepwise change of the wavelength at a constant interval sometimes creates a "ghost" response due to contamination by the Fourier components of the stepper motor drive.

The instrumentation shown in Fig. 2.4 enables us to detect a change in the reflectance of the order of 10^{-5} of the total reflectance at a metal electrode at $f = 14$ Hz with a lock-in amplification time constant of 3 s.

In another configuration of instrumentation, irradiation of white light onto the electrode surface and detection of reflected light by the use of a multi-channel detector (such as CCD or photo-diode array) through a grating monochromator makes multi-channel measurement possible [21].

The intensity of ordinary laboratory light sources, typically a halogen lamp, is not satisfactory to establish the measurement over a wide wavelength region with a high S/N ratio, especially for small organic molecules possessing absorption bands in the UV range of 200–320 nm. To overcome this limitation, Henglein and coworkers adopted synchrotron radiation using an optically thin-layer cell to measure the ER spectra [22, 23]. They were able to extend the wavelength region down to below 180 nm and succeeded in the observation of 257 nm and 197 nm absorptions of pyridine adsorbed on an Au(110) electrode surface [22] and the influence of adsorbed pyridine on the surface reconstruction and the optical properties of Au(100) [23].

2.7
ER Measurements for Redox-active Thin Organic Films

In this section, we are concerned with a mirror-like electrode surface covered with a redox-active thin organic film. Assume that the redox interconversion of the species in the film causes detectable change in the optical properties. In particular, at least one of the compound's oxidation states (both or either of the reduced and oxidized forms) exhibits optical absorption. We first assume that the reflectance at the modified electrode is a linear (first-order) function of the superficial fraction of a chromophore in a given oxidation state. Note that this assumption does not necessarily have a strict rationale in optical theory. We will later return to this point and reconsider it.

The redox active film of interest is assumed to undergo electrochemically reversible interconversion between oxidized and reduced states as a function of electrode potential as described by the Nernst equation. If so, then the plot of reflectance R_{dc} as a function of the electrode potential E can be represented schematically as Fig. 2.5. When $\Delta E_{ac} \ll RT/n_{app}F$, the linear response approximation is valid: A sine wave potential modulation should give rise to a sinusoidal change of the reflectance, as shown in Fig. 2.5, where R is the gas constant, T is the absolute temperature, n_{app} is the apparent number of electrons involved in the redox equilibrium relationship, and F is the Faraday constant.

Now we address a question: is the first assumption that the reflectance is a linear function of the superficial fraction ratio of two oxidation states really applicable? If the answer is "yes", R_{dc} at a given potential E is written as

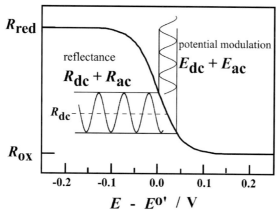

Fig. 2.5 The relationship between the reflectance of an electrode covered with an electroactive thin organic film and the potential modulation. The solid line represents Eq. (9) with a Nernstian equilibrium:

$f_{ox} = \{1+\exp[-n_{app}F\ (E-E^{0\prime})/RT]\}^{-1}$ where f_{ox} is the Nernstian superficial fraction of oxidized form in the film and $E^{0\prime}$ is the formal potential.

$$R_{dc}(E) = R_{ox}f_{ox} + R_{red}(1 - f_{ox}) \tag{9}$$

where f_{ox} is the superficial fraction of oxidized forms among the total amount of adsorbed species of interest, i.e. $f_{ox} = \Gamma_{ox}/\Gamma_t$ (Γ_t is assumed to be independent of E), R_{ox} and R_{red} are, respectively, the reflectance of fully oxidized state and fully reduced states. To discuss whether Eq. (9) is valid or not, we should come back to the optical theory representing the reflection as a function of optical properties of the surface adlayer. Applying the continuum approximation for the adsorption layer, Fresnel's reflection theory can be used in a layered model of the electrode/solution interface to express reflectance [12, 19]. A typical three-phase model is shown in Fig. 2.6. Phases 1, 2, and 3 are, respectively, solution, thin organic film of the electroactive species, and electrode. Using this stratified three-phase layered optical model with optically anisotropic phases 2 and 3, if a thin layer approximation for phase 2 [that is, the thickness of phase 2 (d) is much smaller than the wavelength of the incident light (λ)] can be applied, a quantity $(\Delta R/R)_p$ is described as follows. Here, $(\Delta R/R)_p$ is defined as

$$(\Delta R/R)_p = (R_p - R_0)/R_0 \tag{10}$$

where R_p is the reflectance for the p-polarized light and R_0 is the reflectance of the film-free electrode. The optical theory with the thin layer approximation leads to an expression:

$$(\Delta R/R)_p = (8\pi dn_1 \cos\theta/\lambda)\text{Im}[\{(\hat{\varepsilon}_{2t} - \hat{\varepsilon}_{3t})/(\hat{\varepsilon}_1 - \hat{\varepsilon}_{3t})\} W] \tag{11}$$

incident light (wavelength: λ)

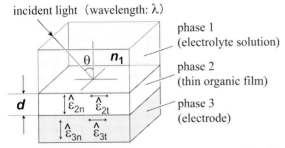

phase 1
(electrolyte solution)

phase 2
(thin organic film)

phase 3
(electrode)

Fig. 2.6 The stratified three-phase layered optical model with optically anisotropic phases 2 and 3, where phase 1 is the electrolyte solution, phase 2 is the thin organic film of thickness d, and phase 3 is the electrode. For the optical constants, see text.

where

$$W = \frac{1 - [\hat{\varepsilon}_1/(\hat{\varepsilon}_{2t} - \hat{\varepsilon}_{3t})][(\hat{\varepsilon}_{2t}/\hat{\varepsilon}_{3n}) - (\hat{\varepsilon}_{3t}/\hat{\varepsilon}_{2n})]\sin^2\theta}{1 - [(\hat{\varepsilon}_1^2 - \hat{\varepsilon}_{3t}\hat{\varepsilon}_{3n})/\{\hat{\varepsilon}_{3n}(\hat{\varepsilon}_1 - \hat{\varepsilon}_{3t})\}]\sin^2\theta} \tag{12}$$

where n_1 is the refractive index of the solution phase (phase-1), θ is the incident angle at the phase-1/phase-2 interface, $\hat{\varepsilon}$ is the complex dielectric constant, the numerical subscripts represent the phase (electrode substrate is phase-3), the subscript "t" means a parallel-to-surface component, and the subscript "n" means a normal-to-surface component.

Taking a close look at these equations, one can immediately recognize that $(\Delta R/R)_p$ is never a linear function of the optical properties of the phase 2. The change in the oxidation state changes the optical constants with the subscript "2". Even if a daring approximation is made, $(\Delta R/R)_p$ cannot appear to be a linear function of any optical constant of the phase-2.

However, if it is *experimentally* verified that R_{dc} is a linear function of f_{ox}, we can accept the apparent validity of Eq. (9).

Using Eq. (9), we can rewrite the reflectance under potential modulation as

$$R_{dc} + R_{ac} = R_{ox}(f_{dc} + f_{ac}) + R_{red}(1 - f_{dc} - f_{ac}). \tag{13}$$

Therefore,

$$R_{ac} = f_{ac}(R_{ox} - R_{red}). \tag{14}$$

Two examples of the experimental verification of Eq. (14) are shown below.

The first one is an adsorption monolayer of MB on a basal plane pyrolytic graphite (BPG) electrode surface [9, 24, 25]. Fig. 2.7a shows the experimental set-up and Fig. 2.7b shows the plot of the change $\Delta R_{dc}(E)$ at constant electrode potentials. The experimental $\Delta R_{dc}(E)$ can actually fit to a Nernst equation:

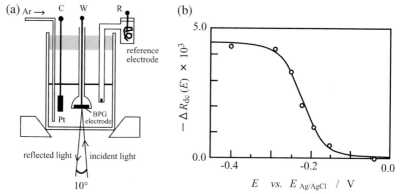

Fig. 2.7 (a) Experimental set-up for a near-normal incidence UV-visible reflectance measurement at constant potential for a basal plane pyrolytic graphite (BPG) electrode. (b) Plot of the change of dc reflectance as a function of electrode potential at $\lambda=737$ nm at a BPG electrode on which an MB monolayer was absorbed from a phosphate buffer solution (30 mM, pH 7.0). The solid line is the least-squares fit to Eq. (15). Reproduced with permission from Ref. [24]. Copyright 1992 American Chemical Society.

$$R_{dc}(E) = R_{ox} + (1 - f_{ox})(R_{red} - R_{ox}) = R_{ox} + \frac{R_{red} - R_{ox}}{1 + \exp\left[\frac{n_{app}F(E - E^{0'})}{RT}\right]} . \tag{15}$$

The values of n_{app} and the formal potential $E^{0'}$ obtained by the least square fitting calculation of the experimental data to Eq. (15) at 737 nm are in good agreement with those obtained from the voltammetric measurement [24].

The second example is a sub-monolayer adsorption film of hemin (iron proto-porphyrin (IX)) (0.7 ML [ML=amount of monolayer adsorption]) on a basal plane of a highly oriented pyrolytic graphite (HOPG) electrode surface [26]. Figure 2.8a shows the results of the simultaneous measurements of current i and $\Delta R_{dc}(E)$ at 440 nm during a cyclic potential sweep at an electrochemically reversible condition. At this condition, the Nernstian redox equilibrium at the electrode/adsorption layer is satisfied regardless of E. The reasons why $\lambda=440$ nm was used include: (a) a negative-going ER band of the real part reaches a maximum at 440 nm (Fig. 2.9) and (b) ER signals due to the Stark effect (see also Section 2.10.2) for both oxidized and reduced forms of hemin give almost identical intensity. A steep rise and subsequent decrease in $\Delta R_{dc}(E)$ synchronizes with the occurrence of the faradaic current peaks (Fig. 2.8a). The linear change in $\Delta R_{dc}(E)$ in the double-layer charging potential regions has been ascribed to the Stark effect on the adsorbed hemin [27]. Since the $\Delta R_{dc}(E)$ vs t curve represents a mirror image with respect to $t=15$ s, the data analysis was restricted to the cathodic scan. The linear portions of the $\Delta R_{dc}(E)$ vs t curve before and after the current peak exhibit an almost identical slope. We use the extrapolated straight line of the linear portion of the $\Delta R_{dc}(E)$ vs t curve at $t<6$ s as the base-

(a)

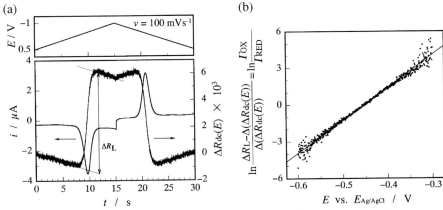

(b)

Fig. 2.8 Result of simultaneous measurement of current and change of reflectance during cyclic potential scan at 100 mV s^{-1} for a sub-monolayer of hemin on an HOPG electrode in 0.1 M Na$_2$B$_4$O$_7$ aqueous solution. In part (a), potential, current, and dc reflectance change (ΔR_{dc}) at a wavelength of 440 nm were plotted against time. Part (b) shows the Nernst plot using the reflectance data. Reproduced with permission from Ref. [26]. Copyright 2005 Society for Applied Spectroscopy.

line. We define ΔR_L as being the gap between the linear portion of the $\Delta R_{dc}(E)$ vs t curve at 12 s$<t<$15 s and the baseline (Fig. 2.8 a). Assuming that $\Delta(\Delta R_{dc}(E))$, which is the value of $\Delta R_{dc}(E)$ with respect to the baseline, is directly proportional to the amount of hemin interconverted from oxidized to reduced form, we can write

$$[\Delta R_L - \Delta(\Delta R_{dc}(E))]/\Delta(\Delta R_{dc}(E)) = \Gamma_{ox}/\Gamma_{red} \,. \tag{16}$$

As shown in Fig. 2.8 b, a plot of $\ln[\{\Delta R_L - \Delta(\Delta R_{dc}(E))\}/\Delta R_{dc}(E))]$ as a function of E yielded a straight line. The least-squares fitting calculation gave a slope of

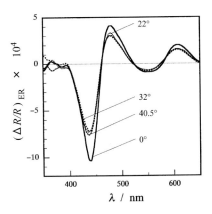

Fig. 2.9 ER spectra (real part) measured with p-polarized incident light for an HOPG electrode on which a submonolayer (0.7 ML) of hemin is adsorbed with four different incident angles (θ). Solution: 0.1 M Na$_2$B$_4$O$_7$. $E_{dc}=-$449 mV vs Ag/AgCl-sat'd KCl. $\Delta E_{ac}=113$ mV. $f=14$ Hz.
Reproduced with permission from Ref. [26]. Copyright 2005 Society for Applied Spectroscopy.

26.91 V^{-1}. At –0.453 V, $\ln[\{\Delta R_L - \Delta(\Delta R_{dc}(E))\}/\Delta(\Delta R_{dc}(E))]$ became 0. We can have a Nernst equation:

$$E = E^{0\prime} + (RT/n_{app}F)\ln(\Gamma_{ox}/\Gamma_{red})$$
$$= E^{0\prime} + (RT/n_{app}F)\ln[\{\Delta R_L - \Delta(\Delta R_{dc}(E))\}/\Delta(\Delta R_{dc}(E))] \qquad (17)$$

where $E^{0\prime}$ is the formal potential for the hemin Fe(III)/Fe(II) couple. From the slope and intercept cited above, $E^{0\prime} = -0.453$ V and $n_{app} = 0.70$ are obtained. These values of $E^{0\prime}$ and n_{app} obtained from the voltammetric reversible response were consistent with those obtained from the $\Delta R_{dc}(E)$ vs t curve. This fact enables us to conclude that the reflectance response is Nernstian and that the change in $\Delta R_{dc}(E)$ is directly proportional to the amount of hemin interconverted between oxidized and reduced forms (Eq. (9)).

2.8
Interpretation of the Reflection Spectrum

A reflection spectrum contains plenty of information about the structure of the interface and the state of the molecules at the interface. However, full understanding and interpretation of the spectrum is often difficult and has remained the subject of debate.

The modulation spectrum usually represents the difference spectrum between two limiting states under the potential modulation when the rate of the surface process can be regarded as being infinite in the linear response system. In the case of potential modulation, the spectrum can be written using Eqs. (11) and (12) for p-polarized light, and a similar representation for s-polarized light can be given [2].

As far as the ER signal obtained by p-polarized light incidence is concerned, we need to divide the amplitude of the ac reflectance by the time average of the reflectance in order for the ER signal to be represented by the use of R_p. Therefore, the ER spectrum, i.e. the spectrum of $(\Delta R/R)_{ER}$, is expressed for a redox interconversion as

$$(\Delta R/R)_{ER} = (R_p^{ox} - R_p^{red})/[0.5(R_p^{ox} + R_p^{red})]. \qquad (18)$$

Using Eq. (10), Eq. (18) is rewritten as

$$(\Delta R/R)_{ER} = [(\Delta R/R)_p^{ox} - (\Delta R/R)_p^{red}]/[1 + 0.5\{(\Delta R/R)_p^{ox} + (\Delta R/R)_p^{red}\}]. \qquad (19)$$

Generally speaking, two limiting cases exist for experimental results for the redox reactions of adsorbed species on electrode surfaces.

In one case, the ER spectrum obtained is in accordance with the difference absorption spectrum. For example, when taking a look at a redox reaction by potential modulation, the ER spectrum is the absorption spectrum of the reduced form from which that of the oxidized form is subtracted. One of the typical systems is cytochrome c_3 adsorbed on either a bare or modified Au electrode (Fig. 2.10) [28]. The ER spectrum is typically almost the same as the difference spectrum of the solution phase species as far as the protein structure remains in its native form. This is also the case for cytochrome c adsorbed on bare or modified electrode surfaces [29–33]. Another example of the typical systems is the self-assembled monolayer of an alkanethiol derivative bearing a redox-active center such as bipyridinium (viologen unit) (Fig. 2.11) [34–36].

In the other limiting case, the ER spectrum obtained is apparently very different from the difference absorption spectrum. The reflection spectrum in general is not only determined by the imaginary part of the complex refractive index but is also governed by the real part of it [2]. A typical case has been shown in Fig. 2.9. A simulation was examined for this typical system [26]. The simulation procedure was as follows: (a) The isotropic extinction coefficient (i.e. the imaginary part of the isotropic complex refractive index) was calculated from the absorption spectrum of a dilute aqueous solution of hemin for both oxidized and reduced forms. (b) The imaginary parts of the anisotropic complex indices were calculated at various presumed orientation angles. (c) The real part of the refractive index was calculated from its imaginary part using the Kramers-Kronig transformation by a numerical integration based on Maclaurin's formula. (d) Using the anisotropic complex refractive index obtained above and the literature value of the optical constant of HOPG, the ER spectral curve for p-polarized light was calculated using Eqs. (11), (12), and (19). Figure 2.12 shows typical simulated ER spectral curves. In comparison to the experimental curves in Fig. 2.9, it is clear that the simulation failed in the reproduction of the experimental spectrum.

A similar simulation of the reflection spectrum has been conducted previously by Kim and coworkers [37], indicating that the reflection spectrum of p-polarized light incidence is far different from the difference absorption spectrum, while that of s-polarized light incidence exhibits some similarity to the difference absorption spectrum.

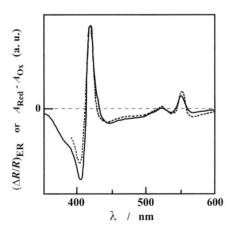

Fig. 2.10 ER spectrum (real part) for cytochrome c_3, a tetra-heme protein, adsorbed on a bare Au electrode surface at the macroscopic formal potential (solid line) and difference absorption spectrum (dotted line), where A_{red} is the absorption spectrum of the reduced form and A_{ox} is that of the oxidized form. The two spectra are normalized at the positive-going Soret absorption band peak for comparison.

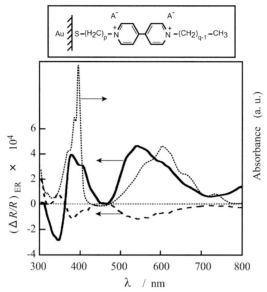

Fig. 2.11 ER spectrum of a viologen thiol ($p=11$, $q=5$, $A^-=PF_6^-$ [see upper scheme]) monolayer on an Au electrode in 0.1 M KPF$_6$ solution. The solid and broken lines represent, respectively, the real and imaginary part ($E_{dc}=-501$ mV vs Ag/AgCl-sat'd KCl. $\Delta E_{ac}=28.3$ mV. $f=14$ Hz. $\theta=45°$ with non-polarized light). The dotted line represents the absorption spectrum of the reduced form of viologen in an aqueous solution in an arbitrary absorbance scale.

Note that, in the monolayer of the viologen thiol at these conditions, the dominant state of the reduced form in the monolayer is the dimer of the viologen radical cation, while the solution spectrum is of the monomer form. The wavelength shift of the ER spectrum in reference to the absorption spectrum is due to the dimer formation. In fact, the ER spectrum matches the dimer absorption spectrum.

In order to interpret the reflectance spectrum, modeling of the interface is the key issue. For example, in the simulation above, we tacitly made some assumptions. One is that the change of the optical properties of the substrate and refractive index of the solution immediately adjacent to the film surface are independent of potential and the presence of the film. The use of the Fresnel model with optical constants is based on the assumption that the phases in the three-strata model are two-dimensionally homogeneous continua. However, if the adsorbed molecule is a globular polymer which possesses a chromophore at its core, a better model of the adsorption layer would be a homogeneously distributed point dipole incorporated in a colorless medium. To gain closer access to the interpretation of the spectrum, a more precise and detailed model would be necessary. But this may increase the number of adjustable parameters and may demand a too complex optical treatment to calculate mathematically. Moreover, one has to pile up approximations, the validity of which cannot easily be confirmed experimentally.

As a conclusion here, when the ER spectrum does not match with the difference absorption spectrum, the precise interpretation of the reflection spectrum

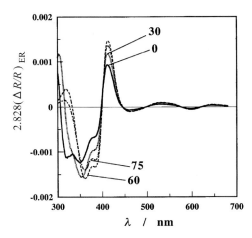

Fig. 2.12 Results of the simulation of ER spectra with *p*-polarized incident light for various orientation angles of the porphyrin plane of hemin with respect to the surface normal when $\theta = 32°$. The numbers in the figure designate the orientation angles. Reproduced with permission from Ref. [26]. Copyright 2005 Society for Applied Spectroscopy.

is not easy. It would be astute to avoid the interpretation problem while extracting other useful information from the ER signal. At the same time, challenges to understand the spectral curve are still of importance to develop this spectroelectrochemical method.

2.9
Reflection Measurement at Special Electrode Configurations

The electrode substrate for the reflection measurement is not necessarily in the form of a flat plate in a quiescent electrolyte solution but sometimes in the form of a sphere or in a hydrodynamic condition. We here consider three typical examples of *in-situ* UV-visible reflectance measurement in special electrode configurations.

The first example is the measurement at a mercury (Hg) electrode. After the invention of polarography, an enormous body of work on the adsorption of organic molecules has been made at a mercury electrode surface. However, mercury is a liquid metal and its hanging drop changes its shape in response to the change of the surface tension, thus to the electrode potential. It is a difficult task to measure the reflection change at a mercury drop electrode surface, since exclusion of the perturbation due to the change of the shape of the electrode is critical.

So far, five types of mercury electrode have been used in specular reflection measurements: (i) hanging mercury drop electrode (HMDE) [38, 39], (ii) mercury film deposited on a platinum or gold substrate [40–42], (iii) mercury pool electrode with or without an amalgamated platinum ring guide [43, 44], (iv) mercury drop bottom electrode placed on the optical window [45–47], and (v) mercury drop bottom electrode placed on an underlying ion-conductive optically transparent polymer film [48]. To avoid the difficulty due to mechanical vibration and shape change, the mercury drop bottom electrode would be useful.

Barker tried to measure the ER signal using an HMDE, though the experimental performance using the HMDE was not satisfactory [38]. Even when the mercury drop bottom electrode placed on the optical window is used for IR reflection measurement by Blackwood and coworkers, the potential control at the bottom surface/electrolyte solution thin layer seems to fail because of too high uncompensated resistance required to control the potential [46, 47].

The use of a mercury drop bottom electrode placed on an underlying Nafion film in ER measurement was demonstrated for the reaction of heptyl viologen incorporated in a Nafion film (Fig. 2.13) [48]. A mercury drop was placed on a Nafion 117 film of thickness 0.175 mm. The transparent nature of the Nafion as well as the high conductivity for cations through its film made it actually possible to measure the ER spectrum of the redox reaction of heptyl viologen with a perpendicular incidence of the light to the mercury electrode surface through the cell bottom window and the Nafion film.

The UV-visible reflectance measurement for a mercury electrode not in contact with polymer but solely with electrolyte solution may be still in demand. The challenge to make the measurement with a mercury drop possible may be quite valuable.

The second example of the special configuration is an electrode at a hanging meniscus (H-M) configuration. Perpendicular incidence of the light to the electrode surface and detection of the reflected light at an H-M configuration can be easily achieved by the use of a half-mirror. Figure 2.14 a shows the experimental set-up for an HOPG electrode at an H-M configuration [49]. At this configuration, the phase transition of heptyl viologen on an HOPG electrode was monitored by the ER method. The amplitude of the potential modulation was large enough to observe the ER signal originating from the redox reaction of a monolayer level amount of heptyl viologen (i.e. monolayer condensation of the reduced form and oxidative transition to gas-like adsorption layer of the oxidized form). In this case, the ER spectrum (Fig. 2.14 b) is far different from the difference absorption spectrum of the viologen (see also the dotted line in Fig. 2.11). The reflection spectrum, therefore, was simulated according to the procedure described in the previous section at $\theta = 0°$. It was assumed that the radical cation (one-electron reduced form of viologen) moiety was lying flat on the electrode surface with an edge-on orientation. The absence of the dimer of the reduced radical cation form is also presumed. The result was added in Fig. 2.14 b using a dotted line. Although the experimental curve did not perfectly match the simulated one, the whole spectral shape could be reproduced by the simulation. Note that this is in sharp contrast to the case of the hemin adsorption layer of an HOPG electrode (see Section 2.8), indicating that modeling is sharply dependent on the system of interest.

The third example is the reflection measurement at a rotating disk electrode (RDE). Scherson and his coworkers have developed near-normal incidence UV-visible reflection-absorption spectroscopy at RDEs [50–52]. Both $(\Delta R/R)_{dc}$ and $(\Delta R/R)_{ER}$ have been measured under hydrodynamic conditions. The use of an RDE enables them to quantitatively control the diffusion layer concentration profile of the solution phase species, especially the species generated electro-

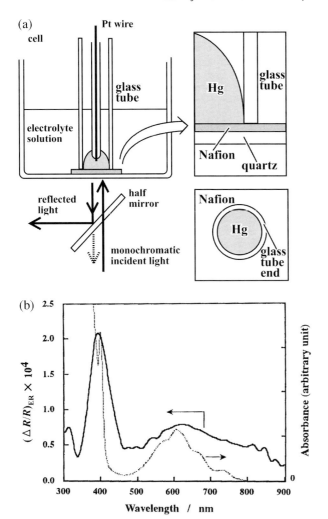

Fig. 2.13 (a) Schematic view of the cell used to measure the normal incidence ER spectrum at the bottom of a mercury drop electrode placed on an underlying Nafion film. (b) Solid line: ER spectrum (real part) for heptyl viologen at a mercury electrode on Nafion film equilibrated with 1 mM heptyl viologen + 1 M KBr ($E_{dc} = -0.5$ V vs Ag/ AgCl-sat'd KCl, $\Delta E_{ac} = 71$ mV, $f = 8$ Hz) at a normal incidence. Dotted line: Absorption spectrum of reduced form of heptyl viologen in an aqueous solution containing excess of sodium hydrosulfite. Reproduced with permission from Ref. [48]. Copyright 1998 Chemical Society of Japan.

chemically. Another benefit is the low uncompensated solution resistance of the system compared to the optically transparent thin-layer electrode cell method in the observation of the electro-generated species.

Two examples of the use of UV-visible reflectance spectroscopy at an RDE are of particular significance:

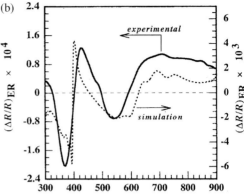

Fig. 2.14 (a) Optics for ER measurement for an HOPG at a hanging-meniscus configuration at a gas/solution interface. (b) Solid line: ER spectrum (real part) for an HOPG electrode in contact with 1.0 mM heptyl viologen + 0.3 M KBr solution at normal incidence with $E_{dc}=-313$ mV vs Ag/ AgCl-sat'd KCl. $\Delta E_{ac}=28.3$ mV. $f=14$ Hz. $\theta=45°$. Broken line: Simulated ER spectrum by the use of the anisotropic optical constant of HOPG. Reproduced with permission from ref. [49]. Copyright 2001 American Chemical Society.

1. The proportionality of the absorbance change due to the electrogeneration of the species with the Levich current and square root of the rotation frequency was confirmed under totally diffusion-controlled conditions [51].
2. Not only the change in absorbance in response to the concentration change of solution phase species, but also the change in the reflectance in response to the electrode surface oxidation could be monitored simultaneously [52].

Note that, although it is out of scope of this chapter, the ER signal of solution phase species partially or totally controlled by a diffusion process at a stationary electrode has been formulated [53–57].

2.10
Estimation of the Molecular Orientation on the Electrode Surface

The light absorption by a molecule is the result of the resonance interaction between the electric field vector of the light and the transition electric dipole of the molecule. This fact tells us that, if the orientation of the molecular electric dipole responsible for the light absorption is aligned and ordered on an electrode surface and if one can control the electric field of the light at the position of the dipole, the optical signal includes information on the molecular orientation. This allows one to estimate the molecular orientation.

So far, two ways of probing the orientation by ER measurement have been proposed and applied. One of them is based on the direct calculation of the interaction between the electric field of the light and the molecular electric dipole.

The other is the use of the Stark effect. The Stark effect arises from the interaction of transition electric dipole with the static electric field at the electrified interface, resulting in a change of the absorption spectrum. Details are given in the following two subsections.

2.10.1
Estimation of the Molecular Orientation on the Electrode Surface using the Redox ER Signal

Let us consider the ER signal due to the redox reaction of a surface-confined dye molecule. For simplicity, we assume that, for an electrode/adsorption layer incorporating a chromophore/solution interface, the absorption of the oxidized form is negligibly small, i.e. the oxidized form is colorless. We also assume that the electric dipole moment of the reduced form is of a single linear dipole and that it has a unique director angle φ with respect to the surface normal while its azimuthal angle is two-dimensionally isotropic (Fig. 2.15 a). The angle φ of the director vector with respect to the surface normal represents the molecular orientation.

Under potential modulation, the two events along with the light path determine the ER signal: absorption by chromophores with an apparent absorbance of A_a and reflection of the light with an apparent reflectance R_s. Then, we have

$$R_t = R_{dc} + R_{ac} = (1 - A_a) R_s . \tag{20}$$

Under the potential modulation of Eq. (3), we write A_a and R_s as

$$A = A_0 + \Delta A \exp[j(\omega t + \phi_A)] \tag{21}$$

$$R_s = R_{s0} + \Delta R_s \exp[j(\omega t + \phi_R)] . \tag{22}$$

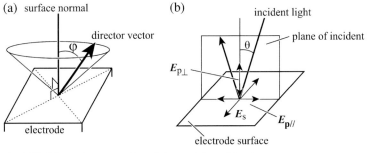

Fig. 2.15 Schematic presentation of the model used for the estimation of molecular orientation. Definition of orientation angle φ (a) and direction of electric field oscillation for p- and s-polarized incident light (b) are shown.

Combining these three equations, and omitting dc parts and high harmonic terms, we have

$$(R_{ac}/R_{dc})_\omega = -\frac{\Delta A}{1 - A_0} \exp[j(\omega t + \phi_A)] + \frac{\Delta R}{R_{s0}} \exp[j(\omega t + \phi_R)]. \tag{23}$$

Now, let us assume that one can experimentally extract the first term of the right hand side of Eq. (23). Since the amplitude of the ER signal now has the form $\Delta A/(1-A_0)$ for both p- and s-polarized light, the intensity ratio of the faradaic ER signal is written as

$$p/s = \frac{1 - A_0^s}{1 - A_0^p} \times \frac{\Delta A^p}{\Delta A^s}. \tag{24}$$

For a monolayer of dye molecules, A_0 may be much smaller than unity. Then, we have an approximated expression of $p/s = \Delta A_p/\Delta A_s$ [34]. The absorption by the chromophore in the thin film is determined by the square of the scalar product of the two vectors: the electric dipole moment of the chromophore and the standing wave electric field of the light at the position of the chromophore. Therefore, ΔA is proportional to the square of the scalar product of reduced forms (remember that the oxidized form is presumed to be colorless).

The p/s ratio of the faradaic ER signal is derived [34] to be

$$p/s = \frac{2E_{p\perp}^2 \cos^2 \varphi + E_{p//}^2 \sin^2 \varphi}{E_s^2 \sin^2 \varphi} \tag{25}$$

where $E_{p\perp}^2$ and $E_{p//}^2$ are the perpendicular-to-surface and parallel-to-surface components, respectively, of the mean square electric field of the surface standing wave of the p-polarized light at the position of the chromophore (Fig. 2.15 b) and E_s^2 is the parallel-to-surface component of the mean square electric field of the surface standing wave of the s-polarized light there. Equation (25) appears to be in the same form as a reported equation describing the dichroic ratio in the linear dichroism theory for an absorption band [58–60].

When the electric dipole moment of the chromophore is circularly in-plane polarized, the ratio p/s is given [34] as

$$p/s = \frac{2E_{p\perp}^2 \sin^2 \varphi + E_{p//}^2 (1 + \cos^2 \varphi)}{E_s^2 (1 + \cos^2 \varphi)} \tag{26}$$

where φ in this case should be the angle value obtained by subtracting from 90° the tilt angle of the chromophore plane with respect to the surface normal. This equation is also of the same form as an equation previously given [60]. The electric field of the standing wave of the light as a function of the distance from the electrode surface can be calculated using the classical optical theory in Fresnel's reflection model [19]. Equations (25) and (26) have been used to obtain the adsorption orientation angle, respectively, of viologen thiol monolayer on an Au electrode [34, 35] and of hemin sub-monolayer on an HOPG electrode [27].

The experimental results of the measurements for viologen monolayers are used as an example. The ER spectra for monolayers of two different viologen thiols, one for which $p=q=4$ and the other for which $p=q=5$ (see molecular structure in Fig. 2.11) on an Au electrode were measured at a various incident angles in 0.3 M KPF$_6$ solution. The spectral curves obtained were almost the same as that in Fig. 2.11 regardless of the incident angle. The obtained p/s ratio was plotted as a function of θ in Fig. 2.16 (circles). The broken lines in Fig. 2.16 were the working curves obtained by the use of Eq. (25) for various presumed φ values. In the lower part of Fig. 2.16, the mean square electric field amplitudes of the light at the position of viologen moiety $\langle E^2 \rangle$ were shown. In these calculations, the optical constants of Au from literature were used, and the refractive

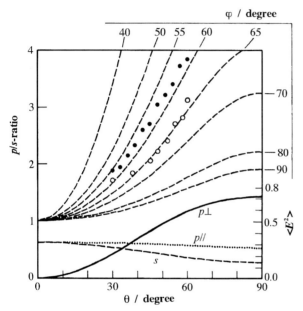

Fig. 2.16 Broken lines in upper part: Working curves of the plots of p/s of ER signal vs incident angle θ for various values of φ based on the three-phase model. (Note that the value of θ here is not the incident angle at the electrode surface but at the interface of an air/optically-flat glass cell window which is parallel to the electrode surface. This applies only to this figure. The true value at the electrode surface can be calculated by the use of refractive indices of the glass and the solution.) In the lower part of the figure, solid, dotted, and dashed lines represent, respectively, $E_{p\perp}^2$, $E_{p//}^2$, and E_s^2 relative to the mean square of the electric field of incident light traveling in the bulk of the solution. The experimental data are shown by circles: closed circle, viologen thiol of $p=q=5$ [see scheme in Fig. 2.11]; open circle, viologen thiol of $p=q=4$. ER measurements were conducted for monolayers of two viologen thiols on an Au electrode in 0.3 M KPF$_6$ solution at $\lambda=603$ nm. Optical constants used in the calculation of the working curves: real and imaginary parts of the dielectric constant of Au were, respectively, -9.41 and 1.19; real and imaginary parts of the refractive index of the monolayer were, respectively, 1.4 and 0.13, $n_1=1.333$, and $d=1.3$ nm.

index of the adsorption layer was calculated by the same procedure in Section 2.8 for both oxidized and reduced states. The experimental points were in good agreement with the working curves so that the line of $\varphi = 58°$ is in accordance with the experimental data for the molecule of $p = q = 5$, and $\varphi = 65°$ for $p = q = 4$.

2.10.2
Estimation of the Molecular Orientation on the Electrode Surface using the Stark Effect ER Signal

The Stark effect is the change of absorption spectrum of a dye molecule due to the interaction between the static electric field and the transition electric dipole. The strength of the static electric field at an electrode/solution interface at a certain condition may exceed 10^6 V m^{-1}, which is strong enough for the Stark effect to be observed. As for the linear Stark effect, a shift in the absorption spectral band is observed and can be expressed as

$$\Delta\varepsilon = hc/\Delta\lambda = k\Delta\mu F_E \cos\alpha \tag{27}$$

where $\Delta\varepsilon$ is the band energy shift, $\Delta\lambda$ is the wavelength shift, k is a constant, and $\Delta\mu$ is the change in the static dipole moments in the ground and excited states of the chromophore ($\Delta\vec{\mu} = \vec{\mu}_{excited} - \vec{\mu}_{ground}$), F_E is the static electric field at the position of the chromophore, and α is the angle between the direction of charge displacement and F_E (Fig. 2.17 a) [27, 61–65]. To mention just one example of the estimation of the orientation of adsorbed molecule using the Stark effect, Schmidt and Plieth analyzed the ER spectrum of p-aminonitrobenzene (p-ANB) on a Pt electrode surface [66]. Fig. 2.17 b represents the ER spectra with p- and s-polarized incident light. The bipolar shape spectral curve for p-polarized

(a)

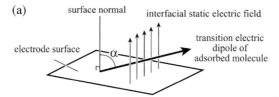

(b)

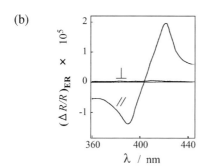

Fig. 2.17 (a) Scheme to explain the linear Stark effect. (b) ER spectrum of a smooth Pt electrode in 10 μM p-ANB + 0.5 M Na$_2$SO$_4$ solution at $E_{dc} = 0.4$ V vs SHE, $\Delta E_{ac} = 200$ mV, $f = 33$ Hz, and $\theta = 65°$. ⊥: s-polarized incident light. //: p-polarized incident light. Part (b) is reprinted from Ref. [66]. Copyright 1986 with permission from Elsevier.

light is typical for the Stark effect. When the absorption band is of a bell-shaped peak, the difference between the wavelength-shifted band by $\Delta\lambda$ and the original band is a bipolar curve. The s-polarized incident light gives a much weaker signal. Since the electric dipole of p-ANB is parallel to the molecular longitudinal axis, the adsorption orientation of the p-ANB molecule should be upright (i.e. the longitudinal axis is vertical with respect to the electrode surface) in light of Eq. (27). Using the experimental linear relationship of $\Delta\varepsilon$ and the electrode potential together with dipole moments of ground and photo-excited states, Schmidt and Plieth calculated the change of static electric field at the position of a p-ANB molecule accompanied by the change of potential as being 8×10^7 V cm^{-1} [66].

2.11
Measurement of Electron Transfer Rate using ER Measurement

As described in Section 2.5, a potential-modulated spectroscopic signal is well suited to track the dynamics of chemically reversible electrode surface processes. The use of the dynamic response of the ER signal to obtain the kinetics of an electrode surface process was first examined by Yeager [67]. The method to obtain the electron transfer rate constant for adsorbed electroactive species k_s using an ER signal has been developed after the simulation and application by Sagara and coworkers [68]. In the following subsections, the ER measurement of electron transfer rate of a thin organic film is described.

2.11.1
Redox ER Signal in Frequency Domain

For kinetic measurements using the ER signal, Eq. (14) is important, although its validity should always be confirmed by experiments. This equation ensures a quantitative relationship between ac faradaic current and the ac reflectance signal, insofar as we can assume that $\Delta E_{ac} \ll RT/n_{app}F$ (linear response approximation). Remember that ac impedance is an operator in the frequency domain, relating potential modulation input to the ac current output. Thus, demonstration of the relationship between potential modulation input (E_{ac}) and ac reflectance signal output (R_{ac}) is very useful.

When a redox reaction is only the origin of the ER signal and the change of reflectance is directly proportional to the amount of the redox couple interconverted between oxidized and reduced state, R_{ac} can be written [9, 56, 69] as

$$R_{ac} = jKi_f/\omega \tag{28}$$

where K is a constant that scales the magnitude of the ER signal with respect to the current signal (thus, K is the function of optical properties of the thin organic film of interest), and i_f is the ac current due to the faradaic process of interest. This equation explicitly indicates that the reflectance change is the integrated quantity of the faradaic ac current and also that R_{ac} has $-90°$ phase difference from i_f.

We now use an equivalent circuit representing an electrode/solution interface where the electrode surface is covered by an electroactive monolayer. The simplest circuit is shown in Fig. 2.18. We assume that the molecules in a Langmuir monolayer undergo an n-electron transfer reaction in response to E_{ac} and that the ER signal is exclusively due to this faradaic process [69]. The faradaic process of the surface-confined species at the formal potential $E^{0\prime}$ is represented by a series connection of a constant capacitance associated with the redox reaction of the adsorbed species C_a and a charge transfer resistance R_{ct}, where C_a is written for a Nernstian process as

$$C_a = nn_{app}F^2\Gamma_t/4RT \tag{29}$$

where n is the number of electrons involved in the redox reaction. The form of R_{ct} depends on the choice of the rate equation. In the equivalent circuit, the uncompensated solution resistance and the double-layer capacitance are designated, respectively, as R_s and C_d. As shown in Fig. 2.18, the series connection of C_a and R_{ct} is assumed to be in parallel with the double-layer capacitance, C_d, which is also represented by a constant capacitance. As detailed in a previous report [69], $(\Delta R/R)_{ER}$ at the formal potential $E^{0\prime}$ is given as

$$Re[(\Delta R/R)_{ER}]/\Delta E_{ac}^{0-p} = -KC_a(1 - \omega^2 R_s R_{ct} C_a C_d)/\xi \tag{30}$$

$$Im[(\Delta R/R)_{ER}]/\Delta E_{ac}^{0-p} = KC_a\omega(R_{ct}C_a + R_sC_a + R_sC_d)/\xi \tag{31}$$

where

$$\xi = (1 - \omega^2 R_s R_{ct} C_a C_d)^2 + \omega^2(R_{ct}C_a + R_sC_a + R_sC_d)^2. \tag{32}$$

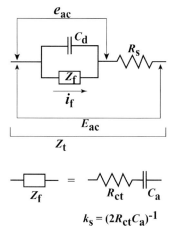

Fig. 2.18 An equivalent circuit representing an electrode/solution interface. The electrode surface is covered by a monolayer of a redox-active species. e_{ac}: ac potential across the faradaic unit of equivalent circuit, C_d: double-layer capacitance, R_s: uncompensated solution resistance, Z_f: impedance representing solely the electron transfer reaction process of the monolayer, i_f: ac current due to the faradaic process, Z_t: total impedance of the whole system, k_s: heterogeneous electron transfer rate constant of the monolayer of electroactive species, R_{ct}: charge transfer resistance, C_a: capacitance associated with the redox reaction of the adsorbed species.

When using a Buter-Volmer type rate equation, R_{ct} at $E^{0'}$ is written as

$$R_{ct} = 2RT/nn_{app}F^2 k_s \Gamma_t \tag{33}$$

and, thus, the rate constant k_s is given by

$$k_s = (2R_{ct}C_a)^{-1} . \tag{34}$$

Note that when the value of R_s can be set to zero, k_s can be obtained directly from an equation we derived previously [68]:

$$k_s = \frac{\omega}{2} \left| \frac{\mathrm{Re}[(\Delta R/R)_{ER}]}{\mathrm{Im}[(\Delta R/R)_{ER}]} \right| . \tag{35}$$

In order for the value of k_s to be determined, it should be possible to discriminate the experimental ω-dependence of $(\Delta R/R)_{ER}$ from that calculated assuming that k_s is infinite. Infinite k_s corresponds to the case of electrochemically reversible reaction in the whole frequency range used for the measurement. The phase of R_{ac}, ϕ, can be written as

$$-\cot\phi = \frac{\mathrm{Re}[(\Delta R/R)_{ER}]}{\mathrm{Im}[(\Delta R/R)_{ER}]} = \frac{1 - \omega^2 R_s R_{ct} C_a C_d}{C_a \omega (R_{ct} C_a + R_s C_a + R_s C_d)} . \tag{36}$$

Therefore, the plot of $-\omega \cot \phi$ against $-\omega^2$ gives a straight line when all the equivalent circuit elements have ω-independent values [69]. The slope and intercept to the $-\omega \cot \phi$ axis are, respectively,

$$\frac{R_s R_{ct} C_a C_d}{C_a (R_{ct} C_a + R_s C_a + R_s C_d)} \quad \text{and} \quad \frac{1}{C_a (R_{ct} C_a + R_s C_a + R_s C_d)} .$$

From these values and Eq. (34), k_s can be determined.

The main advantage of using the spectroelectrochemical signal for kinetic measurement is that the signal is due solely to the oxidation state interconversion and can easily be extracted in many cases. Usually, the simultaneously occurring process, for example the double layer charging, does not produce a detectable ER signal. Even when the ER signal due to the change of the metal surface electron density (viz. electroreflectance effect in the narrow sense) overlaps, it can easily be subtracted graphically from the ER spectrum because it is almost linear to the electrode potential, and its change with potential is not significant within the potential range of the redox reaction. However, it should be emphasized that *the phase of the redox ER signal is strongly affected by the double-layer charging process unless R_s is zero*, as shown by Eqs. (30)–(32). This fact should be correctly taken into account in the data analysis of the frequency dependence of an ER signal.

2.11.2
Examples of Electron Transfer Rate Measurement using ER Signal

Two reported examples of rate measurements by the use of modulation optical signals are described here.

Figure 2.19 shows the results of the ER measurements obtained at a glassy carbon electrode on which hemin is adsorbed [69]. Figure 2.19a and b shows the complex plane plot of an ER signal at $E^{0\prime}$. Figure 2.19c is the plot of $-\omega \cot \phi$ as a function of $-\omega^2$. k_s was found to be 4.9×10^3 s^{-1} using the values of R_s and

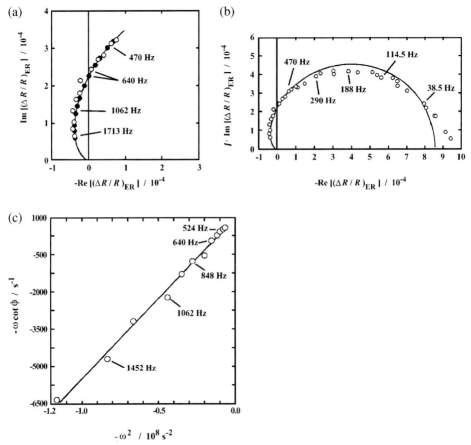

Fig. 2.19 (a) and (b): Complex plane plot of ER signal for an adsorption layer of hemin on a glassy carbon electrode in 0.5 M NaF + 30 mM phosphate buffer solution (pH 6.85) at a wavelength of 433 nm with $E_{dc} = -325$ mV vs Ag/AgCl/sat'd-KCl and $\Delta E_{ac} = 10.25$ mV. Part (a) is the expanded view of the high-frequency region of part (b).

Part (c) shows a plot of $-\omega \cot \phi$ as a function of $-\omega^2$. In parts (a) and (b), open circles are experimental values and closed circles and solid line are simulated by the use of best fit parameters. In part (c), the solid line is the least-squares best fit. Reproduced with permission from Ref. [69]. Copyright 1995 American Chemical Society.

(a)

(b)

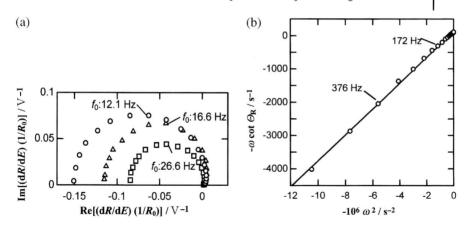

Fig. 2.20 (a) Complex plane plots of ER signals for monolayers of three different manganese porphyrin derivatives on Au electrodes in DMSO. The measurements were all conducted at their formal potentials at a wavelength of 460 nm with $\Delta E_{ac} = 50$ mV(rms). The lengths of alkyl chain linkers between the porphyrin group and the mercapto group were C2 (square), C6 (triangle), and C12 (circle). Part (b) shows the plot of $-\omega \cot \phi$ as a function of $-\omega^2$ for the compound of C12. Reprinted from Ref. [71]. Copyright 1999 with permission from Elsevier.

C_d obtained from the separate ac impedance measurement followed by using Eqs. (34) and (36). This value is approximately three times the reciprocal of the cell time constant (τ_{cell}^{-1}). The cell time constant is expressed as $R_s C_d$, which corresponds to the exponential decay time constant of current upon application of a potential step for an electrochemical cell in the absence of redox reaction. It is an index representing the capability of the cell to respond to the external perturbation of the electrode potential [70]. Generally speaking, when k_s is greater than τ_{cell}^{-1}, the sign of the real part is inverted ($f > 650$ Hz in Fig. 2.19a) in the high frequency region. The upper limit of measurable k_s value will be discussed later.

Yamada and coworkers applied the ER method to determine the electron transfer rate constant for manganese porphyrin self-assembled monolayers on Au electrodes in DMSO solution [71, 72]. Figure 2.20 shows the results of the ER measurements reported [71]. Figure 2.20a is the complex plane plot of the ER signal for three compounds of different alkyl chain lengths. Note that the ER signal here is presented after normalization by ΔE_{ac}, in contrast to Fig. 2.19a. The plot of $-\omega \cot \phi$ as a function of $-\omega^2$ for the compound of C12 is shown in Fig. 2.20b, from which they obtained $k_s = 65$ s^{-1}. They also obtained a rate constant of 56 s^{-1} from the ac impedance measurement. This good agreement provides us with an actual proof of the harmony of the kinetic analysis between ac spectroscopic signal and ac current measurements.

2.11.3
Improvement in Data Analysis

In the method reviewed above, a few drawbacks can be pointed out.

1. In the analysis using the plot of $-\omega \cot \phi$ as a function of $-\omega^2$, only a few points at higher frequencies are highly weighted, especially when k_s is greater than τ_{cell}^{-1}, as shown in Fig. 2.19. The regression error associated with such an analysis using only a few points may be considerable.

2. The double layer charging capacitance was assumed to be a constant. This assumption of a constant capacitor appears to be a rough approximation in some cases. As pointed out by Gaigalas and his colleagues [73–75] and MacDonald [76], the use of a constant phase element, for example, may better represent the electrode interface of interest, although the physical meaning of constant phase element is quite ambiguous.

3. Usually, we measure a set of ac impedance data in a wide frequency range. But only two constant values, R_s and C_d, are used in the analysis. This means that a large part of ac impedance data, which may include a considerable amount of information about the interfacial kinetics, is unused.

These problems are resolved by a refined method proposed by Gaigalas [73–75]. The same proposal was described later by Yamada and Finklea and coworkers [21, 71, 72].

The point of the refinement was to express the ac potential applied to the faradaic impedance unit, e_{ac}, using R_s, ac impedance, and the total ac current at the interface. As detailed in the next subsection, this method can exclude the effect of R_s. In other words, the deviation of the phase and amplitude of e_{ac} from E_{ac} due to the double-layer charging can be correctly considered. This procedure was first proposed by Gaigalas and coworkers in the measurement of the rate constant of adsorbed cytochrome c [74, 75]. After they excluded the effect of R_s, they used Eq. (35) to obtain $k_s \approx 850 \text{ s}^{-1}$ when τ_{cell}^{-1} was ca. 300 s^{-1} [74]. They pointed out the importance of the consideration of the potential distribution at the electrode interface. They also developed potential-modulated fluorescence measurements [77]. Brevnov and Finklea used a similar procedure to that proposed by Gaigalas and experimentally confirmed that the ER signal after the exclusion of the effect of R_s is in line with the faradaic ac admittance data [21]. Note that they obtained a k_s value of 300–500 s^{-1} from the ac admittance data, while τ_{cell}^{-1} estimated by this author according to the description in their paper was greater than 2000 s^{-1}. Yamada also used the same procedure as Gaigalas before using Eq. (35) [71].

The refinements described here are all incorporated into the analysis described in the next section.

2.11.4
Combined Analysis of Impedance and Modulation Spectroscopic Signals

The refined method of analysis of ω-dependent ER signal data set to obtain k_s described in the last part of the previous section is in principle the combined use of ac impedance and the modulated spectroscopic signal. A similar calculation was reported by Yamada and Finklea and their colleagues [21, 71]. We describe below in detail the procedure for a concerted use of these two potential modulation techniques in the kinetic analysis.

We now use an equivalent circuit shown in Fig. 2.21. As opposed to Fig. 2.18, the faradaic impedance Z_f and double-layer non-faradaic impedance Z_d are used without any specified constituent represented by constant resistance and capacitance elements. That is, it is unnecessary to specify each of the elements involved in these two partial impedance units. We assume that we have a data set of Z_t including the frequencies at the high limit from which we can determine the value of R_s (this frequency is usually called the "high frequency limit" in the ac impedance analysis protocol). Then, the ac potential difference across Z_f, e_{ac}, is given by

$$e_{ac} = E_{ac}\left(1 - \frac{R_s}{Z_t}\right) \tag{37}$$

where Z_t is the total impedance of the whole equivalent circuit.

In Eqs. (30)–(32), R_{ac} was expressed in reference to the ac potential modulation of E_{ac}. In the light of Eq. (37), R_{ac} can now be alternatively analyzed with reference to the ac potential difference modulation of e_{ac}. In latter analysis, the effects of R_s and double-layer charging upon R_{ac} have already been excluded. This exclusion is independent of the form of Z_d. Therefore, a complex quantity of R_{ac}/e_{ac} has a phase Φ that is nothing else than the phase of the optical signal in reference to e_{ac}. Thus, if Z_f has a form such that the rate constant is represented by Eq. (34), Φ satisfies the following equation:

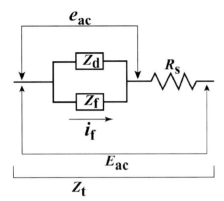

Fig. 2.21 An equivalent circuit representing an electrode/solution interface, where the electrode surface is covered by a monolayer of a redox active species, used to explain the combined analysis of ac impedance and ER signal. Z_d: impedance representing the double-layer charging process.

$$|\cot \Phi| = \frac{k_s}{\pi f} . \tag{38}$$

Taken together, the analysis procedure in the combined use of ac impedance and potential modulation spectroscopic signal can be summarized as follows [78]: (i) Using the high frequency limit of ac impedance, obtain R_s. (ii) Using R_s and Z_t measured at the same set of frequencies as that of optical signal, calculate e_{ac} using Eq. (37). (iii) Divide R_{ac} by e_{ac} and obtain the phase Φ. (iv) Plot $|\cot \Phi|$ as a function of f^{-1}. (v) When a straight line crossing at (0, 0) point is obtained, its slope is equal to k_s/π.

It is clear that the problems in the use of the plot of $-\omega \cot \phi$ as a function of $-\omega^2$ listed in the previous section are well resolved.

Now we show a practical example of the application of above-described method [79]. For a viologen thiol ($p=11$, $q=5$, $A^- = PF_6^-$; see Fig. 2.11) monolayer on an Au electrode, the ER spectral curve was found to be independent of f in the range 1–10 kHz. We used $\Delta E_{ac}^{O-P} = 20$ mV, assuming that this amplitude satisfies the linearity of the response.

Figure 2.22 shows the ac impedance, Z_t, measured with $\Delta E_{ac}^{O-P} = 20$ mV. The ac impedance with this modulation amplitude was in accord with that measured at $\Delta E_{ac}^{O-P} = 5$ mV, rationalizing the linear approximation. From Fig. 2.22, R_s was obtained to be 230 Ω. The plot of the imaginary part of Z_t as a function of the reciprocal of f was linear, with a crossing at the point (0, 0). Using the slope of this plot and R_s, τ_{cell}^{-1} was estimated to be 750 s^{-1}, if the electron transfer resistance can be assumed to be nearly zero (i.e. the reaction is electrochemically reversible).

Figure 2.23 shows the frequency dependence of the ER signal at a wavelength of 572 nm. At lower frequencies, the trajectory of the complex plane plot looks like a semi-circle. However, taking a close look at the higher frequency region, the trajectory deviates from a semi-circle, indicating that the electron transfer

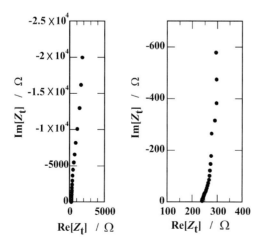

Fig. 2.22 Complex plane plot of ac impedance (Z_t) for a monolayer of viologen thiol ($p=11$, $q=5$, $A^- = PF_6$ in the upper scheme in Fig. 2.11) on a polycrystalline Au electrode in 0.1 M KBr solution with $E_{dc} = -313$ mV vs Ag/AgCl/sat'd-KCl and $\Delta E_{ac} = 20$ mV. The highest frequency was 10 kHz and the lowest was 1 Hz.

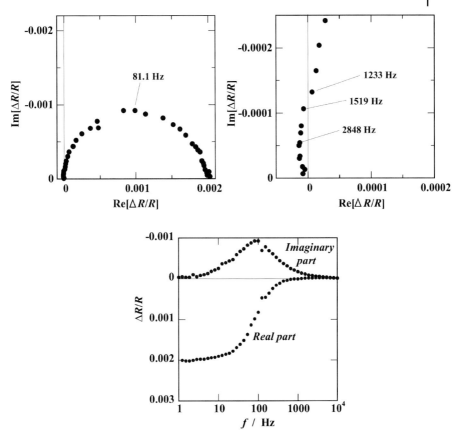

Fig. 2.23 Complex plane plots of ER signals for the monolayer of viologen thiol at a wavelength of 572 nm with $E_{dc}=-313$ mV vs Ag/AgCl/sat'd-KCl and $\Delta E_{ac}=20$ mV. In the lower part, a plot of the ER signal vs f is shown.

reaction is not electrochemically reversible at high frequencies. The frequency at the top of the semi-circle was 81.1 Hz, and was used to estimate τ_{cell}^{-1}, which was 750 s^{-1}, in accordance with that obtained from the ac impedance.

Following the analysis procedure described in the preceding paragraphs, we calculated e_{ac}, and obtained cot Φ as a function of f^{-1}. The result is shown in Fig. 2.24. In the range of $f > 1$ kHz, we obtained a straight line crossing the (0, 0) point. The slope was 1780 s^{-1}, from which we obtained $k_s = 5,590$ s^{-1}. In turn, the reciprocal of the time constant 750 s^{-1} is now firmly ascribed to τ_{cell}^{-1}. The rate constant is approximately one order of magnitude greater than the reciprocal of the cell time constant that was obtained. The plot deviates from a straight line at frequencies lower than 1 kHz. The reason for this deviation is currently unclear.

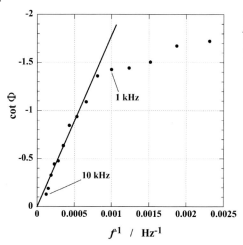

Fig. 2.24 Plot of cot Φ as a function of f^{-1} for the ER signal shown in Fig. 2.23. The solid line is the best fit for $f > 1.2$ kHz.

2.11.5
Upper Limit of Measurable Rate Constant

The upper limit of the rate constant that can be obtained by potential modulation spectroscopy is of interest. It has been frequently stated that a rate constant greater than τ_{cell}^{-1} is not accessible at all. However, τ_{cell}^{-1} is not a strict upper limit. In the potential modulation method, including ac impedance measurements, the upper limit of the modulation frequency *at which one can distinguish faradaic component from non-faradaic components* actually limits the measurable rate constant [80]. As expressed by Eq. (28), the optical signal is proportional to the product of i_f and the reciprocal of ω. Higher frequency measurement is required to impose the kinetics onto the optical signal when a greater value of k_s is to be measured. However, the signal becomes smaller in proportion to the reciprocal of ω at higher frequencies. These oppositing two factors determine the upper measurable limit of the k_s. Roughly speaking, in order for a k_s value 10 times greater to be measured at the same precision, a 10 times higher S/N ratio and sensitivity are required. To obtain k_s of the order of 10^5 s^{-1}, we may need the ER measurement in the frequency range of 2–20 kHz with extremely high sensitivity (probably with slightly higher sensitivity than presently available in practical experiments). Therefore, the measurable k_s value cannot exceed approximately 10 times the value of τ_{cell}^{-1}.

2.11.6
Rate Constant Measurement using an ER Voltammogram

In the previous subsection, the analysis of an ER signal in a frequency domain is restricted to the use of the experimental signal at the formal potential of the redox reaction of interest. If we can take a similar approach using the potential

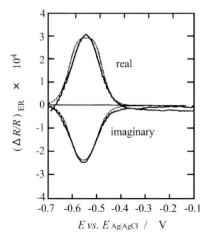

Fig. 2.25 ER voltammogram of cytochrome c_3 adsorbed on an Au(111) electrode in 10 mM phosphate buffer solution, pH 7.0. Solid lines: experimental curves ($\Delta E_{ac} = 84.9$ mV, $f = 8.17$ Hz, $\lambda = 420$ nm). Dotted lines: best fit line by nonlinear least squares calculation. The fitting calculation gives rise to $k_s = 200$ s^{-1} and $\Gamma_t = 1.6 \times 10^{-11}$ mol cm^{-2}. The value of Γ_t is consistent with that obtained from voltammetric measurement.
Reprinted from Ref. [81]. Copyright 1996 with permission from Elsevier.

dependence of the ER signal in combination with the modulation frequency dependence, we may have more detailed information on the kinetics. For example, the transfer coefficient in the rate equation, interaction between the neighboring adsorbed molecules, and the distribution of the adsorption state can be included explicitly in the analysis. For the case of $k_s \ll \tau_{cell}^{-1}$, the simulation of the ER voltammogram and the method of the analysis was first described in detail by Sagara and coworkers [68]. The cases in which the linear response approximation ($\Delta E_{ac} \ll RT/n_{app}F$) does not apply were also treated in detail in the simulation. Feng and coworkers simulated an ER voltammogram taking the effect of R_s into consideration [81]. Using these results, they succeeded in the measurement of the electron transfer rate constant of cytochrome c_3 adsorbed on an Au(111) electrode surface (Fig. 2.25) [81]. Gaigalas and his coworkers also made a kinetic analysis of an ER voltammogram for putidaredoxin adsorbed on an Au electrode [55].

2.12
ER Signal Originated from Non-Faradaic Processes – a Quick Overview

Many of the dynamic processes occurring on the electrode surface take place without the accompanying electron transfer process. Such processes can be represented by adsorption-desorption or change of electric dipole orientation. We need to gain sensitive access to the non-faradaic processes to track the non-faradaic dynamics of molecular assemblies on electrode surfaces. The following non-faradaic processes can be the targets of the ER measurements.
1. Adsorption-desorption processes: This process is the interconversion of molecular species between the soluble and the surface-confined state. This may be detectable when the two states exhibit a difference in light absorption. The soluble state gives rise to an absorption spectral feature, while the adsorbed

state exhibits a reflection spectral feature. These two features are distinguishable in many cases. Otherwise, the adsorption-desorption process can still be detectable when an ER signal from the underlying electrode surface varies with this process or when this process also interconverts the orientation of the molecules between ordered and random. We have actually succeeded in detecting adsorption-desorption on an HOPG electrode for MB [25] and rhodamine dyes [82].

2. Change of orientation of the adsorbed species: As described in Section 2.10.1, ER spectroscopy provides us with an opportunity to estimate the orientation of the molecules confined on an electrode surface using polarized incident light at various incident angles. Detection of the ER signal due to the change of orientation may be possible.

3. Change of the optical constant such as the refractive index of the environment: Based on the optical theory of reflection, the reflection spectrum depends on the refractive index of the medium.

4. Electrochemical Stark effect: As described already in Section 2.10.2, the Stark effect is based on the interaction of the interfacial static electric field with the transition electric dipole or molecular polarizability. The Stark effect may give rise to the first (or sometimes second) derivative of the absorption spectrum, depending on the type of interaction with the electric field. It is important to note that the ER signal due to the Stark effect should have the same frequency dependence as the ac change of the static electric field insofar as orientation change does not take place simultaneously, because the Stark effect is a field effect. In fact, this has been experimentally confirmed by frequency domain analysis [82].

2.13
ER Signal with Harmonics Higher than the Fundamental Modulation Frequency

The basic approximation used in the above discussion of the ER signal is the linearity of response to the potential perturbation (see Section 2.5). This approximation is not always applicable. Especially when the optical signal is very weak, one should use a ΔE_{ac} value much greater than $RT/n_{app}F$ to obtain the signal with a reasonably high S/N ratio. Simulation of the ER signal for such a case was reported, although R_s was presumed to be zero [68].

Formulation of ac current response when $\Delta E_{ac} \gg RT/n_{app}F$ has been reported by Engblom and his colleagues [83, 84]. Analysis of the ER signal in reference to this formulation should be possible. Another feature of the non-linear response is the appearance of the second and higher harmonic signals [see Eq. (4)]. The ER measurement of higher harmonics is not a difficult task. Indeed, simultaneous measurement of the second-harmonic ER voltammogram and ac voltammograms for soluble redox species has been made [85]. The measurement was conducted at a polycrystalline Au electrode in an aqueous solution of methyl viologen. The 90° phase difference relationship was clearly observed in the bipolar-shaped voltammetric curves between the second harmonic ER signal and the second harmonic ac current. In the future, when the kinetic description of the second and higher harmonic ER signals will be established, higher harmonic measurements will also be useful to track the dynamics on the electrode surfaces.

2.14
Distinguishing between Two Simultaneously Occurring Electrode Processes

At the electrified interfaces, multiple redox processes may sometimes take place simultaneously. For example, two redox-active species or states possessing nearly the same formal potentials undergo simultaneous reactions. The spectroscopic signal can allow us to identify each of the species that cause the signal. Therefore, on the basis of a characteristic spectral feature, one can see the two different redox processes separately. If the two exhibit explicit differences in the spectral curve, we can distinguish two simultaneously occurring electrode processes [86]. Another way to distinguish them is to use the difference in the kinetics of the two or more simultaneously occurring processes. This is possible using a dynamic spectroelectrochemical measurement sensitive to the reaction rate. The potential modulation method is well suited, because the kinetic discrimination can be made in the frequency domain as described in the previous section.

Distinction by the spectral characteristics, i.e. wavelength dependence, is well known. When two processes are occurring simultaneously and if the spectrum of one of them is known, deconvolution of the experimental spectral curve enables us to distinguish them. The details of this general method are not given here, but the latter method is described using a practical example of the discrimination of two simultaneously occurring redox processes in the frequency domain.

The phase-shifting technique is easy and useful to separate two electrode surface processes, whether faradaic or non-faradaic. Assume that R_{ac} has two components:

$$R_{ac} = \Delta R_{ac1} \exp[j(\omega t + \phi_{a1})] + \Delta R_{ac2} \exp[j(\omega t + \phi_{a2})]. \tag{39}$$

By making a counterclockwise rotation of the complex plane coordinate by an angle of ϕ_{a2} (Fig. 2.26), the ac reflectance R'_{ac} on the new coordinate plane is written as

$$R'_{ac} = \Delta R_{ac1} \exp[j(\omega t + \phi_{a1} - \phi_{a2})] + \Delta R_{ac2} \exp[j\omega t]. \tag{40}$$

The second term now has only a real component. This means that the imaginary part of R'_{ac} solely contains the spectral information of the process ΔR_{ac1} in the first term in the right hand side of Eq. (40). This method is called the phase shifting technique [87, 88].

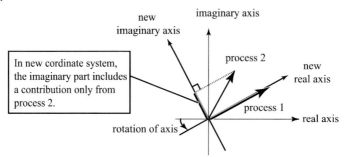

Fig. 2.26 Schematic representation of the principle of the phase-shifting technique by rotation of the complex plane coordinate.

The value of ϕ_{a2} can be found from the ER signal at a given wavelength where ΔR_{ac1} is zero. The phase shift can be done during the experiment. At the wavelength where ΔR_{ac1} is zero, the phase adjustment to make the imaginary part zero is applied to the lock-in amplifier with detection of the signal from the real sample. This procedure is equivalent to the rotation of the complex plane coordinate. At this condition, the imaginary part of the signal at any wavelength corresponds to the spectrum of ΔR_{ac2}.

The phase shift procedure can be applied even after the measurement, called an internal phase shift procedure [88]. Select a wavelength, measure the phase angle, and apply an operator for the rotation to all the data points.

An example of the use of the phase shifting technique is demonstrated in Fig. 2.27 [87]. The redox process of adsorbed cytochrome c_3, whose signal is much weaker than that of the simultaneously occurring solution redox process of methyl viologen, is successfully extracted at an Ag electrode.

The discussion here points to a very important criterion for judging whether an ER signal observed experimentally is of a single component or a sum of multiple components. If a spectral curve shape, regardless of its intensity and sign, is identical for real and imaginary parts at any ω, the ER signal is of a single component. Then, the cross point(s) of real and imaginary parts, if any, should always be on the zero-ER signal line. At metal and graphite electrodes, we sometimes observe the redox ER signal superimposed on the electroreflectance signal in the narrow sense, Stark effect signal, or adsorption-desorption signal [25, 27, 65, 89]. Therefore, the above-mentioned criterion and the method to distinguish multiple processes are highly important. Note that the electroreflectance effect signal in the narrow sense is usually largely weakened when modifying a metal electrode with a self-assembled monolayer of long-chain alkanethiol or a derivative [89].

(a)

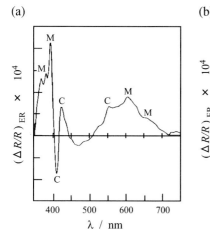

(b)

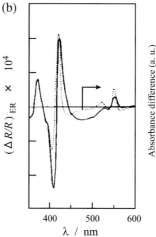

Fig. 2.27 Phase-shifted ER spectra for an Ag electrode with irreversibly adsorbed cytochrome c_3 in 0.5 mM methyl viologen aqueous solution (0.1 M K_2SO_4, pH 7.0) with $E_{dc}=-0.5$ V vs Ag/AgCl/sat'd-KCl, $\Delta E_{ac}=50$ mV, $f=14.2$ Hz, $\theta=45°$ (nonpolarized light). (a) Real part of the phase-shifted spectrum ("M" and "C" designate, respectively, the absorption bands of methyl viologen and cytochrome c_3). (b) Imaginary part of the phase-shifted spectrum (solid line) obtained at the same time as the real part (a). Broken line represents the difference absorption spectrum of cytochrome c_3 in aqueous solution (the spectrum of the oxidized form was subtracted from that of the reduced form). Partly reprinted from Ref. [87]. Copyright 1994 with permission from Elsevier.

2.15
Some Recent Examples of the Application of ER Measurement for a Functional Electrode

The high sensitivity of the ER method benefits bioelectrochemists in the detection of the redox reaction of the electron transfer proteins. Even for an adsorption monolayer of proteins, the superficial density of the electroactive center is much smaller than that of small molecules, especially when the molecular weight of the protein is several kilo-Daltons. The redox reaction of adsorbed protein buried in the double-layer charging current in the voltammogram can be detected by the ER method. Ikeda and coworkers succeeded in the clear observation of the ER spectrum and ER voltammogram of a heme c in an adsorbed protein (alcohol dehydrogenase) of ca. 140 kD containing hemes and PQQ, while direct redox reaction could not be detected by cyclic voltammetry [90].

In the development of the electrode surface modification method, nano-regulation of the surface films with inorganic nanoclusters and molecular-level tracking of the dynamic phase transition are highlighted. For the *in-situ* characterization of such electrode surfaces covered with a monolayer-level organic film or organic molecule-nanocluster hybrid film, UV-visible reflectance spectroscopy can be one of the indispensable spectroelectrochemical research tools. In the following, practical applications of UV-visible reflectance measurement for two typical systems are demonstrated.

Nano-regulated immobilization of metal nanoparticles onto a thin organic film on the electrode surface has attracted attention recently. The research targets include the emergence of novel functions originating from the intrinsic properties of the particles. The clarification of the interfacial structure of such an electrode surface as well as catalytic activities is important. UV-visible reflectance spectroscopy has been used to characterize Au particles on thin organic films [91–93]. Figure 2.28b shows the ER spectrum of an Au electrode covered with an aminoalkane thiol monolayer on which citrate-stabilized Au nanoparticles (diameter 11 nm) were immobilized [92]. After comparison to the plasmon absorption band of the Au colloidal solution (Fig. 2.28a) and in the light of the electro-optical properties of the metal particles, it was concluded that the ER

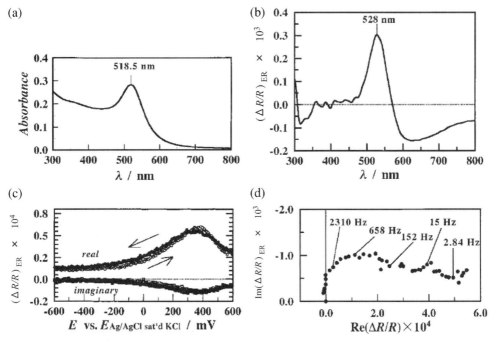

Fig. 2.28 Absorption spectrum of citrate-stabilized Au particles of average diameter 11 nm (a) and the results of ER measurements (b–d) for the Au particles immobilized on a 1-mercapto-11-amino-undecane-modified Au electrode in 0.1 M phosphate buffer (7.0). (a) Absorption spectrum of 24 nM Au solution. (b) ER spectrum (real part) with $E_{dc}=0.1$ V vs Ag/ AgCl/sat'd-KCl, $\Delta E_{ac}=99$ mV, $f=14$ Hz, and $\theta=32°$ (non-polarized light). (c) Cyclic ER voltammogram with $\lambda=530$ nm, $\Delta E_{ac}=99$ mV, $f=14$ Hz, $\theta=32°$ (non-polarized light), and a potential sweep rate of 2 mV s^{-1}. (d) Complex plane plot of ER signal with $E_{dc}=0.3$ V, $\lambda=528$ nm, $\Delta E_{ac}=99$ mV. Reproduced with permission from Ref. [92]. Copyright 2002 American Chemical Society.

spectrum originated from the blue shift of the plasmon band due to electron injection onto the particles at more negative potentials. Fig. 2.28c represents the ER voltammogram, showing that the ER signal exhibits a maximum at a potential near the pzc (potential of zero charge) of bulk Au. Fig. 2.28d represents the modulation frequency dependence of the ER signal. The tailing of the plot to lower frequencies was attributed to the presence of a very slow process giving rise to the delay of the ER signal relative to the ac interfacial potential modulation. The ER spectrum of an Au nanoparticle-immobilized Au electrode in the presence of a dye in close proximity to the particles was also reported [93]. An anomalous ER signal presumably due to the particle-dye electronic interaction was found.

Figure 2.29 shows the results of the measurements of $(\Delta R/R)_{dc}$ in response to two different potential step perturbations for an HOPG electrode at a hanging-meniscus configuration onto an aqueous solution of heptyl viologen [94]. Heptyl viologen shows a faradaic phase transition between a gas-like expanded adsorption layer of the oxidized form and a two-dimensional condensed monolayer of the one-electron reduced form. The latter has absorption bands in the wavelength region of 350–700 nm. Fig. 2.29a shows the results of simultaneous measurement of the current transients (upper part) and $(\Delta R/R)_{dc}$ transients (lower part) in response to a single potential step. The reflectance change at 635 nm was measured at the normal incidence and averaged. The hump of the transient current corresponds to the faradaic condensation process. The time integral of this current was found to be in good agreement with the change of $(\Delta R/R)_{dc}$. This fact reveals that the interfacial fraction of the condensed phase on the electrode surface can be directly monitored by the reflectance signal. The $(\Delta R/R)_{dc}$ transients in Fig. 2.29b and c were obtained as the response to the double potential step. At the first potential step, nucleation of the condensation was initiated. At the second potential step, the growth or diminishment of the condensed phase was tracked, and the finally reached fraction was obtained. For example, the two lines labeled 50 or 100 ms show a decrease of the condensed phase fraction as the decay after the second step (Fig. 2.29b). On the other hand, further growth was clearly recorded at 50 ms in Fig. 2.29c.

These phase changes with time are difficult to monitor by dc current measurements, because isolation from non-faradaic current and measurement with high enough S/N ratio were difficult to attain. In contrast, the reflectance exclusively represents the faradaic process. Difficulty in obtaining the time integration of the current could also be avoided.

(a)

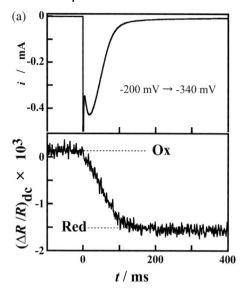

(b)

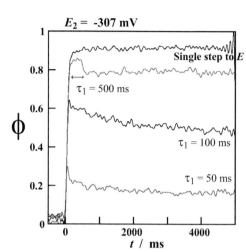

(c)

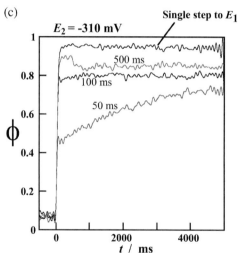

Fig. 2.29 Results of the single or double potential step measurements of current and reflectance signal for an HOPG electrode at a H-M configuration on 1 mM HV^{2+} 2 Br$^-$ +0.3 M KBr. The electrode area was 1.44 cm^2 and the temperature was 23 °C.
(a) Simultaneously measured single potential step transient of current (upper) and reflectance signal at a wavelength of 634.5 nm. Initial potential was –0.20 V vs Ag/AgCl/sat'd-KCl and the first step potential

was $E_1 = -0.34$ V. (b) and (c) Typical transients of ϕ (the fraction of condensed phase in this figure) obtained from the reflectance signal in response to a double potential step at 24 °C with an initial potential of –0.20 V, the first step potential of –0.34 V, and the secondary step potentials at –307 mV (b) and –310 mV (c).
Reprinted from Ref. [94]. Copyright 2004 with permission from Elsevier.

2.16
Scope for Future Development of UV-visible Reflection Measurements

2.16.1
New Techniques in UV-visible Reflection Measurements

In the measurement of surface processes on a single-crystal metal electrode, a single facet is used as the electrode. A single facet on the small single crystal prepared at the end of the metal wire by a melting-cooling-annealing procedure can be a target of voltammetric measurement using a hanging meniscus configuration or a sample for scanning-probe microscopic measurement. Using a laser light source, UV-visible reflection measurement can be carried out for a single facet. For the optical set-up, the use of a half mirror and a microscopic objective lens is helpful. If the contact of the single crystal and electrolyte solution is restricted to the facet of interest, it may behave as a microelectrode, providing us with an opportunity to conduct spectroelectrochemical measurements in a very short time. Scherson and his coworkers have recently demonstrated the measurement of $(\Delta R/R)_{dc}$ at $\lambda = 633$ nm on a single crystal Pt(111) microfacet using a rapid cyclic potential scan in H_2SO_4 and $HClO_4$ aqueous solutions [95]. The time resolution of this optical probe is remarkable, making it possible to monitor adsorption dynamics at the interface not readily accessible by other techniques. The same group has recently reported the time-resolved reflectance spectroscopy of a polycrystalline Pt microelectrode using a laser source and a beam-splitter/microscope objective arrangement [96].

By the use of collimated laser light divided into double beams, one can measure the reflectance at two spots on one single facet, as shown schematically in Fig. 2.30 [97].

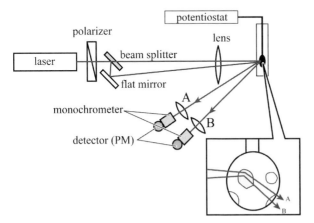

Fig. 2.30 Schematic diagram of a set-up of optics for double-beam UV-visible reflectance measurement at the two spots on a single facet of a small single crystal metal electrode.

The position and distance of the two spots are variable. This approach has merit for the following reasons: One can see whether the process of interest is of mean-field type or nucleation-growth. In the former case, the reflection signals at the two spots as the response to a single potential step should be a identical function of time. Additionally, one can see the two-dimensional uniformity of the kinetics or one can estimate the rate of the surface diffusion process.

Weightman and coworkers have recently applied reflection anisotropy spectroscopy to a solid electrode/solution interface [98–100]. This spectroscopic technique has been used for the characterization of semiconductor or metal surfaces in ultrahigh vacuum since the pioneering work by Aspnes [101, 102]. The application of this method in the UV-visible wavelength region to electrochemical interfaces has recently been demonstrated by Weightman and coworkers. The main merit is the surface sensitivity based on the difference in symmetry between bulk and surface. This method seems to be powerful in the elucidation of surface anisotropy of the optical constant, though the modeling of the interface may be challenging, as in the case of the interpretation of the ER spectrum.

2.16.2
Remarks on the Scope for Future Development of UV-visible Reflection Measurements

The uses of the UV-visible reflectance spectroscopy for the characterization of thin organic films on the electrode surface are reviewed. The potential modulation method is the main focus. Instead of providing a comprehensive description with the background of optics, recent applications are detailed.

In comparison to other spectroelectrochemical techniques, the reflection of UV-visible light is more often negatively commented on, so that: (a) modeling of the electrode/solution interface for the understanding of optical aspects, thus the interpretation of the spectrum, is not easy, (b) it is sometimes difficult to selectively see a surface process in distinction from bulk processes, and (iii) the amount of information obtainable on the details of the molecular structure is much less than in the case of vibration spectroscopy. However, UV-visible reflectance has a number of strong aspects that are not applicable to other methods and thus should always be regarded as of paramount importance in the area of spectroelectrochemistry. We emphasize that high sensitivity and time resolution enable us to follow the dynamics. Local measurements can be made with a small beam spot. Especially in response to recently growing demands for following the dynamics on the functional electrode surface, these merits, together with the ease and adaptability of the optical set-up will become more and more significant. As briefly described in Sections 2.15 and 2.16.1, a wealth of information on the interfacial processes can be made available by novel types of devices. Of course, much more effort at the modeling of the electrode/solution interface for the application of optical theory is required in future, and the UV-visible reflectance technique would appear to be one of the best methods available, for example, in the determination of molecular orientation in a thin organic film and two-dimensional anisotropy in a wide range of types of electrified interface.

Acknowledgments

Thanks are due to Yasuhiko Tanaka for his help in assembling this chapter. This work was partly supported by PRESTO, the Japanese Science and Technology Corporation, and a Grant-in-Aid for Scientific Research from MEXT of the Japanese government.

References

1 M. Shikata, *Rev. Polarogr.*, (1954) 15.
2 J.D.E. McIntyre, in: R.H. Müller (Ed.), *Advances in Electrochemistry and Electrochemical Engineering*, Vol. 9, John Wiley and Sons, Chap. 2 (1973).
3 W.N. Hansen, A. Prostak, *Phys. Rev.*, 160 (1967) 600.
4 W.N. Hansen, A. Prostak, *Phys. Rev.*, 174 (1968) 500.
5 R.H. Müller, in: R. Verma, J.R. Selman (Eds.), *Techniques for Characterization of Electrodes and Electrochemical Processes*, John Wiley & Sons, 1991, p. 31.
6 W. Plieth, W. Kozlowski, T. Twomey, in: J. Lipkowski, P.N. Ross (Eds.), *Adsorption of Molecules at Metal Electrodes*, VCH Publishers, 1992, p. 239.
7 D.M. Kolb, in: R.J. Gale (Ed.), *Spectroelectrochemistry: Theory and Practice*, Plenum Press, New York, 1988, p. 87.
8 W. Plieth, in: C. Gutiérrez, C. Melendres (Eds.), *Spectroscopic and Diffraction Techniques in Interfacial Electrochemistry*, Kluwer Academic Publishers, Netherlands, 1990, p. 223.
9 T. Sagara, *Recent Res. Dev. Phys. Chem.*, 2 (1998) 159.
10 T. Sagara, V. Zamlynny, D. Bizzotto, A. McAlees, R. McCrindle, J. Lipkowski, *Isr. J. Chem.*, 37 (1996) 197.
11 B. Hapke, *Theory of Reflectance, and Emittance Spectroscopy*, Cambridge Univ. Press, New York (1993).
12 M. Born, E. Wolf, *Principles of Optics*, 7th expanded edition, Cambridge University Press, Cambridge (1999).
13 I. Balslev, *Solid State Comm.*, 3 (1965) 213.
14 J.T. Lue, *J. Phys. E.: Sci. Instrum.*, 12 (1979) 833.
15 S. Kim, D.A. Scherson, *Anal. Chem.*, 64 (1992) 3091.
16 S. Nakabayashi, A. Kira, *J. Phys. Chem.*, 95 (1991) 9961.
17 B. Beden, C. Lamy, in: R.J. Gale (Ed.), *Spectroelectrochemistry: Theory and Practice*, Plenum Press, 1988, p. 189.
18 R.J. Nichols, in: J. Lipkowski, P.N. Ross (Eds.), *Adsorption of Molecules at Metal Electrodes*, VCH Publishers, 1992, p. 347.
19 W.N. Hansen, *J. Opt. Soc. Am.*, 58 (1968) 380.
20 B. Pettinger, in: J. Lipkowski, P.N. Ross, (Eds.), *Adsorption of Molecules at Metal Electrodes*, VCH Publishers, 1992, p. 293.
21 D.A. Brevnov, H.O. Finklea, *J. Electrochem. Soc.*, 147 (2000) 3461.
22 F. Henglein, J. Lipkowski, D.M. Kolb, *J. Electroanal. Chem.*, 303 (1991) 245.
23 F. Henglein, D.M. Kolb, L. Stolberg, J. Lipkowski, *Surf. Sci.*, 291 (1993) 325.
24 T. Sagara, J. Iizuka, K. Niki, *Langmuir*, 8 (1992) 1018.
25 T. Sagara, K. Niki, *Langmuir*, 9 (1993) 831.
26 T. Sagara, H. Murase, M. Komatsu, N. Nakashima, *Appl. Spectrosc.*, 54 (2000) 316.
27 T. Sagara, M. Fukuda, N. Nakashima, *J. Phys. Chem. B*, 102 (1998) 521.
28 T. Sagara, *Res. Proj. Rev., Nissan Sci. Found.*, 15 (1992) 159.
29 C. Hinnen, R. Parsons, K. Niki, *J. Electroanal. Chem.*, 147 (1983) 329.
30 C. Hinnen, K. Niki, *J. Electroanal. Chem.*, 264 (1989) 157.
31 T. Sagara, H. Murakami, S. Igarashi, H. Sato, K. Niki, *Langmuir*, 7 (1991) 3190.
32 T. Sagara, H. Sato, K. Niki, *Bunseki Kagaku*, 40 (1991) 641.
33 Z.Q. Feng, S. Imabayashi, T. Kakiuchi, K. Niki, *J. Electroanal. Chem.*, 394 (1995) 149.

34 T. Sagara, N. Kaba, M. Komatsu, M. Uchida, N. Nakashima, *Electrochim. Acta*, 43 (1998) 2183.

35 T. Sagara, H. Maeda, Y. Yuan, N. Nakashima, *Langmuir*, 15 (1999) 3823.

36 T. Sagara, H. Tsuruta, N. Nakashima, *J. Electroanal. Chem.*, 500 (2001) 255.

37 S. Kim, Z. Wang, D. A. Scherson, *J. Phys. Chem. B*, 101 (1997) 2735.

38 G. C. Barker, *J. Electroanal. Chem.*, 39 (1972) 480.

39 M. M. J. Pieterse, M. Sluyters-Rehbach, J. H. Sluyters, *J. Electroanal. Chem.*, 91 (1978) 55.

40 A. Bewick, J. Robinson, *J. Electroanal. Chem.*, 71 (1976) 131.

41 F. Kitamuta, T. Ohsaka, K. Tokuda, *J. Electroanal. Chem.*, 353 (1993) 323.

42 R. O. Lezna, S. A. Centeno, *Langmuir*, 12 (1996) 591.

43 A. Bewick, H. A. Hawkins, A. M. Tuxford, *Surf. Sci.*, 37 (1973) 82.

44 M. W. Humphreys, R. Parsons, *J. Electroanal. Chem.*, 82 (1977) 369.

45 W. McKenna, C. Korzenieski, D. Blackwood, S. Pons, *Electrochim. Acta*, 33 (1988) 1019.

46 D. Blackwood, S. Pons, *J. Electroanal. Chem.*, 247 (1988) 277.

47 D. Blackwood, C. Korzenieski, W. McKenna, J. Li, S. Pons, in: M. P. Soriaga (Ed.), *Electrochemical Surface Science – Molecular Phenomena at Electrode Surfaces*, American Chemical Society, Washington DC (1988), Chap. 23, p. 338.

48 T. Sagara, H. Hiasa, N. Nakashima, *Chem. Lett.*, 1998, 783.

49 T. Sagara, S. Tanaka, Y. Fukuoka, N. Nakashima, *Langmuir*, 17 (2001) 1620.

50 M. Zhao, D. A. Scherson, *J. Electrochem. Soc.*, 140 (1993) 729.

51 P. Shi, D. A. Scherson, *Anal. Chem.*, 76 (2004) 2398.

52 Y. V. Tolmachev, D. A. Scherson, *Electrochim. Acta*, 49 (2004) 1315.

53 A. S. Hinman, J. F. McAleer, S. Pons, *J. Electroanal. Chem.*, 154 (1983) 45.

54 M. Zhao, D. A. Scherson, *J. Electrochem. Soc.*, 140 (1993) 1671.

55 A. K. Gaigalas, V. Seipa, V. Vilker, *J. Colloid Interface Sci.*, 186 (1997) 339.

56 T. Sagara, H. Murase, N. Nakashima, *J. Electroanal. Chem.*, 454 (1998) 75.

57 I. C. Stefan, Y. V. Tolmachov, D. A. Scherson, *Anal. Chem.*, 73 (2001) 527.

58 S. Frey, L. K. Tamm, *Biophys. J.*, 60 (1991) 922.

59 R. P. Sperline, Y. Song, H. Freiser, *Langmuir*, 8 (1992) 2183.

60 R. P. Sperline, Y. Song, H. Freiser, *Langmuir*, 10 (1992) 37.

61 G. Grewer, M. Lösche, *Makromol. Chem., Makromol. Symp.*, 46 (1991) 79.

62 W. J. Plieth, P. Schmidt, P. Keller, *Electrochim. Acta*, 31 (1986) 1001.

63 W. J. Plieth, P. Gruschinske, H.-J. Hensel, *Ber. Bunsenges. Phys. Chem.*, 82 (1978) 615.

64 J. R. Platt, *J. Chem. Phys.*, 34 (1961) 862.

65 T. Sagara, T. Midorikawa, D. A. Shultz, Q. Zhao, *Langmuir*, 14 (1998) 3682.

66 P. H. Schmidt, W. J. Plieth, *J. Electroanal. Chem.*, 201 (1986) 163.

67 R. Adzic, B. Cahan, E. Yeager, *J. Chem. Phys.*, 58 (1973) 1780.

68 T. Sagara, S. Igarashi, H. Sato, K. Niki, *Langmuir*, 7 (1991) 1005.

69 Z.-Q. Feng, T. Sagara, K. Niki, *Anal. Chem.*, 67 (1995) 3564.

70 A. J. Bard, K. R. Faulkner, *Electrochemical Methods, Fundamentals and Application*, John Wiley & Sons, New York, 2001, p. 15.

71 T. Yamada, M. Nango, T. Ohtsuka, *J. Electroanal. Chem.*, 528 (2002) 93

72 T. Yamada, T. Hashimoto, S. Kikushima, T. Ohtsuka, M. Nango, *Langmuir*, 17 (2001) 4634.

73 L. Li, C. Meuse, V. Silin, A. K. Gaigalas, *Langmuir*, 16 (2000) 4672.

74 A. K. Gaigalas, T. Ruzgas, *J. Electroanal. Chem.*, 465 (1999) 96.

75 T. Ruzgas, L. Wong, A. K. Gaigalas, V. Vilker, *Langmuir*, 14 (1998) 7298.

76 J. Ross MacDonald, W. R. Kenan, *Impedance Spectroscopy: Emphasizing Solid Materials and Systems*, John Wiley & Sons, New York, 1987.

77 A. K. Gaigalas, L. Li, T. Ruzgas, *Curr. Top. Colloid Interface Sci.*, 3 (1999) 83.

78 T. Sagara, *Rev. Polarogr.*, 48 (2003) 153.

79 T. Sagara, H. Tsuruta, unpublished results.

80 E. Laviron, In: A. J. Bard (Ed.), Electroanalytical Chemistry, Vol. 12, Marcel Dekker, New York, 1982, p. 53 (especially, the statements on p. 101).

81 Z. Q. Feng, S. Imabayashi, T. Kakiuchi, K. Niki, *J. Electroanal. Chem.*, 408 (1996) 15.

82 T. Sagara, H. Murase, unpublished result.

83 S. O. Engblom, J. C. Myland, K. B. Oldham, *J. Electroanal. Chem.*, 480 (2000) 120.

84 D. Lelievre, S. Saur, E. Laviron, *J. Electroanal. Chem.*, 117 (1981) 17.

85 T. Sagara, T. Takeuchi, unpublished result.

86 T. Sagara, S. Takagi, K. Niki, *J. Electroanal. Chem.*, 349 (1993) 159.

87 T. Sagara, H. X. Wang, K. Niki, *J. Electroanal. Chem.*, 364 (1994) 285.

88 T. Sagara, H. Kawamura, K. Ezoe, N. Nakashima, *J. Electroanal. Chem.*, 445 (1998) 171.

89 T. Sagara, H. Kawamura, N. Nakashima, *Langmuir*, 12 (1996) 4253.

90 T. Ikeda, D. Kobayashi, F. Matsushita, T. Sagara, and K. Niki, *J. Electroanal. Chem.*, 361 (1993) 221.

91 D. Bethell, M. Brust, D. J. Schiffrin, C. Kiely, *J. Electroanal. Chem.*, 409 (1996) 137.

92 T. Sagara, N. Kato, N. Nakashima, *J. Phys. Chem. B*, 106 (2002) 1205.

93 T. Sagara, N. Kato, A. Toyota, N. Nakashima, *Langmuir*, 18 (2002) 6995.

94 T. Sagara, K. Muichi, *J. Electroanal. Chem.*, 567 (2004) 193.

95 I. Fromondi, A. L. Cudero, J. Feliu, D. A. Scherson, *Electrochem. Solid-State Lett.*, 8 (2005) E9.

96 I. Fromondi, P. Shi, A. Mineshige, D. A. Scherson, *J. Phys. Chem. B*, 109 (2005) 36.

97 B. Pozniak, D. A. Scherson, *J. Am. Chem. Soc.*, 126 (2004) 14696.

98 D. S. Martin, P. Weightman, *Surf. Interface Anal.*, 31 (2001) 915.

99 C. I. Smith, A. J. Maunder, C. A. Lucas, R. J. Nichols, P. Weightman, *J. Electrochem. Soc.*, 150 (2003) E233.

100 C. I. Smith, G. J. Dolan, T. Farrell, A. J. Maunder, D. G. Fernig, C. Edwards, P. Weightman, *J. Phys.: Condens. Matter*, 16 (2004) S4385.

101 D. E. Aspnes, A. A. Studna, *Appl. Opt.*, 14 (1975) 220.

102 J. D. E. McIntyre, D. E. Aspnes, *Surf. Sci.*, 24 (1971) 417.

3

Epi-fluorescence Microscopy Studies of Potential Controlled Changes in Adsorbed Thin Organic Films at Electrode Surfaces [*)]

Dan Bizzotto and Jeff L. Shepherd

3.1
Introduction

Optical microscopy is a technique that is widely used to study nature at a level of detail that is smaller than what is possible using the unaided eye. Microscopy has played an important role in research for investigating small objects in biological systems, and in the characterization of materials for the design and creation of new electronic devices. In electrochemical research, optical microscopy is typically used for characterizing the quality of a solid metal electrode surface during the preparation and polishing. In addition, optical microscopy has been used to monitor the electrolytic metal deposition and growth process or the growth of other materials through a faradaic process. It is interesting to note that in an early volume of this series (Vol. 9, 1973), Simon [1] described the use of optical microscopy in the study of electrochemical processes, and in a section on fluorescence he noted that the use of fluorescence microscopy in electrochemical studies has yet to be published and certain possibilities exist for its use in electrochemistry.

The technique of fluorescence microscopy is mature, and the limits of this optical method have been described in detail especially for use in research on biologically relevant systems. The reader is referred to a number of texts for more details [2–4]. Up to now, few fluorescence microscopy studies exist for electrochemical interfaces. The pH dependence of the fluorescein and quinine fluorescence was used to image regions where oxygen evolution and oxygen reduction were taking place through changes in the local pH [5, 6], but there are few examples of fluorescence from species adsorbed or bonded to metal electrodes under electrochemical control. Gaigalas [7–11] studied the potential-induced changes in the fluorescence intensity of an adsorbed layer of highly fluorescent molecules or of fluorescein-labeled single-strand oligonucleotides. Buttry [12, 13] measured the magnitude of the electric field in the electrical double layer using fluorophores im-

---

[*)] Please find the structures and abbreviations used in text at the end of this chapter.

Advances in Electrochemical Science and Engineering Vol. 9.
Edited by Richard C. Alkire, Dieter M. Kolb, Jacek Lipkowski and Philip N. Ross
Copyright © 2006 WILEY-VCH Verlag GmbH & Co. KGaA, Weinheim
ISBN: 3-527-31317-6

mobilized in alkylthiol self-assembled monolayers (SAMs) on Ag electrodes. Fox [14, 15] studied the fluorescence from various lengths of alkylthiol SAMs on gold surfaces showing the increase in fluorescence with an increase in the length of alkyl chain tethering the fluorophore to the metal surface.

Molecular luminescence from species near a metal surface is influenced by the distance separating the molecule from the metal. Reduction of the lifetime of the excited state results in very weak fluorescence, and changing the separation influences the lifetime and resulting fluorescence yield. This distance-dependent quenching process is described in detail in Section 3.3, relying upon a number of theoretical calculations by Chance [16] and Barnes [17, 18]. Other examples of luminescence from species on or near electrode surfaces are found in the electrogenerated chemiluminescence (ECL) literature. In the early works, ECL from an adsorbed molecular layer was demonstrated [19–21]. The authors concluded that the lifetime of the excited state rivals the rate of generation of the radical luminescent species, resulting in a measurable fluorescence.

Fluorescence analysis of such quenching surfaces does not provide much of an advantage for the investigation of the potential-dependent nature of the metal surface, but provides a very useful tool for investigating the region near the electrode surface, a region which may extend past the end of the diffuse part of the double layer which is not unambiguously accessible using electrochemical techniques. Fluorescence was used to investigate the adsorption/desorption process for adsorbed molecules that are lipid-like in nature. This review will focus on this unique interface, where the fluorescent molecule is adsorbed onto the electrode surface and does not exist in the bulk of the electrolyte. This can be accomplished by deposition from a floating monolayer present at the electrolyte surface, or self-assembly onto the metal before introduction into the electrochemical cell, or through a reaction at the electrode surface generating a new fluorescent molecule. In all examples from our work described below, the fluorescing species is confined to the region on or near the electrode surface. This is an important consideration, since the small fluorescence measured from the electrode/electrolyte interface would be negligible in comparison to that measured if the fluorescent species were present throughout the electrolyte.

Electrochemical measurements give an average picture of the changes occurring at the electrode surface. *In-situ* analysis of the interface using microscopy enables investigation of the degree of heterogeneity, which is difficult to extrapolate from the electrochemical studies alone. This is clearly demonstrated in the selective removal of a SAM from various areas of a polycrystalline Au bead electrode. In addition to the advantages inherent in investigating heterogeneous processes, microscopy simplifies the alignment of difficult samples for *in-situ* spectroelectrochemical analysis. One example presented in this review is the analysis of the potential-induced changes in a lipid layer adsorbed onto an Hg drop. The fluid nature of this metal surface makes analysis using traditional reflection set-ups impossible and requires the use of thin Hg films deposited on planar metal surfaces. Microscopy allows for *in-situ* analysis of the Hg drop with minor modifications to the standard electrochemical methodologies.

The electrochemical systems studied by the fluorescence method are based upon the adsorption of lipid-like compounds similar to the molecules that make up the cell membrane. The vast literature of methods and a variety of fluorophores are available for our use in the study of electrochemical systems. A brief review of the use of fluorescence microscopy in the study of biological systems is presented, because a number of the probes used for staining the biological structures are relevant for the electrochemical work presented. Moreover, the methods used in biological imaging to encourage fluorescence and to improve contrast are relevant for the work on electrode surfaces.

3.2
Fluorescence Microscopy and Fluorescence Probes

Fluorescence microscopy is used extensively in the study of biological systems, namely cell and tissue structure, enabling the visualization of structures within the cell through the creation of an enhanced contrast. An effective methodology involves the use of fluorophore probes that target specific components of the system under study. Currently, a large number of these probes are available from commercial sources like Molecular Probes, and a variety of motifs exist for interaction with cellular components. Fluorophores consisting of long hydrocarbon tails can mix within lipid membranes without a specific chemical binding interaction [22–25]. These molecules may preferentially concentrate within a specific lipid domain [26], providing contrast for fluorescence imaging of lipid membranes. Other examples include probes that use the affinity and specificity of antibody-antigen interactions in the labeling of these cellular components. Probes exist that are sensitive to potential or ion concentrations or the presence of ion channels; others are used to indicate enzyme activity by creating substrates that fluorescence upon reaction. The characteristic under study can be observed using an appropriate fluorescent probe, and the spatial distribution of this activity can be determined with microscopy.

In all examples, the choice of fluorophore is crucial for obtaining meaningful results. For example, the spectra, quantum yield, and fluorescence intensity of the fluorophore can be significantly affected by the environmental surroundings. The intensity of the dye can be affected by photobleaching, resulting in a continual decrease in fluorescence intensity with time. Furthermore, dye-dye interactions can affect both the absorption and emission spectra. These limitations create great difficulties in correlating fluorescence intensity to the concentration of the component of interest. Usually, two fluorescent probes are used, and their intensities are ratioed, providing a degree of compensation [27, 28].

In biological studies, little attention is given to the quenching due to the presence of a nearby metal film, as this rarely occurs in the samples studied. The use of fluorescence as a tool for the study of electrode surfaces requires a review of the effect of a metal surface on the rate of molecular fluorescence near a metal electrode.

3.3
Fluorescence near Metal Surfaces

The influence of a metal surface on the rate of a luminescence process must be considered for *in-situ* electrochemical measurements. This topic has been treated extensively both in theory [16, 29–31] and through experimental studies [32, 33] having its roots in Förster energy transfer. A molecule in an electronically excited state will return to the ground state through a number of processes, either non-radiative or radiative. Fluorescence resonance energy transfer (FRET) is a non-radiative relaxation process and was first realized by Förster, who described the transfer of energy from a donor molecule (D) to an acceptor molecule (A) through a long-range resonance transition dipole-dipole interaction. FRET occurs when there is a significant spectral overlap between the emission band of the donor and the absorption band of the acceptor. The donor-acceptor pair must be close enough (10–100 Å) and oriented in the appropriate manner for efficient energy transfer. This radiationless transfer of electronic excitation energy results in quenching of donor fluorescence through energy transfer to the acceptor, which in turn will relax to the ground state via radiative or non-radiative routes. Since the molecules need not be in contact, this process is considered "long range", and is inversely proportional to the sixth power of the separation between donor and acceptor. Förster described the energy transfer rate (k_T) through the following relation:

$$k_T = \left(\frac{1}{\tau_S}\right)\left(\frac{d_o}{d}\right)^6$$

where τ_S is the measured fluorescence lifetime of the sensitizer in the absence of the acceptor, and d is the distance between the centers of the chromophores. At the Förster critical distance d_o, 50% of the excitation energy is transferred to the acceptor, and the probability of relaxation by energy transfer equals the probability of relaxation in the absence of the acceptor. In Förster's derivation, the energy transfer is believed to occur between a single donor acceptor pair, and therefore a transition dipole-dipole type interaction was used. The Förster critical distance is dependent upon the sensitizer fluorescence quantum yield in the absence of the acceptor, the absorption of the absorber layer, the wavelength of sensitizer fluorescence, the refractive index of the medium, and a term that depends on the relative directions of the transition dipole moments of donor and acceptor. Currently FRET is a widely used method for determining structural information on molecular length scales [34].

Moving from discrete interactions between molecules to larger superstructures, Kuhn and coworkers used Langmuir-Blodgett techniques to construct surfaces modified by monolayers of D and A which were separated by non-fluorescent spacer molecules in a multilayer sandwich arrangement [30]. The measured fluorescence intensity depended on the inverse fourth power of the separation between D and A, obeying the following:

$$\left(\frac{I_d}{I_\infty}\right)_S = \left(1 + \left(\frac{d_o}{d}\right)^4\right)^{-1}$$

The fluorescence yield is less strongly dependent upon the distance separating D and A layers, which is explained through dimensionality considerations for the sheet-like arrangement of D and A dipoles [31]. Replacing the acceptor layer of molecules by a metal surface results in a fluorescence quenching behavior similar to that observed by Kuhn. Luminescence near a metal surface has been theoretically described by Chance [16], explaining the pioneering results of Drexhage [35], who measured the radiative lifetime of an Eu complex near a metal surface separated by spacer layers using the Langmuir-Blodgett approach. Barnes used this theory and measured fluorescence lifetimes to study the role of thickness and corrugation of the metal mirror and the environment in which emission occurs as a function of separation from the metal surface [17, 18]. In the near field ($<1/4$ of the wavelength), significant quenching of the excited state occurs because of transfer of energy to the metal substrate. If the transfer occurs to the bulk of the metal substrate, the quenching dependence takes on a d^{-3} character, while if the transfer occurs to the metal surface (surface plasmons), then quenching is dependent on d^{-4} [16]. At distances far from the electrode, comparable to the wavelength of emitted light, constructive and destructive interferences of the emitted radiation and that reflected from the metal surface produce an oscillation in the fluorescence intensity. This variation in intensity with separation distance from a metal has been extensively studied and is now becoming useful in the creation of waveguide structures to improve the performance of LEDs and to increase the quantum yield of inefficient fluorophores [36–39]. In this review, we take advantage of the quenching property of the metal electrode, which proves useful for exploring the potential-dependent changes of an adsorbed organic layer in the region near the electrode surface.

3.4
Description of a Fluorescence Microscope for Electrochemical Studies

The optical microscope is designed to present a magnified image of the sample to the eye. This is accomplished by a microscope objective lens, which is the primary magnifying element. The resolution of the microscope is defined by the objective, and further magnification can result in an increased blurring of the image and a decrease in image quality or contrast. The proper use of a compound microscope is discussed in many texts [2, 3], and the methods used depend upon the nature of the sample and the type of information required. A large number of contrast techniques have been developed (e.g., dark-field microscopy, phase contrast, differential-interference contrast), with fluorescence microscopy playing an important role. Most fluorescence microscopy is now implemented in the epi configuration, where the objective is used for both the focusing of excitation radiation and the collection of the emitted light. In this way, the microscope objective serves a dual purpose and avoids significant complications in the alignment of the focusing and collection optics, which must coincide at the area of interest. This epi configuration is used to great advantage in fluorescence microscopy, since the region in focus is optimally illuminated and any luminescence from this in-focus region is also optimally collected. This arrangement is particularly convenient for studying electrode surfaces, since the excitation illumination originates from the objective and does not require a separate optical pathway. In addition, the objective is typically corrected for the refraction through a cover glass, which is nominally 0.17 mm thick.

The spectroelectrochemical cell positioned on the inverted microscope is shown in Fig. 3.1. The cell used in these experiments was created with a cover glass as the optical window along ports to accommodate the calomel reference electrode, the gold counter electrode, and ports for Ar purge gas, which is used to remove oxygen. The illuminating excitation radiation is provided through proper filtering of the broadband output from an Xe or Hg arc lamp. This light is collimated, passed through an aperture, and then directed toward the wavelength-discriminating optical element, the filter cube. Two bandpass filters and

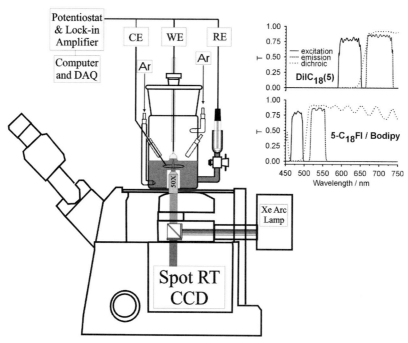

Fig. 3.1 Schematic of the epi-fluorescence experimental setup modified for *in-situ* electrochemical measurements. The transmission characteristics (from Chroma) are shown on the right for the filter cubes which were used for the two fluorescence probes. The filter cubes were mounted in a turret which allowed exchange without disrupting the electrochemistry. In some experiments the CCD camera was replaced by an S2000 fiber optic spectrometer.

a dichroic mirror are housed within the filter cube. The excitation bandpass filter typically allows a 50-nm bandpass of wavelengths through to the dichroic mirror, which reflects the appropriate band of excitation wavelengths toward the objective, which focuses the light onto the sample. The reflected and fluorescence light is collected by the objective. The red-shifted fluorescence is transmitted through the dichroic mirror to the emission filter, which passes a specified bandpass of wavelengths to the detector. These filter cubes are very efficient and are produced with various filters and dichroic mirrors so as to accommodate the large number of fluorophores available. The transmission characteristics of a filter cube used in one example detailed in this chapter is shown in Fig. 3.1. The fluorescence is directed toward a solid state CCD array detector, which is thermo-electrically cooled, decreasing the dark noise to low levels. This type of detector is typically more sensitive to the longer wavelengths. The optimum field of view is obtained by matching the image size to the CCD size, this being accomplished by an optical coupler. The maximum resolution in the digital image is achieved by considering the optical lateral resolution, the CCD pixel size, and the Nyquist sampling theorem.

3.4.1
Microscope Resolution

In the design of a microscope, magnification is achieved through the use of an objective that produces an image of the object in the so-called intermediate image plane. Viewing through an eyepiece magnifies this intermediate image, resulting in a magnification which is the product of the objective and eyepiece magnifying power. The limit on the optical magnification is given by the maximum possible resolution. Resolving or distinguishing between two features in a sample is dependent on the light-gathering power of the objective, or its numerical aperture (NA). Based upon Huygens's principle, light from each "point" in the object propagates through the objective, which behaves as a circular aperture, producing an Airy diffraction pattern (or Airy disc). High NA objectives enable efficient collection of more diffraction orders, thereby decreasing the size of the Airy disc. The point spread function (PSF) is another way of describing resolution and characterizes the propagation of a point source through the microscope. The maximum resolution achievable is given by the system's ability to resolve two Airy discs or to distinguish between the PSFs for two point sources that are close together. This concept of resolution is combined with contrast, and together they make up a more useful measure of the resolving power of a microscope. Two point sources can be resolved if the Airy discs do not overlap. Once the discs overlap, the contrast between the maxima of the PSF is still possible as long as the two maxima can be recognized. A more precise definition for diffraction-limited resolution is given by the Rayleigh criterion, where the maximum of one Airy disc overlaps with the first minimum in the second Airy disc (overlap of 26%), resulting in a lateral resolution (R) of

$$R = \frac{1.22\lambda}{2 \cdot NA}$$

High NA objectives will achieve the best resolution, with the use of an intermediate fluid (immersion oil) that replaces air, increasing the refractive index in the gap between the glass and the objective. This allows more diffraction orders to be collected, decreasing the size of the Airy disc. Typically, the high magnification objectives have large NA (~1) as well as a very small working distance (wd), which is the distance between the objective and the focal plane. For example, a 50× objective may have an NA=0.8, a wd of 0.1 mm, and a calculated lateral resolution for λ=500 nm of 0.38 µm. *In-situ* measurements as described here are done in a regular electrochemical environment rather than a thin-layer environment (single crystal surfaces are easily damaged), so ultra-long working distance objectives are employed at the expense of resolution (e.g. 50×, NA=0.5, wd=10 mm; lateral resolution for λ=500 nm is 0.6 µm).

The depth of focus or axial resolving power (R_{ax}) is also important and depends on the characteristics of the objective.

$$R_{ax} = \frac{n\lambda}{(NA)^2}$$

Parallel to the optical axis, R_{ax} characterizes the minimum distance between two point sources oriented along the optical axis, which is normal to the electrode surface. The depth of field is related to R_{ax} and is a measure of the thickness of the region in focus, from the closest object plane to the furthest object plane, which is typically 1–5 µm. A significant drawback to the use of high NA objectives is the collection of scattered or out-of-focus light that comes from regions that are not in the focal plane. This light significantly contributes to blurring of the fluorescence image. Confocal fluorescence microscopy uses appropriately placed pinholes, which significantly decrease the blurring of fluorescence images. Currently, software deconvolution techniques are available that can eliminate the blurring through the determination of the PSF of the optical system. More complex processes collect images from a number of focal planes above and below the one of interest. The blurring contribution is removed through a computational procedure, and a 3D image of the sample is determined [40].

3.4.2
Image Analysis

Techniques for image analysis are well established as detailed in a book by Russ [41]. The raw images must be manipulated to correct for dark charge and for non-uniform illumination across the field of view. This is required for particle-size and number analysis as well as extracting the "intensity" from the images. Figure 3.2 schematically illustrates the image analysis procedure used for the

(a)

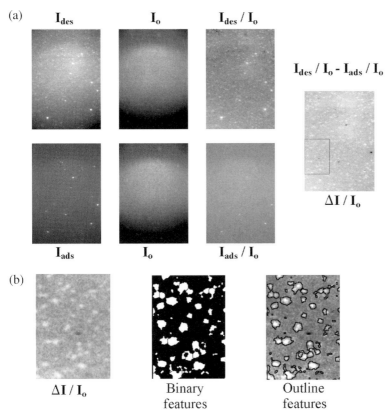

I_{des} I_o I_{des} / I_o

$I_{des} / I_o - I_{ads} / I_o$

$\Delta I / I_o$

I_{ads} I_o I_{ads} / I_o

(b)

$\Delta I / I_o$ Binary features   Outline features

Fig. 3.2 Image analysis procedure used for the images presented in this review. (a) The fluorescence images acquired at the adsorption (I_{ads}) and desorption (I_{des}) potentials are divided by the image which represents the distribution of the incident intensity (I_o) creating flat field corrected images which are used for further analysis. The change in fluorescence is determined by subtraction, yielding I/I_o. A section of this image was expanded (b) to demonstrate the particle analysis procedure. Creation of binary features was accomplished through edge-enhancing techniques followed by watershed filtering. The features outlines are clearly representative of the features in the pictures. Further computation gives particle density and average sizes.

fluorescence images measured. Microscopy images suffer from a number of systematic errors that are easily corrected using techniques that have been in extensive use in astrophotography [42]. The array photodetector (CCD) is subject to dark current, which can be subtracted through measurement in the absence of illumination. The fluorescence images are also subject to non-uniform illumination from the excitation light source. The excitation from an Xe arc lamp is collected, focused, and passed through an aperature, an image of which is focused onto the sample after passing through the wavelength-selecting filters contained within the filter cube. The spatial distribution of excitation light intensity will influence the spatial distribution of the fluorescence. Correction is simply applied by measuring a "fluorescence image" of the electrode before deposition of the fluorescent monolayer. The leakage of the excitation wavelengths through the dichroic mirror and emission filter is sufficient to establish the spatial distribution of the excitation intensity. The raw fluorescence images are then divided by the intensity of the excitation light after correction for the dark signal. This procedure also allows for correction of the small differences in the excitation light intensity or filter cube transmission over time. This procedure also introduces some noise into the fluorescence images. Multiple images were averaged and smoothed using a Gaussian Filter. The resulting images are effectively normalized with respect to the incident intensity I/I_o creating a flat field image.

For use in the study of electrode surfaces, our interest is in the potential-dependent changes in the adsorbed layer, which can be observed through creation of difference images, in a fashion similar to electroreflectance measurements. The flat-field images are then subject to removal of the adsorption potential image creating I/I_o images for analysis. This ensures that the analysis that follows concentrates on the potential-dependent changes.

An average histogram is calculated for each I/I_o image. The particle analysis is performed using standard techniques detailed by Russ [41]. The image is edge enhanced and thresholded to give outlined features, which are then filled. The image is watershed filtered to separate touching particles, and the particle statistics such as average number and mean diameter of particles are computed. An example of this image analysis procedure and the resulting image and particle analysis is given in Fig. 3.2. The analysis does accurately find and measure each particle, yielding average number of particles and their mean size.

3.5
Electrochemical Systems Studied with Fluorescence Microscopy

The imaging of potential-dependent changes from an organic-modified metal electrode using fluorescence techniques is limited because of the metal-mediated quenching of fluorescence. As mentioned in the introduction, the combination of a technique sensitive to the region near the electrode coupled with the interfacially sensitive electrochemical measurements provides a complementary study of the changes in the adsorbed organic monolayer. The examples chosen are taken from recent work which has focused on the characterization of the organically modified electrode surface. Modification of metal surfaces with organic molecules takes two basic forms, physically adsorbed and chemisorbed layers. The use of Langmuir-Blodgett (LB) deposition techniques has allowed the modification of these surfaces by transferring onto the metal water-insoluble molecules organized on the electrolyte surface. Chemisorbed variations of this process have been popularized by the so-called self-assembly of organic molecules bearing thiol moieties, which undergo an oxidative reaction with gold (for example), forming a covalent linkage with the surface, yielding a very stable monolayer. This process has been extensively described in the literature [43–47].

The physically adsorbed monolayer of lipid-like molecules is less well known and does not form interfaces that are as robust as the SAM. These molecules are present on the metal surface as a consequence of the LB deposition procedure and because of the reduction of surface energy achieved by remaining on the metal surface. The electrochemical nature of these films has been characterized using the traditional approaches that have occupied electrochemical technique development for the past 50 years, focusing on the measurement of current-potential curves, the measurement of the capacitance of the interface, and the surface energetics through the electrocapillary equation. Over the past two decades, these techniques, usually limited to Hg drop electrodes, have been expanded to include well-defined metal surfaces through single-crystal studies. The adsorption of lipid-like molecules onto metal single-crystal electrodes has been studied by electrochemical techniques and was recently reviewed [48]. Electrochemistry provides a very sensitive measure of the quality of an adsorbed film, and in addition enables control over the surface energetics of the metal/solution (M|S) interface. This control allows for the investigation of a large range of stable and metastable arrangements of the adsorbed molecule on the metal electrode surface.

Overall, electrochemical characterizations are very sensitive to the nature of the interface and the changes that occur due to potential perturbation. It is important to remember that these measurements give us a picture of the average state of the adsorbed layer, measurements which tend to hide some of the important nuances in the potential-dependent behavior of the adsorbed layer. Determining these average changes in the adsorbed layer in tandem with the simultaneous fluorescence measurement that includes spatial resolution can be accomplished through various *in-situ* techniques. *In-situ* optical microscopy allows non-invasive probing of the film and enables questions about its nature to be studied without significantly impacting on the character of the layer. While this technique does not provide a molecular picture of the changes in the interface, as does STM or AFM, the perturbation is less invasive, so less robust systems can be investigated without undo impact.

The systems that demonstrate the capabilities of fluorescence imaging on electrochemical interfaces include physically adsorbed organic layers that are doped with fluorescent lipophilic probe molecules, in a manner similar to the biological studies, characterization of the electrochemical dimerization process which produces a fluorescent compound only at the electrode surface, imaging of a lipid-coated Hg drop, and the investigation of the heterogeneous reductive desorption of a fluorescence SAM from a polycrystalline gold bead. The examples are chosen to illustrate the unique information obtained with fluorescence imaging, while electrochemistry controls the interfacial characteristics.

3.5.1
Adsorption of C$_{18}$OH on Au(111)

The system most studied using electrochemistry/fluorescence microscopy is the adsorption of lipid-like molecules onto Au and Hg electrodes, which has been recently reviewed [48]. The electrode surface is modified by bringing the metal into contact with an organized monolayer of the surfactant molecules floating on the electrolyte surface at the naturally formed equilibrium spreading pressure. The transfer to the nearly uncharged metal surface is accomplished at 0 V/SCE, forming a well-defined adsorbed layer characterized by capacitance values typical for a monolayer of a low-dielectric adsorbed film. As the potential or charge on the electrode surface is made more negative, the adsorbed film becomes porated and is replaced by water at extreme negative values of charge. This process can be described as the transition between multiple electrocapillary curves that occur at defined values of the electric potential. These transitions are easily measured with differential capacitance. As an example, the cyclic voltammogram (CV) and capacitance of Au(111) coated by a film of octadecanol transferred from the electrolyte surface are shown in Fig. 3.3. At potentials near to values for the uncharged metal/electrolyte interface (pzc=0.255 V/SCE), the adsorbed organic layer has a capacitance of nearly 6 µF cm^{-2}. As the potential is swept negatively, poration of the monolayer occurs, and this change is represented by the pseudo-capacitance peaks which are prevalent in the CV at −0.250 V/SCE. At more negative potentials, the capacitance increases to 14 µF cm^{-2}, which is just below the value characteristic for a water-covered surface. At the negative potential scan limit (−0.800 V/SCE) the capacitance in the absence and in the presence of the surfactant are the same, suggesting the complete desorption of the surfactant layer from the metal surface. The desorbed molecules

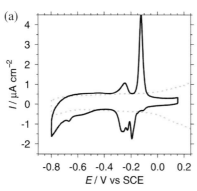

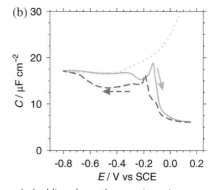

Fig. 3.3 Cyclic voltammogram (a) and capacitance (b) measured for an octadecanol-coated Au(111) electrode. The dotted line was measured in the absence of the surfactant. The solid line represents the positive-going potential scan, while the dashed line shows the negative going scan for the capacitance. The electrolyte was 0.05 M KClO$_4$; the CV was measured at 20 mV/s. A series RC circuit was assumed in determining the capacitance using a 25 Hz, 5 mV rms sine wave perturbation.

can be readsorbed onto the electrode surface by reversing the potential sweep direction, making the electrode less negatively charged. On the positive potential scan, a slight decrease in capacitance is noted preceding the pseudo-capacitance peaks, and at the positive potential limit the original low capacitance layer is re-established, indicating the readsorption of the organic molecules onto the electrode surface. Hysteresis is noted between the desorption and readsorption processes, suggesting a mechanistic difference in these processes. The recovery of the original adsorbed layer indicates no loss of the lipid-like molecules to the bulk of the electrolyte, which suggests that at desorption potentials the molecules are near to but not on the electrode surface. This differs from the Frumkin model for surfactant adsorption, which has been devised for the adsorption of surfactants that are soluble in the electrolyte. This model assumes that, at equilibrium, desorption into the diffuse layer or into the bulk of solution occurs, so that the chemical potential of the species near the electrode and in the bulk of the electrolyte are equal [49]. For the systems using insoluble or Langmuir-Blodgett-type surfactants, the desorbed molecules may be in a metastable state, incongruent with the Frumkin model.

The fluorescence imaging technique will be useful in characterizing the desorbed organic film, since the molecules may be far enough from the metal electrode surface so as not to be significantly affected by metal-mediated quenching. Since octadecanol is not fluorescent, 2 mol% of a carbocyanine fluorophore DiIC$_{18}$(5)(1,1'-dioctadecyl-3,3,3',3'-tetramethylindodicarbocyanine perchlorate) and 2 mol% of 5-C$_{18}$Fl (5-octadecanoylaminofluorescein) were added to the solution used to create the floating monolayer of octadecanol. These fluorophores are not FRET pairs since their spectral overlap is small, so can be used for the characterization of a given adsorbed layer. In this regard, the interface can be imaged twice by separately collecting the fluorescence from each fluorescent probe molecule from the same region of the electrode surface. Furthermore, the use of two probes will indicate how uniformly dispersed the probe is within the octadecanol monolayer. The same potential range as shown in Fig. 3.3 is used, and fluorescence images from the dye-containing octadecanol/Au(111) interface was recorded in 25-mV increments. The imaging starts at positive potentials, stepping in the negative direction to the negative potential limit (−0.8 V/SCE), then reversing the potential steps back to +0.15 V/SCE. Fluorescence from 5-C$_{18}$Fl was measured followed by the same imaging sequence for the fluorescence from DiIC$_{18}$(5). The images for each fluorophore were normalized according to the procedure in Section 3.4.2 and then pseudo-colored consistently with their natural fluorescence wavelengths (green for 5-C$_{18}$Fl and red for DiIC$_{18}$(5)). The colored images were combined, creating one image of the interface, revealing green, red, or a mixed pseudo-colored fluorescence.

The modified images of the interface are shown in Fig. 3.4 superimposed upon the capacitance curve recorded during the optical measurement. At positive potentials (images A and B), fluorescence from both dye molecules are quenched because of the proximity of the surfactant and metal. At potentials negative of −

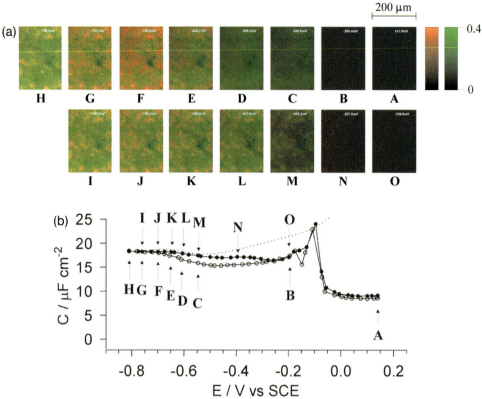

Fig. 3.4 (a) The fluorescence images measured for an octadecanol-coated Au(111) electrode which was labeled with 2 mol% of both DiIC$_{18}$(5) and 5-C$_{18}$Fl. The images are pseudo-colored and combined to show the distribution of each fluorescent probe using green for 5-C$_{18}$Fl and red for DiIC$_{18}$(5). The labeled images in (a) correspond to the labeled potentials in (b).

0.500 V/SCE, an increase in the fluorescence from 5-C$_{18}$Fl is noticed, which occurred slightly before the fluorescence from DiIC$_{18}$(5) observed at –0.625 V/SCE. It will be shown that this is a result of a difference in the metal-mediated quenching profiles for the two dye molecules. The distribution of desorbed octadecanol is clearly seen at more negative potentials (images E to G), indicating the increased separation between the surfactant and metal. Fluorescent features from both probes are present in these images, revealing heterogeneity in the morphology of the desorbed layer structure that remains consistent over the time spent at the desorption potentials. The fluorescence intensity increases to a maximum at the negative potential limit (image H), and the yellow-orange color in the image results from a fluorescence from both probes. This is a clear indication that the probe molecules are evenly mixed in the octadecanol matrix and therefore faithfully report on the potential-induced behavior of the surfactant. The fluorescence remains intense during the initial stage of the readsorption potential scan (images I to L), reaffirming the hysteresis observed in the capacitance measurement. At the positive potential limit (image O), the fluorescence from both probes are quenched because of the readsorption of the surfactant onto the metal surface. These fluorescence imaging results confirm the potential control over the desorption and readsorption process for these insoluble surfactant molecules. Moreover, fluorescence shows that the desorbed molecules stay near to but not on the electrode, and, surprisingly, the desorbed species do not diffuse from the metal surface. This suggests that the layer is held in place by forces that have been proposed to be due to the ordering of the water layer between the electrode and the desorbed organic molecules [50].

The intensity and structure of the desorbed surfactant was analyzed following the procedure in Section 3.4.2 for the normalized gray-scale images. Figure 3.5 shows the dependence of fluorescence intensity, number of image features, and their mean area on potential. Consistent with the pseudo-colored images, the intensity of $5\text{-}C_{18}Fl$ begins to increase at potentials less negative than those for $DiIC_{18}(5)$. For both probe molecules, a continual increase in the fluorescence intensity occurs with increasing negative potentials. A slight deviation in this

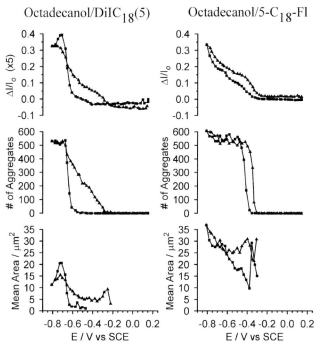

Fig. 3.5 Results of image analysis for octadecanol adsorbed onto Au(111). The monolayer contained two fluorescent probes, $DiIC_{18}(5)$ and $5\text{-}C_{18}Fl$, which resulted in fluorescence images as shown in Fig. 3.4.

The intensity, number of aggregates and the mean area of the aggregates were determined through the image analysis procedure described previously.

trend between the two curves is observed negative of −0.700 V/SCE. While the intensity of 5-C_{18}Fl continues to increase up to a maximum value, the fluorescence intensity of $DiIC_{18}(5)$ decreases. This decrease is ascribed to the association of $DiIC_{18}(5)$ molecules forming non-fluorescent aggregates, which decrease the measured fluorescence. Interestingly, as detailed in Ref. [51], the readsorption of the dye molecules results in destabilization of the dye aggregate, re-establishing the monomer fluorescence upon subsequent desorption. This process illustrates the use of electrode potential to control the characteristics of dye-dye intermolecular interactions that influence photophysical characteristics. For the positive scan of potentials, the fluorescence intensity remains greater than the negative potential scan for both probe molecules, again consistent with the hysteresis observed in the images and capacitance plot.

Particle analysis of the fluorescence images indicated that potential of the metal surface significantly influenced the number of particles that were detected, but not their mean area. Interpretation of such analysis is difficult because of the uncertainty in the degree of quenching experienced by various parts of the desorbed layer. Assuming the desorbed molecules are all separated from the electrode surface to the same extent, then the image analysis can be used to gain insight into the desorption process. The monolayer desorbs as uniformly small particles, the number of which is dictated by the potential. Because of limitations in the optical resolution, it is unclear if these aggregates are single particles or a collection of smaller aggregates. This analysis does depend upon the dye molecule used as the probe species, with $DiIC_{18}(5)$ showing more hysteresis in the adsorption/desorption process than 5-C_{18}Fl, as well as differences in the potentials at which the aggregates are first observed. The increase in fluorescence intensity of 5-C_{18}Fl at less negative potentials than $DiIC_{18}(5)$ can be explained through the difference in the metal-mediated quenching-distance profiles for the two fluorescent probe molecules. The quenching-distance profiles can be measured using a gold-coated glass slide covered by steps of various thicknesses of SiO_2, creating the stepped surface as shown schematically in Fig. 3.6. Deposition of a fluorescent dye-containing octadecanol layer onto the stepped surface allowed for a controlled variation of the separation from a gold film. The fluorescence intensity was measured for each of the steps and plotted as a function of the separation in Fig. 3.6 b. The fluorescence from 5-C_{18}Fl is quenched less at small separations in comparison to $DiIC_{18}(5)$. Following the description in literature [30, 31], the experimental data was fitted to the equation proposed by Kuhn and confirmed by Barnes (Section 3.3), with the result shown in Fig. 3.6 c. This template can be used to provide an estimate of the gap between the desorbed surfactant and the Au(111) electrode (taking into account the differences in the refractive indices of water and SiO_2). The fluorescent images of the desorbed surfactant presented in Fig. 3.4 are approximately 10% of the intensity recorded from the floating monolayer prior to deposition onto the electrode surface. Difficulties exist with this estimation, since fluorescence measured from the floating monolayer cannot be directly compared to the electrochemical interface, since this measurement does not have the advantage of a

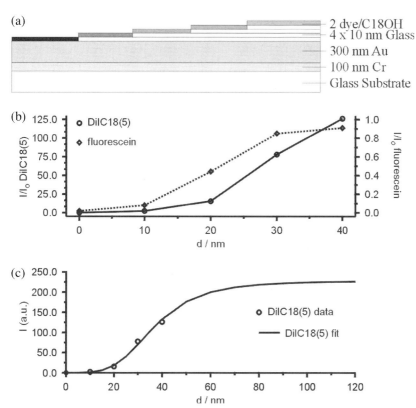

Fig. 3.6 (a) Schematic of the stepped surface used to determine the quenching characteristics for the two dyes used in this work. A monolayer of octadecanol containing 2 mol% of both dyes was deposited onto the stepped surface using the horizontal touch technique. b) The measured intensity or quenching curve as a function of distance from the metal surface. c) The solid line represents a fit to the data using the equation in the text.

mirror-reflecting fluorescence that is emitted upwards, back towards the objective. Even with this limitation, the upper limit for a value of this separation is estimated at less than 20 nm, which is based on the calculated value for d_o (37 nm) determined from the fits. This distance is further than the edge of the diffuse layer but is still near the electrode surface, where the forces holding it in place are effective and not washed out by random Brownian motion.

The advantage provided by fluorescence imaging for these unique systems is clear. For example, investigations by probe microscopy (e.g., AFM) would not easily observe these desorbed structures, since the probe would disrupt the organization or act as a broom sweeping these desorbed molecules away. The role fluorescence has played in the confirmation of the mechanism which was imagined through electrochemical measurements is significant. The identification of the heterogeneity of the desorbed layer is also quite important and is usually not considered when performing other averaging optical measurements that do not have the ability to discriminate or detect the heterogeneity present in the electrochemical interface.

3.5.2
The Adsorption and Dimerization of 2-(2′-Thienyl)pyridine (TP) on Au(111)

The adsorption/desorption of the lipid-like surfactants was monitored using fluorescence which was simplified because of a lack of significant background since the molecules were only present on the electrolyte surface or metal/electrolyte interface and not present in the bulk of the electrolyte. In this example, the adsorbing molecule, TP, which is a bidentate ligand composed of thiophene and pyridine covalently linked through the 2 and 2′ carbons, is present in the bulk of the electrolyte as is usual in the traditional adsorption measurements [52]. The electrochemical characterization of TP adsorption onto Au(111) is shown in Fig. 3.7a,b for two different positive potential sweep limits. Using quantitative measurements, it was found that TP adsorbs onto the Au(111) electrode and establishes a maximum surface concentration of 5×10^{-10} mol cm^{-2} at −0.2 V/SCE, similar to that observed for bi-pyridine [53]. At potentials positive of −0.2 V/SCE, an oxidative process was observed, which was believed to be the dimerization of two TP molecules to form PTTP (5,5′-bis(2-pyridyl)-2,2′-bithienyl) facilitated by the pre-organized layer of monomers created through adsorption onto a well-defined single-crystal surface.

Confirmation of this dimerization reaction was possible using *in-situ* fluorescence spectroscopic measurements (Fig. 3.7c,d). TP fluoresces at 360 nm and PTTP (synthesized separately) fluorescence is red-shifted to 435 and 450 nm because of the increased conjugation. The dimer was electrochemically created by adsorbing TP at the potential of high surface coverage (−0.2 V/SCE), then stepping the potential to 0.2 V/SCE, where the dimer was created. The modified adsorbed layer was then desorbed and the fluorescence image and spectra were measured. Clear evidence of the dimer was observed even in the presence of a large fluorescence background, presumably from the weak fluorescence from TP in the electrolyte. The fluorescence spectrum (measured at the desorption potential) for an interface with an adsorbed layer at −0.2 V/SCE was characteristic of TP. These spectra represent the change in fluorescence intensity due to the change in potential, and so in this case the spectrum represents an excess of TP that was present in the surface region because of adsorption. The fluorescence spectrum measured for an adsorbed layer at 0.2 V/SCE showed a peak at 460 nm, which is characteristic of PTTP. To confirm this finding, PTTP was adsorbed onto the electrode surface, and fluorescence of this monolayer, recorded at the desorption potential, was compared to the electrochemically created PTTP, as shown in Fig. 3.7d. The spectra are very similar and compare well with the solution-phase fluorescence spectrum of PTTP. The fluorescence measurements were needed to conclusively demonstrate the electrochemical formation of the TP dimer.

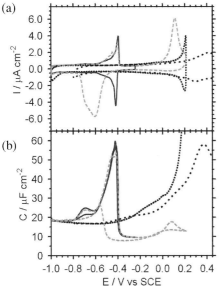

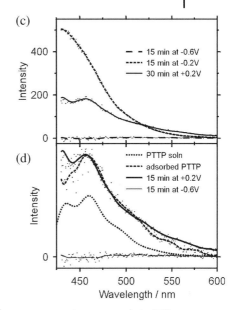

Fig. 3.7 (a) Cyclic voltammograms (20 mV/s) of Au(111) in 0.05 M KClO4 (. . . .), after addition of NaOH, pH = 10 (.....) in the presence of 1 mM TP for two positive potential limits: −0.25 V/SCE (——) and 0.2 V/SCE (– – – –); (b) differential capacitance measured under the same conditions as the CVs presented in (a). The capacitance was measured using 25 Hz, 5 mV rms sine wave perturbation and calculated assuming a series RC circuit;

(c) fluorescence spectra measured at −1 V/SCE for electrode surface conditioned at various adsorption potentials: −0.6 V/SCE (– – –), −0.2 V/SCE (- - - -), +0.2 V/SCE (——); (d) fluorescence spectra of PTTP in CHCl₃ solution (.....), fluorescence spectra measured at −1 V/SCE for an electrode surface modified by adsorbed PTTP (– – – –), in the presence of TP at +0.2 V/SCE (——) and at −0.6 V/SCE (— —).

3.5.3
Fluorescence Microscopy of the Adsorption of DOPC onto an Hg Drop

The adsorption and potential-induced changes in an adsorbed monolayer of dioleoyl phosphatidylcholine (DOPC) onto an Hg surface has been studied and used to model half of the lipid membrane for many years [54–56]. The changes in the lipid monolayer are represented in the capacitance measurement as sharp pseudo-capacitance peaks that are strongly correlated to the adsorbed monolayer order. A few approaches have been attempted in modeling these changes [57, 58], though no direct experimental data was available other than electrochemical measurements, which give an average picture of the interface.

The measurement of fluorescence from an Hg drop coated with a liquid-like lipid monolayer of DOPC containing a small amount of dye was performed following the procedure described for the Au single-crystal electrodes. The *in-situ* application of microscopic analysis during the electrochemical measurements enabled the direct imaging of the potential-dependent changes in the adsorbed layer on an Hg drop. The image of the Hg drop is composed of an in-focus region of the bottom of the drop with the rest of the Hg surface above the focal plane and therefore out of focus (Fig. 3.8 inset). In all images presented, the in-focus region was outlined. As mentioned previously, fluorescence that may originate from regions not in the focal plane will still contribute to the measured fluorescence. Another important difference between a smooth metal like Hg

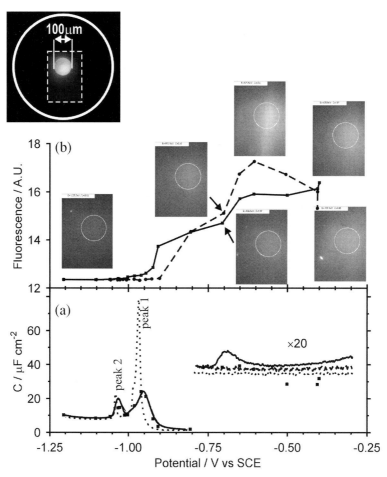

Fig. 3.8 Fluorescence study of DOPC adsorbed onto an Hg drop. (a) Capacitance of the pure DOPC (dotted line), DOPC + 2 mol% 5-C_{18}Fl (dashed line is the positive-going scan, solid line is the negative going scan) measured using a 75 Hz, 2 mV rms perturbation with capacitance calculated assuming a series RC circuit. (b) The average fluorescence intensity and selected images of the Hg surface (0.2×0.3 mm). The inset above shows a bright-field image of the Hg drop. The in-focus region is 100 μm in diameter. The extent of the image is given by the dashed line, while the thick solid ring outlines the full diameter of the Hg drop.

and a rougher Au surface is the significantly different quenching-distance profiles. The quenching of the excited state near a smooth surface is two or three
orders of magnitude less than that near rough surfaces [59]. Therefore, fluorescence may be observed when the distance separating the fluorophore and the
Hg surface is much smaller than that in the case of the rougher Au surface.
This difference will be evident in this section since fluorescence maybe observed even if the fluorophore is only slightly separated from the Hg electrode
and may not need to be desorbed.

Spectroelectrochemical studies of an Hg surface is hampered by the nature of
the liquid-metal surface. The curvature and fluctuations in the curvature due to
the changes in surface tension with potential (electrocapillarity) can have serious consequences in traditional reflectance measurements. Imaging of the Hg
drop using fluorescence microscopy was attempted in order to provide some
experimental data to understand the nature of these potential-induced changes.
The DOPC-coated Hg drop surface was created by pushing the Hg drop
through the floating lipid monolayer [60]. The membrane-labeling lipophilic fluorescent dye (5-C_{18}Fl) used previously was introduced as a component in the
DOPC monolayer. The adsorbed layer is characterized by a low capacitance
(1.8 μF cm^{-2}) at potentials around the potential of zero charge (pzc = -0.4 V/
SCE) and by two narrow pseudo-capacitance peaks (peak 1 at -0.95 V/SCE and
peak 2 at -1.1 V/SCE) as shown in Fig. 3.8a. The presence of the small amount
of 5-C_{18}Fl was enough of a perturbation that the minimum capacitance was
slightly elevated; the pseudo-capacitance peaks were broadened and smaller
when compared to pure DOPC. Overall, the characteristics assigned to a pure
DOPC monolayer adsorbed onto Hg were all represented in this adsorbed layer
even with the addition of a foreign component. The potential-induced changes
are completely reversible if the potential scan is limited to -1.2 V/SCE with the
peaks being retraced and the re-establishing of the minimum capacitance layer.
As shown in Fig. 3.8b, these limited-potential scans resulted in a constant fluorescence intensity for the minimum-capacitance potential region, decreasing
once the potential became negative of the phase transitions represented by
peaks 1 and 2. Originally, the peaks were thought to signify the poration of the
monolayer (shown via facile metal ion reduction [61–63]) and the displacement
of some of the lipid molecules into a second layer. Figure 3.8b shows a decrease
in the fluorescence intensity due to potential-induced lipid reorganization,
which requires the fluorophore to be closer to the metal surface at the negative
potential region. The DOPC monolayer in the low-capacitance configuration
may force the dye molecule away from the Hg surface. The dye probe, which
preferentially resides in the headgroup region, would be farther from the electrode than the lipid molecules. The defects created through potential excursion
to -0.8 V/SCE would allow the dye molecules to become, on average, closer to
the metal surface. The reversible nature of this process also supports this
"squeezing out" hypothesis. The static images shown in Fig. 3.8 do not reveal
the mobile nature of the fluorescence features, in contrast to that observed for
the adsorbed layer on Au. The mobile fluorescent features made image analysis

impossible and may be related to the fluidity of the Hg surface and the induced tangential flow of the interface, which is due to the changes in interfacial tension known to contribute to non-idealities in the electroanalytical polarography [64]. Poration of the monolayer at circa −1 V/SCE significantly diminished the mobility, and the fluorescence features became frozen in place. The fluorescence features were once again mobile when the potential was returned to the minimum-capacitance potential region. This example demonstrates the enhanced capability provided by microscopy for examining these unique interfaces, not possible with methods that measure average properties.

The DOPC monolayer is thought to be desorbed or displaced from the Hg surface at very negative potentials. This desorption was found to be strongly dependent upon the cation in the electrolyte [65]. Fluorescence imaging at very negative potentials indicated displacement of the DOPC and a large increase in fluorescence, as expected based upon our experience with the Au system. The correspondence of the capacitance values in the absence and in the presence of DOPC at these negative potentials along with these fluorescence measurements imply desorption from the interface where the separation from the metal surface is explicitly demonstrated. This behavior was also observed for the Au/octadecanol/dye system. In that case, the organic layer remained near the electrode and did not diffuse away, yielding images that were very static (except for photobleaching), allowing particle size distribution analysis. As mentioned, the fluorescence images acquired for the DOPC-coated Hg drop were not static, making image analysis difficult. These fluorescence measurements constituted the first direct measurement of the displacement of the DOPC from the Hg surface and will prove very useful in the investigation of a hybrid liposome/monolayer system described next.

3.5.4
Fluorescence Microscopy of Liposome Fusion onto a DOPC-coated Hg Interface

Previous electrochemical studies of the adsorption and lysis of DOPC liposomes onto an Hg electrode proposed the existence of an adsorbed hybrid layer composed of lipid monolayer with regions that were composed of liposomes fused with the adsorbed monolayer [66]. This layer was a result of liposome interaction with a porated DOPC monolayer, which can occur by setting the potential more negative than the pseudo-capacitance phase transition potentials. The liposomes become incorporated or fuse with the adsorbed layer. Return of the potential to the minimum-capacitance region resulted in an adsorbed lipid layer with a minimum capacitance intermediate between a monolayer and bilayer, indicative of a liposome that is partially incorporated into the adsorbed lipid layer, forming a so-called hemi-liposomal adsorbed layer. This was proposed through the indirect electrochemical investigations. Fluorescence microscopy was used to demonstrate the fusion of liposomes to a previously formed dye-containing DOPC monolayer transferred from the electrolyte surface. The liposomes were free of the dye so as to eliminate the substantial background that would have

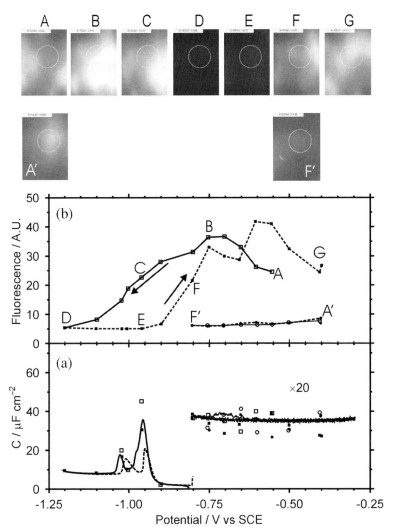

Fig. 3.9 Fluorescence study of an adsorbed layer of DOPC containing 2 mol% of 5-C_{18}Fl on an Hg drop before and after interaction with dye-free DOPC liposomes.
(a) Capacitance of an Hg surface coated with DOPC+2 mol% 5-C_{18}Fl (dashed line is the positive-going scan, solid line is the negative-going scan) measured using a 75 Hz, 2 mV rms perturbation with capacitance calculated assuming a series RC circuit. (b) The average fluorescence intensity and selected images of the surface which correspond to the labels in (b). The images A′ and F′ were taken before the interaction with the liposomes and are represented with differing gray-scale palettes.

existed for the liposomes containing unquenched dye in the bulk of the electrolyte, hindering the observation of interfacial behavior. A true fusion event would result in the transfer of dye from the adsorbed monolayer to the hemi-liposomal feature. Through lateral diffusion, a highly fluorescent interface would result, since the dye molecules in the hemi-liposomes would be furthest from the Hg and would not be quenched to the degree expected for an adsorbed dye molecule.

The *in-situ* capacitance, average fluorescence intensity, and fluorescence images are shown in Fig. 3.9 for a dye-containing lipid-coated Hg surface in electrolyte containing liposomes. The formation of the hybrid layer is possible only after a potential excursion to values more negative than the pseudo-capacitance peaks. Before fusion, little change in the fluorescence intensity was measured for potentials positive of the peaks. A slight decrease in the fluorescence intensity at the negative limit was observed, in agreement with the adsorbed layer created by transfer of DOPC from the electrolyte surface. The images A' and F' show a relatively constant level of fluorescence. Scanning the potential negative of the capacitance peaks resulted in a significant change in the observed fluorescence due to the fusion of liposomes with the dye-containing DOPC monolayer. The images and fluorescence intensity curve labeled A–G in Fig. 3.9 were acquired after poration of the monolayer and therefore after allowing interaction with liposomes. After liposome fusion, transfer of dye from the monolayer into the hemi-liposomal feature occurred, indicating that the interaction of the liposomes with the DOPC monolayer was intimate, resulting in an exchange of the probe dye molecules. The large increase in fluorescence and evidence of true fusion is significant and further supports our model of an adsorbed liposome hybrid layer.

3.5.5
Fluorescence Imaging of the Reductive Desorption of an Alkylthiol SAM on Au

The desorption of an alkylthiol SAM involves the reduction of the Au-S bond and the removal of the resulting alkylthiolate from the electrode surface. These two processes may occur sequentially or at the same time. In both cases, the passage of electrochemical current is observed resulting from the faradaic reduction and as a result of changes in the interfacial capacitance. Usually these processes overlap and are difficult to resolve using electrochemistry. The reductive desorption of alkylthiol SAM has been studied extensively, and the potential for this process depends strongly on the surface crystallography of the gold surface [67]. For polycrystalline gold electrodes, the reductive desorption process can take place over a large potential region, and electrochemical measurements can only give an estimate of the processes occurring at the electrode surface. Fluorescence microscopy was used to investigate the reductive desorption of a BODIPY-modified C_{10} alkylthiol from a gold-bead electrode measured in basic solution [68]. The SAM, created *ex-situ*, resulted in a low-capacitance film at 0 V/SCE. From electrochemical measurements in 0.1 M NaOH, reductive desorption began at −1.25 V/SCE. The desorption of the fluorescent thiol did not occur uniformly over the electrode surface. The desorption of the molecule resulted in a fluorescent signal that was localized to specific regions on the electrode surface, as shown in Fig. 3.10a. Desorption was observed to occur from a

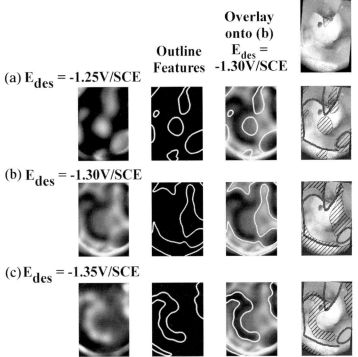

Fig. 3.10 Fluorescence imaging of a SAM-coated Au bead during reductive desorption of fluorescent thiol. Heterogeneity of the removal of thiol as indicated by fluorescence is correlated to the surface morphology (right hand column of figures). Clear evidence of selective removal of the thiol is presented as being due to extent of potential perturbation.

small facet that was present on the Au bead surface and was determined to have (111) surface crystallography using electron back-scattered diffraction (EBSD). The created thiolates diffused away from the electrode surface in contrast to the physically adsorbed surfactant system on Au. The short length of the alkyl chain and charged headgroup contributes to this slight solubility, resulting in diffusion away from the electrode surface. For the same electrode, potential excursions more negative caused further reductive desorption, but from different regions of the electrode. Moreover, the facet appears dark in these fluorescent images, while the region around the facet was fluorescent (Fig. 3.10b). The reductive desorption is strongly dependent upon the character of the surface, and the extent of negative potential excursion determines the region from which the thiol is removed. All the thiol was removed from the gold surface once the potential reached -1.35 V/SCE.

A comparison of the fluorescence images at each of the negative potentials, with the image of the surface constructed from an *ex-situ* analysis of the bead after etching in aqua regia to expose the grain boundaries, shows a distinct correlation. Figure 3.10 compares an optical image oriented appropriately and the potential-dependent fluorescence images. The fluorescence features were outlined and compared, showing the selective nature of the potential desorption process. These outlines were also applied to the optical image, showing a correlation with the visible features on the etched bead surface. From this analysis, it is clear that the reductive desorption process occurs in a heterogeneous fashion across a polycrystalline surface and also results in the presence of the produced thiolate at the electrode surface. Experiments on a longer alkyl chain analog are ongoing and may show that the desorbed molecules remain near the electrode surface available for possible re-adsorption. Fluorescence imaging will contribute to the further understanding of the reductive desorption process.

3.6
Conclusions and Future Considerations

Fluorescence microscopy is proving to be a simple but effective mode of surface analysis that is tailored to interfaces that exclusively contain the fluorophore. The coupling of electrochemical and fluorescence techniques is unique, and we have learned and borrowed much from the use of fluorescence microscopy in the study of complex biological systems. The process was modified so as to adapt to the quenching due to the metal electrode, which turned out to be a two-edged sword. On one hand, the distance-dependent quenching acts as a ruler for measuring the distance the fluorophores are separated from the electrode surface once desorbed, which is advantageous since the required *in-situ* measurement is non-trivial. In addition, the quenching process, which decreases the lifetime of the excited state, tends to save the fluorophore from photobleaching, extending the useful time for analysis. On the other hand, quenching allows examination of only the molecules desorbed from the electrode surface and is blind to the adsorbed surfactant. We have demonstrated that in some cases the type of hydrophobic fluorophore chosen can have a large impact on the measurement. We intend to begin a more systematic study of the characteristics of various classes of fluorophores with an eye to quantifying the distance–fluorescence quenching relationship and using this information in a multivariate methodology that will determine the distribution of the distances the aggregates maintain from the electrode surface at these negative potentials.

Other popular techniques used in biological studies include FRET, which may be applied in these systems as well. The transfer of energy from a donor to an acceptor fluorophore is also distance dependent and can be used to explore the dynamics of the desorbed layer in terms of diffusion and exchange processes.

The imaging of mixed monolayers similarly to what is done on the floating monolayers is also of interest on the electrode surface. Segregation of the dye to various regions of the monolayer, for example, between co-existence areas in lipid monolayer, is clearly observed [69] in the Langmuir trough. It is of interest to pursue in this way the effect of an electric field on the phase behavior of these layers and to determine whether the phase co-existence is also possible for the desorbed monolayer.

The fluorescence of adsorbates on electrode surfaces is not immune to the problems normally encountered in fluorescence microscopy of biological samples. Relating the measured fluorescence to an interesting physical parameter can be difficult because of changes in fluorescence intensity due to scattering in samples, variation caused by photobleaching, and interference from other chromophores. Near a metal surface, we have difficulty in separating the intensity change due to a change in the concentration of the fluorophore from that due to changes in the separation from the electrode surface. The use of fluorescence lifetime imaging would be less affected than the intensity-based measurements which convolute lifetime and concentration. Lifetimes are an intrinsic property of a fluorophore, and measurement of this property has been adopted to avoid problems associated with fluorescence intensity measurements, such as scattering in samples, variation caused by photobleaching, and interference from other chromophores. We are working to modify the instrumentation to allow for fluorescence lifetime measurements in an electrochemical environment in an attempt to solve the problem of how far from the electrode surface are the desorbed surfactants.

Structures and Abbreviations

5-C_{18}Fl

DiIC$_{18}$(5)

TP

BODIPY-C_{10}SH

Acknowledgments

The authors would like to express their thanks to NSERC for funding this work and for scholarship support (JS). Thanks are also extended to the UBC mechanical, electronic, and glassblowing services, whose efforts have been crucial in the development of these experiments. The determination of the surface crystallography by electron-backscattered diffraction (EBSD) was performed by Prof. C. Sinclair, and creation of the stepped surface was accomplished by Dr. M. Beaudoin.

References

1 A. C. Simon (1973): Optical microscopy in electrochemistry. In: Advances in electrochemistry and electrochemical engineering. (Eds: P. Delahay and C. W. Tobias). Wiley, New York, 423–526.

2 S. Bradbury and B. Bracegirdle (1998): Introduction to Light Microscopy. Springer, New York, 123 pages.

3 J. James and H. J. Tanke (1991): Biomedical Light Microscopy. Kluwer Academic Publishers, Dordrecht, 192 pages.

4 D. B. Murphy (2001): Fundamentals of light microscopy and electronic imaging. Wiley-Liss, New York.

5 J. E. Vitt and R. C. Engstrom, Anal. Chem., 69 (1997) 1070.

6 W. J. Bowyer, J. Xie and R. C. Engstrom, Anal. Chem., 68 (1996) 2005.

7 A. K. Gaigalas and T. Ruzgas, J. Electroanal. Chem., 465 (1999) 96.

8 A. K. Gaigalas, L. Li and T. Ruzgas, Curr. Top. Colloid Interface Sci., 3 (1999) 83.

9 L. Li, T. Ruzgas and A. K. Gaigalas, Langmuir, 15 (1999) 6358.

10 L. Li, C. Meuse, V. Silin, A. K. Gaigalas and Y. Zhang, Langmuir, 16 (2000) 4672.

11 L. Wang, V. Silin, A. K. Gaigalas, J. Xia and G. Gebeyehu, J. Colloid Interface Sci., 248 (2002) 404.

12 J. M. Pope, Z. Tan, S. Kimbrell and D. A. Buttry, J. Am. Chem. Soc., 114 (1992) 10085.

13 J. M. Pope and D. A. Buttry, J. Electroanal. Chem., 498 (2001) 75.

14 M. A. Fox, J. K. Whitesell and A. J. McKerrow, Langmuir, 14 (1998) 816.

15 K. W. Kittredge, M. A. Fox and J. K. Whitesell, J. Phys. Chem. B, 105 (2001) 10594.

16 R. R. Chance, A. Prock and R. Silbey (1978): Molecular fluorescence and energy transfer near interfaces. In: Advances in Chemical Physics. Vol. 37 (Ed: I. Prigogine). Interscience Publishers, New York, 1–65.

17 W. L. Barnes, J. Mod. Opt., 45 (1998) 661.

18 S. Astilean and W. L. Barnes, Appl. Phys. B: Lasers and Optics, 75 (2002) 591.

19 X. Zhang and A. J. Bard, J. Phys. Chem., 92 (1988) 5566.

20 Y. S. Obeng and A. J. Bard, Langmuir, 7 (1991) 195.

21 X. H. Xu and A. J. Bard, Langmuir, 10 (1994) 2409.

22 S. Krasne, Biophys. J., 30 (1980) 415.

23 S. Krasne, Biophys. J., 30 (1980) 441.

24 P. J. Sims, A. S. Waggoner, C.-H. Wang and J. F. Hoffman, Biochemistry, 13 (1974) 3315.

25 D. E. Wolf (1988): Probing the Lateral Organization and Dynamics of Membranes. In: Spectroscopic Membrane Probes. Vol. 1 (Ed: L. M. Loew) CRC Press, Boca Raton, Florida (240 pp.)

26 J. M. Crane and L. K. Tamm, Biophys. J., 86 (2004) 2965.

27 J. Zhang, R. M. Davidson, M. D. Wei and L. M. Loew, Biophys. J., 74 (1998) 48.

28 R. B. Silver, Methods Cell Biol., 56 (1998) 237.

29 R. R. Chance, A. H. Miller, A. Prock and R. Silbey, J. Chem. Phys., 63 (1975) 1589.

30 H. Kuhn, D. Möbius and H. Bücher (1972): Spectroscopy of Monolayer Assemblies. In: Physical Methods of Chemistry. Part IIIB Optical, Spectroscopic, and Radioactivity Methods. Vol. 1 (Eds: A. Weissberger and B.W. Rossiter). Wiley-Interscience, New York, 577–578.

31 H. Kuhn and D. Moebius (1993): Monolayer Assemblies. In: Physical Methods of Chemistry, 2nd ed. Vol. IXB (Eds: B.W. Rossiter and R.C. Baetzold). Wiley Interscience, New York, 375–542.

32 K.H. Drexhage (1974): Interaction of light with monomolecular dye layers. In: Progress in Optics. Vol. 12 (Ed: E. Wolf). Interscience, Amsterdam, 163–232.

33 I. Pockrand, A. Brillante and D. Moebius, Chem. Phys. Lett., 69 (1980) 499.

34 H.C. Cheung (1991): Resonance Energy Transfer. In: Topics in Fluorescence Spectroscopy. Vol. 2 (Ed: J.R. Lakowicz). Plenum Press, New York, 128–176.

35 K.H. Drexhage, J. Lumin., 1/2 (1970) 693.

36 T.D. Harris, A.M. Glass and D.H. Olson, Analytical Chemistry Symposia Series, 19 (1984) 49.

37 I. Gryczynski, J. Malicka, Z. Gryczynski, C.D. Geddes and J.R. Lakowicz, J. Fluorescence, 12 (2002) 11.

38 J.R. Lakowicz, J. Malicka, I. Gryczynski, Z. Gryczynski and C.D. Geddes, J. Phys. D: Appl. Phys., 36 (2003) R240.

39 J.R. Lakowicz, Y. Shen, S. D'Auria, J. Malicka, J. Fang, Z. Gryczynski and I. Gryczynski, Anal. Biochem., 301 (2002) 261.

40 J.G. McNally, T. Karpova, J. Cooper and J.A. Conchello, Methods (Orlando, Florida), 19 (1999) 373.

41 J.C. Russ (1995): The Image Processing Handbook, 2nd ed. CRC Press, Boca Raton. 674 pages.

42 S.B. Howell (2000): Handbook of CCD Astronomy. Cambridge University Press, Cambridge, UK.

43 D.S. Everhart, Handbook of Applied Surface and Colloid Chemistry, 2 (2002) 99.

44 A. Badia, R.B. Lennox and L. Reven, Acc. Chem. Res., 33 (2000) 475.

45 T. Ishida, Springer Series in Chemical Physics, 70 (2003) 91.

46 D.K. Schwartz, Annu. Rev. Phys. Chem., 52 (2001) 107.

47 J.C. Love, L.A. Estroff, J.K. Kriebel, R.G. Nuzzo and G.M. Whitesides, Chem. Rev. (Washington, DC, United States), 105 (2005) 1103.

48 D. Bizzotto, Y. Yang, J.L. Shepherd, R. Stoodley, J.O. Agak, V. Stauffer, M. Lathuillière, A. Ahktar and E. Chung, J. Electroanal. Chem., 547 (2004) 165–184.

49 A. Frumkin and B. Damaskin (1964): Adsorption of Organic Compounds at Electrodes. In: Modern Aspects of Electrochemistry. Vol. 3 (Eds: J. O'M. Bockris and B.E. Conway). Butterworths, London, 149–223.

50 J.O. Agak, R. Stoodley, U. Retter and D. Bizzotto, J. Electroanal. Chem., 562 (2004) 135.

51 J.L. Shepherd and D. Bizzotto, J. Phys. Chem. B, 107 (2003) 8524.

52 J. Lipkowski and L. Stolberg (1992): Molecular Adsorption at Gold and Silver Electrodes. In: Adsorption of Molecules at Metal Electrodes (Eds: J. Lipkowski and P.N. Ross). VCH, New York, 171–238.

53 E. Chung, J.L. Shepherd, D. Bizzotto and M.O. Wolf, Langmuir, 20 (2004) 8270.

54 A. Nelson, Biophys. J., 80 (2001) 2694.

55 A. Nelson, Langmuir, 12 (1996) 2058.

56 M.R. Moncelli, L. Becucci, A. Nelson and R. Guidelli, Biophys. J., 70 (1996) 2716.

57 F.A.M. Leermakers and A. Nelson, J. Electroanal. Chem., 278 (1990) 53.

58 A. Nelson and F.A.M. Leermakers, J. Electroanal. Chem., 278 (1990) 73.

59 A.C. Pineda and D. Ronis, J. Chem. Phys., 83 (1985) 5330.

60 R. Stoodley and D. Bizzotto, Analyst, 128 (2003) 552.

61 D. Bizzotto and A. Nelson, Langmuir, 14 (1998) 6269.

62 A. Nelson and H.P. van Leeuwen, J. Electroanal. Chem., 273 (1989) 183.

63 A. Nelson and H.P. van Leeuwen, J. Electroanal. Chem., 273 (1989) 201.

64 J. Heyrovsky and J.K. Kuta (Eds.) (1966): Principals of Polarography. Academic Press, New York. 581 pages. p. 444.

65 F.T. Buoninsegni, L. Becucci, M.R. Moncelli and R. Guidelli, J. Electroanal. Chem., 500 (2001) 395.

66 V. Stauffer, R. Stoodley, J.O. Agak and D. Bizzotto, J. Electroanal. Chem., 516 (2001) 73.

67 S.S. Wong and M.D. Porter, J. Electroanal. Chem., 485 (2000) 135.

68 J.L. Shepherd, A. Kell, E. Chung, C.W. Sinclar, M.S. Workentin and D. Bizzotto, J. Amer. Chem. Soc., 126 (2004) 8329.

69 H.M. McConnell, Annu. Rev. Phys. Chem., 42 (1991) 171.

4

Linear and Non-linear Spectroscopy at the Electrified Liquid/Liquid Interface *

David J. Fermín

4.1
Introductory Remarks and Scope of the Chapter

Electrochemistry at the interface between two immiscible electrolyte solutions is a discipline that occupies a rather special place, combining fundamental aspects of ions and molecules in condensed phase with heterogeneous processes at electrified junctions. Interest in these molecular interfaces ranges from ion-selective electrodes and ion detectors to drug transport in biological membranes and simplified model systems for life-sustaining processes [1–3]. The most widely studied systems involve an aqueous electrolyte containing highly hydrophilic ions in contact with an organic phase featuring large hydrophobic ions. Under this condition, the interface behaves as an ideally polarizable system over a certain potential range. On the other hand, the introduction of a partitioning ion can fix the Galvani potential difference according to the relative concentrations in both phases.

Conventional approaches based on electrochemical techniques, surface tension, and extraction methods have allowed the establishment of thermodynamic and kinetic information concerning partition equilibrium, rate of charge transfer, and adsorption of surfactant and ionic species at the liquid/liquid interface [4–6]. In particular, electrochemical methods are tremendously sensitive to charge transfer processes at this interface. For instance, conventional instrumentation allowed the monitoring of ion transfer across a liquid/liquid interface supported on a single micron-sized hole [7, 8]. On the other hand, the concentration profile of species reacting at the interface can be accurately monitored by scanning electrochemical microscopy [9, 10]. However, a detailed picture of the chemical environment at the junction between the two immiscible liquids cannot be directly accessed by purely electrochemical means. The implementation of *in-situ* spectroscopic techniques has allowed access to key information such as:

---

* Please find lists of Symbols and Abbreviations at the end of this chapter.

Advances in Electrochemical Science and Engineering Vol. 9.
Edited by Richard C. Alkire, Dieter M. Kolb, Jacek Lipkowski and Philip N. Ross
Copyright © 2006 WILEY-VCH Verlag GmbH & Co. KGaA, Weinheim
ISBN: 3-527-31317-6

1. Solvent density, electrostatic potential and polarity profiles at the liquid/liquid boundary,
2. Molecular organization of solvents and specifically adsorbed species,
3. Protonation, aggregation and ion pairing at the interfacial region.

A key requirement for *in-situ* spectroscopic methods in these systems is surface specificity. At liquid/liquid junctions, separating interfacial signals from the overwhelmingly large bulk responses in linear spectroscopy is not a trivial issue. On the other hand, non-linear spectroscopy is a powerful tool for investigating the properties of adsorbed species, but the success of this approach is closely linked to the choice of appropriate probe molecules (besides the remarkably sensitivity of sum frequency generation on vibrational modes of water at interfaces). This chapter presents an overview of linear and non-linear optical methods recently employed in the study of electrified liquid/liquid interfaces. Most of the discussion will be concentrated on the junctions between two bulk liquids under potentiostatic control, although many of these approaches are commonly employed to study liquid/air, phospholipid bilayers, and molecular soft interfaces.

4.2
Linear Spectroscopy

The buried nature of the liquid/liquid junctions introduces tremendous difficulties for *in-situ* spectroscopic analysis of species confined to the interfacial region. UV-visible absorption, luminescence, and Raman signal associated with species at the interface are overwhelmed by the signal arising from the species in the bulk liquids. In this section, different approaches are highlighted in order to increase the linear optical signals arising from probes located at the interface.

4.2.1
Total Internal Reflection Absorption/Fluorescence Spectroscopy

The relationship between the angles of incident (θ_i) and refracted (θ_r) light impinging on a surface is determined by the refractive indexes of both phases according to Snell's law,

$$n_i \sin \theta_i = n_r \sin \theta_r \tag{1}$$

where n_i and n_r are the refractive indexes of the incident and refracted phases, respectively. In the case where $n_i > n_r$, a critical angle of incidence (θ_c) can be obtained in which no refracted light occurs. This condition is commonly referred to as total internal reflection (TIR), where θ_c is defined as

$$\theta_c = \arcsin \frac{n_r}{n_i} . \tag{2}$$

Under TIR conditions, the intensity of the incident electromagnetic field decreases exponentially into the phase of lower refractive index, as illustrated in Fig. 4.1. The intensity of the evanescent field (I) as a function of the distance from the interface in the phase of lower refractive index (z) is determined by

$$I(z) = I_0 e^{-z\frac{4\pi}{\lambda}\sqrt{n_i^2 \sin^2 \theta_i - n_r^2}} . \tag{3}$$

where I_0 and λ are the intensity and wavelength of the incident light. From Eq. (3), the penetration depth of light (ζ) associated with the evanescence field is given by

$$\zeta = \frac{\lambda}{4\pi\sqrt{n_i^2 \sin^2 \theta_i - nr^2}} . \tag{4}$$

Taking as an example the interface between water ($n_w = 1.332$) and 1,2-dichloroethane (DCE, $n_{DCE} = 1.444$), TIR can be accomplished by illumination through the organic phase. According to Eq. (2), the critical angle for this interface is 67.2°, and for a typical angle of 75°, the penetration depth is approximately $\lambda/5$ (see Fig. 4.1). If we consider a charge transfer process generating absorbing/luminescent species at the DCE side of the interface, spectroscopic studies in TIR will be rather sensitive to the changes in concentration in the diffuse layer in the organic side. On the other hand, if the probe is confined in the aqueous side, the spectroscopic signal will involve contributions from species located within 100 nm from the liquid/liquid boundary.

Kakiuchi and coworkers studied the transfer kinetics of ionic fluorescence probes via monitoring the fluorescence signal under TIR illumination [11–14]. A conventional arrangement for this type of measurement is exemplified in Fig. 4.2 [15]. The experiments were performed at the polarized water/DCE interface, with the anionic probe Eosin B (EB) transferring from the aqueous to the organic electrolytes. In this configuration, the most important contributions to

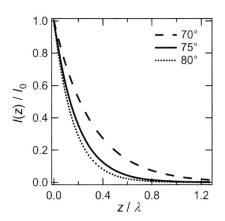

Fig. 4.1 Evanescent wave profile for three different angles higher than the critical angle θ_c. The z axis is associated with the distance from the interface into the phase of lower refractive index. The refractive indexes were taken as $n_r = 1.332$ and $n_i = 1.444$ to simulate the profiles at the water/DCE.

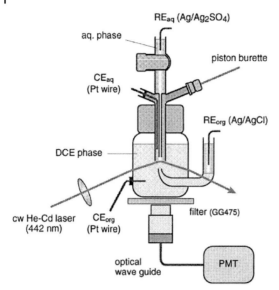

Fig. 4.2 Schematic diagram of an electro-chemical cell arrangement for *in-situ* fluorescence measurements in TIR. The fluorescence is collected in a direction normal to the liquid/liquid interface via an optical wave guide and measured by a photomultiplier tube (PMT). Reprinted with permission from Ref. [15]. Copyright (2000) American Chemical Society.

the changes of the fluorescence intensity under TIR illumination arise from changes in concentration of the probe in the diffusion layer at the organic side of the interface. Considering that the initial concentration of EB in the organic phase is rather small, the fluorescence intensity associated with the species transferred to the organic phase (F^o) is given by

$$F^o(t) = \varepsilon^o \Phi^o I_0 \int_0^\infty c_i^o(z, t) \mathrm{d}z \tag{5}$$

where I_0 is the photon flux of the excitation beam and ε^o, Φ^o and c_i^o are the absorption coefficient, fluorescence quantum yield, and concentration of the probe in the organic phase, respectively. Under potentiostatic conditions, the concentration of the transferring ion corresponds to the temporal integral of the associated faradaic current (i_F)

$$\int_0^\infty c_i^o(z, t) \mathrm{d}z = \frac{1}{z_i F A} \int_0^\tau i_F(\tau) \mathrm{d}\tau \tag{6}$$

where A is the surface area of the liquid/liquid junction and z_i is the charge of the transferring ion. A simple derivation involving Eqs. (5) and (6) shows that changes in the fluorescence intensity are directly proportional to the faradaic current (i_F)

$$\frac{dF^\circ}{dt} = \frac{\varepsilon^\circ \Phi^\circ I_0}{z_i FA} i_F(t).$$ (7)

According to Eq. (7), the derivative of the fluorescence signal measured during a potential cycle will provide a signal proportional to the corresponding cyclic voltammogram. In Fig. 4.3, representative *differential cyclic voltfluorograms* obtained for the transfer of the probe EB at various scan rates are contrasted to the cyclic voltammogram recorded at 200 mV s^{-1} [12]. Both measurements provide consistent values of the formal ion transfer potential. The sensitivity of the spectroscopic signal is also illustrated in Fig. 4.3c, where the initial concentration of EB in the aqueous phase was 5 nM. One of the key advantages of this approach is the specific information concerning the transferring ion. In the case where the electrochemical signal involves several transferring ions in a narrow potential range, the fluorescence signal allows deconvoluting the contribution of each of the probes.

Ding and coworkers have employed a similar approach to study ion [16–18] and heterogeneous electron transfer [19, 20] reactions at the liquid/liquid interfaces. In this case, the reflected light was collected in TIR to estimate the changes in intensity associated with the generation of absorbing species. The

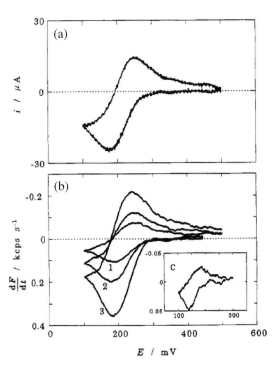

Fig. 4.3 Cyclic voltammogram of the transfer of EB (0.5 mM) at 200 mV s^{-1} (a), and differential cyclic voltfluorograms recorded at 10 mV s^{-1} and EB concentrations of 10 (1), 20 (2) and 50 nM (3) (b). In (c), the concentration of EB was reduced to 5 nM. Reprinted with permission from Ref. [12]. Copyright (1994) American Chemical Society.

absorption spectra illustrated in Fig. 4.4a show the generation of the radical anion tetracyanoquinodimethane (TCNQ⁻) during the voltage-induced heterogeneous reduction by ferrocyanide at the water/DCE interface [19]. The interfacial electron transfer can be expressed as

$$\left[Fe(CN)_6^{4-} \right]_{aq} + [TCNQ]_{DCE} \xrightleftharpoons[k_b]{k_f} \left[Fe(CN)_6^{3-} \right]_{aq} + [TCNQ^-]_{DCE} \tag{8}$$

As indicated by the subscripts "aq" and "DCE", reactants and products remain in their respective electrolyte solutions. The proportionality between the change in absorbance in TIR (A_{TIR}) at a given wavelength and the faradaic current is given by

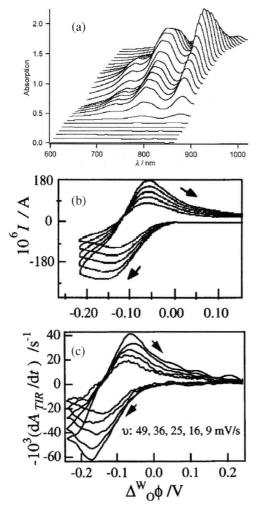

Fig. 4.4 *In-situ* absorption spectra in TIR of TCNQ⁻ generated by heterogeneous reduction of TCNQ by the hexacyanoferrate redox couple at the water/DCE interface (a). From bottom to top, the potential is changed from 0.187 to –0.26 V and reversed back to the initial potential at steps of 40 mV. From Ref. [19]. Reproduced by permission of the Royal Society of Chemistry. Corresponding cyclic voltammograms (b) and differential cyclic voltabsorptiograms recorded at 670 nm and scan rates of 9, 16, 25, 36 and 49 mV s⁻¹. Reprinted from Ref. [20] with permission from Elsevier Science.

$$\frac{dA_{TIR}}{dt} = \frac{2\varepsilon^o}{z_i FA \cos \theta_i} i_F(t) \tag{9}$$

where θ_i is the angle of incident light with respect to the interface normal. As no other faradaic processes take place in the potential range around the formal electron transfer potential of the reaction in Eq. (8), the faradaic current i_F is only related to the heterogeneous electron transfer process. Figure 4b and c also show the direct correlation between the spectroscopic and electrochemical signals [20].

Changes in A_{TIR} associated with the generation of $TCNQ^-$ during a potential step were also employed for monitoring the rate of the heterogeneous electron transfer reaction. Following the development previously published by Kakiuchi et al. [12], it follows that for short times after the potential step

$$A_{TIR}(t) = \frac{2\varepsilon_{TCNQ^-} c_{TCNQ} k_f}{\lambda^2 \cos \theta_i} \left[\frac{2\lambda\sqrt{t}}{\sqrt{\pi}} - 1 \right] \tag{10}$$

where

$$\lambda = \frac{k_f}{\sqrt{D_{TCNQ}}} + \frac{k_b}{\sqrt{D_{TCNQ^-}}} \tag{11}$$

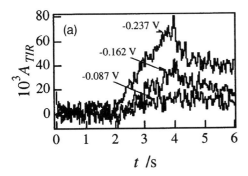

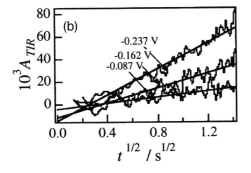

Fig. 4.5 Chronoabsorptograms recorded during the reduction of TCNQ at various potentials (a). The initial potential was held at 0.215 V for 2 s before being stepped to the indicated potential for a further 2 s and stepped back to the initial value. The A_{TIR} vs $t^{0.5}$ curves (b) allow estimation of the rate constant of electron transfer according to Eq. (10). Reprinted from Ref. [20] with permission from Elsevier Science.

D, k_f and k_b are the diffusion coefficient and the forward (reduction of TCNQ) and backward (oxidation of TCNQ$^-$) electron transfer rate constants, respectively. Eqs. (10) and (11) are effectively valid for this system only within the so-called *constant phase approximation*. In these studies, the concentration of the redox species in the aqueous phase is in large excess with respect to TCNQ in the organic phase, and diffusion profiles are only developed in the latter. The absorbance changes upon several potential steps are exemplified in Fig. 4.5 for the process in Eq. (8). The sudden decrease in the absorbance after 4 s takes place as the potential is stepped back to the initial value (–0.215 V). From the slope and intercept of the A_{TIR} versus $\sqrt{t}$ curves in Fig. 4.5 b, the heterogeneous electron transfer rate constant can be effectively estimated. This approach, commonly referred to as *chronoabsorptometry*, has provided valuable kinetic information on interfacial processes such as ion transfer [17, 18] and facilitated ion transfer [21].

4.2.2
Potential-modulated Reflectance/Fluorescence in TIR

Periodic perturbations of the potential across the liquid/liquid boundary induce a modulation of the concentration of species located in the interfacial region. By collecting the spectroscopic signals at the same frequency as that of the potential perturbation, employing phase-sensitive detection, the interfacial sensitivity of the measurements is tremendously enhanced, as the contribution from species in the bulk of the electrolyte solutions can be effectively neglected. Based on this principle, Fermín and co-workers introduced *potential-modulated reflectance* (PMR) and *potential-modulated fluorescence* (PMF) to study a variety of processes including ion transfer [22], electron transfer [20], and the specific adsorption of ionic species [15].

Let us consider a sinusoidal Galvani potential difference ($\phi^w - \phi^o = \Delta_o^w \phi$) of the form

$$\Delta_o^w \phi = (\Delta_o^w \phi)_0 + (\Delta_o^w \phi)_1 \exp(i\omega t) \tag{12}$$

where the sub-indexes 0 and 1 correspond to the dc and frequency-dependent components, respectively. The frequency of potential modulation is represented by ω. In the presence of a fluorescent probe transferring from the aqueous to the organic phase, the potential modulation generates an AC perturbation of the interfacial concentration of the probe. From Eq. (9), it can easily be derived that the frequency-dependent fluorescence signal (ΔF) is related to the AC faradaic current (i_F) by [15]

$$i\omega\Delta F = \frac{4.606 \, \varepsilon^o \Phi^o I_0}{z_i FA \cos \theta_i} i_F(t). \tag{13}$$

Eq. (13) indicates that the periodic optical response is 90° out of phase with the faradaic current. This point is illustrated in Fig. 4.6, where the AC voltammo-

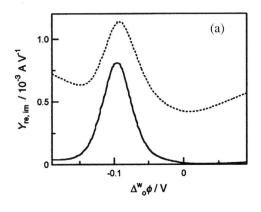

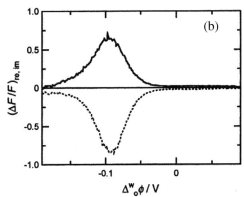

Fig. 4.6 AC admittance (a) and PMF responses (b) in TIR for the transfer of $Ru(bipy)_3^{2+}$ from water to DCE. The continuous and dashed lines correspond to the real and imaginary components of the frequency-dependent signals, respectively. Reprinted with permission from Ref. [15]. Copyright (2000) American Chemical Society.

gram and AC voltfluorogram for the transfer of $Ru(bipy)_3^{2+}$ from water to DCE are contrasted [15]. The real and imaginary components of the admittance exhibit a bell-shaped response with a maximum located at the formal transfer potential of the transferring ion. The AC-fluorescence response shows a similar potential dependence, but the imaginary component has a negative sign. The contribution of the double layer charging to the admittance responses manifests itself by the offset of the imaginary component with respect to the real part. In contrast to the electrochemical responses, the optical signal is not affected by this element in a first approximation. This is due to the fact that the PMF response is directly related to the faradaic current.

The relationship between the AC fluorescence and the faradaic admittance also allows us to study the dynamics of ion transfer responses as a function of the frequency modulation. For a quasi-reversible ion transfer process featuring planar diffusion profiles, the real and imaginary components of the AC fluorescence can be expressed as [15, 23]

$$\Delta F_{\rm re} = \frac{4.606\,\varepsilon^{\circ}\Phi^{\circ}I_0}{z_i FA\cos\theta_i}\left[\frac{\Delta_{\rm o}^{\rm w}\phi_1\sigma\omega^{-1.5}}{(R_{\rm ct}+\sigma\omega^{-0.5})^2+(\sigma\omega^{-0.5})^2}\right] \tag{14}$$

$$\Delta F_{\rm im} = -\frac{4.606\,\varepsilon^{\circ}\Phi^{\circ}I_0}{z_i FA\cos\theta_i}\left[\frac{\Delta_{\rm o}^{\rm w}\phi_1(R_{\rm ct}+\sigma\omega^{-0.5})\omega^{-1}}{(R_{\rm ct}+\sigma\omega^{-0.5})^2+(\sigma\omega^{-0.5})^2}\right] \tag{15}$$

where

$$R_{\rm ct} = \frac{RT}{z^2 F^2 Ak_{\rm tr}c^{\rm w}} \tag{16}$$

$$\sigma = \frac{4RT\cosh^2(zF(\Delta_{\rm o}^{\rm w}\phi - \Delta_{\rm o}^{\rm w}\phi^{\rm o\prime})/2RT)}{z^2 F^2 Ac^{\rm w}(2D^{\rm w})^{0.5}}. \tag{17}$$

$c^{\rm w}$ and $D^{\rm w}$ correspond to the bulk concentration and diffusion coefficient of the transferring ion in the aqueous phase, respectively. Eq. (16) describes the so-called charge transfer resistance familiar from electrochemical impedance spectroscopy. The parameter $\Delta_{\rm o}^{\rm w}\phi^{\rm o\prime}$ is the formal ion-transfer potential, which is determined by the partition coefficient of the ion in both electrolyte phases. Complex representation of the AC fluorescence signal as exemplified in Fig. 4.7 allows us to estimate the apparent ion transfer rate constant $k_{\rm tr}$. In the case of $Ru(bipy)_3^{2+}$, the value of $k_{\rm tr}$ at the formal transfer potential is of the order of 0.1 cm s^{-1} [23]. Nishi et al. employed the same approach for studying the transfer kinetics of Rose Bengal, obtaining rate constants one order of magnitude smaller [24]. In this case, the kinetics of the ion transfer appears to be affected by the rate of adsorption-desorption processes at the water/DCE boundary.

As indicated in Eq. (4), the penetration depth of light in TIR is limited to about 100 nm, and the spectroscopic signals mostly reflect the changes in

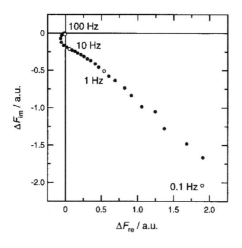

Fig. 4.7 Complex representation of the PMF responses for the transfer of $Ru(bipy)_3^{2+}$ from water to DCE at −0.10 V. Reprinted with permission from Ref. [23]. Copyright (2001) American Chemical Society.

composition of the electrolyte phase of higher refractive index. However, in the case of ionic fluorescent probes featuring high affinity for the liquid/liquid boundary, PMF responses can also provide information on the potential dependence of the surface coverage. Nagatani et al. studied the adsorption properties of the water-soluble porphyrin *meso*-tetrakis(*N*-methypyridyl)porphyrinato zinc(II) (ZnTMPyP^{4+}) at the polarizable water/DCE interface [15, 23]. As exemplified in Fig. 4.8, the admittance and PMF responses of ZnTMPyP^{4+} exhibit more complex features than those observed for Ru(bipy)$_3^{2+}$ (see Fig. 4.6). The imaginary component of the admittance shows a broad signal featuring several shoulders, and the maximum is shifted toward more positive potentials than the real component. On the other hand, PMF responses are observed throughout a wide potential range more negative than the formal transfer potential, i.e. 0.10 V. Recalling that the Galvani potential difference is given by $\phi^w - \phi^o = \Delta_o^w \phi$, the cationic porphyrin remains confined to the aqueous phase at potentials more negative than 0.10 V. The origin of the PMF signal at this potential range has been associated with the specific adsorption of the porphyrin at the interface. In addition to the responses at negative potentials, a well-defined peak and a 180° phase shift of the PMF is observed at potentials more positive than 0.10 V, indicating

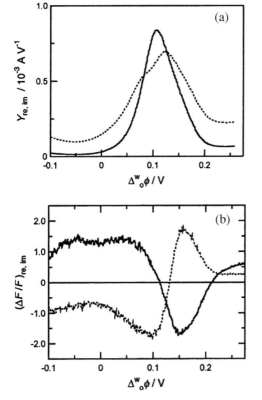

Fig. 4.8 AC admittance (a) and PMF responses (b) in TIR for the adsorption and transfer of ZnTMPyP(tosylate)$_4$ from water to DCE. The continuous and dashed lines correspond to the real and imaginary components of the frequency-dependent signals, respectively. The behavior of the admittance and PMF significantly contrasts with that observed in the presence of Ru(bipy)$_3^{2+}$ because of the specific adsorption of the porphyrin at the interface (see Fig. 4.6). Reprinted with permission from Ref. [15]. Copyright (2000) American Chemical Society.

that other interfacial phenomena take place after the porphyrin is transferred to the organic phase.

The interfacial behavior of ZnTMPyP^{4+} was interpreted within the framework of the so-called two-adsorption planes model [23]. This model is schematically illustrated in Fig. 4.9. Depending on the Galvani potential difference, the adsorption of ZnTMPyP^{4+} can take place either at a plane located at the aqueous side of the interface or at a plane at the organic side. The adsorption equilibria between the bulk concentration in each electrolyte phase and the corresponding adsorption planes are affected by the potential distribution across the interface. The interfacial behavior of the cationic probe initially located in the aqueous phase can be described in terms of three processes as a function of the Galvani potential difference:

1. $\Delta_o^w \phi \leq \Delta_o^w \phi^{o\prime}$; the ZnTMPyP^{4+} species remains confined to the aqueous phase but the surface coverage at the aqueous adsorption plane increases as $\Delta_o^w \phi$ increases,

2. $\Delta_o^w \phi \approx \Delta_o^w \phi^{o\prime}$; the transfer of ZnTMPyP^{4+} takes place and the concentration ratio is given by

$$\Delta_o^w \phi = \Delta_o^w \phi^{o\prime} + \frac{RT}{zF} \ln \frac{c^o}{c^w} \tag{18}$$

3. $\Delta_o^w \phi \geq \Delta_o^w \phi^{o\prime}$; the concentration of the porphyrin in the organic phase will be in equilibrium with the surface coverage at the organic adsorption plane. In contrast to the behavior in 1, the coverage at this adsorption plane will decrease as $\Delta_o^w \phi$ increases. Consequently, the change in the phase of the PMF responses associated with the adsorption at the aqueous and organic adsorption planes comes as a result of the opposite potential dependences of the coverage.

The model developed by Nagatani et al. was based on the assumption that the adsorption and transfer processes are effectively uncoupled and the common steady-state boundary condition is determined by Eq. (18) [23, 25]. Furthermore,

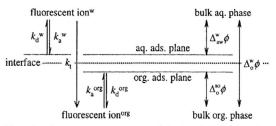

Fig. 4.9 Schematic representation of the liquid/liquid interface featuring two adsorption planes. The various equilibria are affected by the distribution of the Galvani potential difference ($\Delta_o^w \phi$) across the interface. Reprinted with permission from Ref. [23]. Copyright (2001) American Chemical Society.

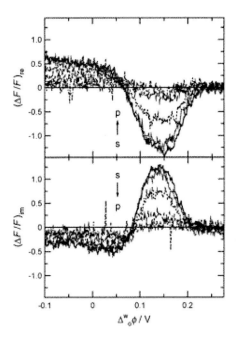

Fig. 4.10 Real and imaginary components of the PMF responses associated with the adsorption of ZnTMPyP^{4+} from the organic side of the water/DCE interface. The interface was illuminated in TIR with a linearly polarized laser beam. The intensity of the PMF responses decreases as the polarization was changed from *s* to *p*. Reprinted with permission from Ref. [15]. Copyright (2000) American Chemical Society.

frequency-dependent studies of the PMF responses also provided information on the kinetics of adsorption at each plane. In the case of ZnTMPyP^{4+}, the adsorption dynamics are characterized by activation energies of the order of 20 kJ mol^{-1}.

PMF studies employing linearly polarized light also allow us to estimate the average orientation of the specifically adsorbed porphyrins. The responses illustrated in Fig. 4.10 show the effect of the angle of light polarization on the normalized PMF signal for the ZnTMPyP^{4+} initially located in the organic phase [15]. Under these conditions, the concentration of the porphyrin in the aqueous phase is negligible, and the PMF signal is dominated by the adsorption process at the organic adsorption plane. Analysis of the PMF intensity as a function of the polarization angle revealed that the average tilting angle of the porphyrin ring at the organic adsorption plane with respect to the surface normal is $78°\pm4°$.

4.2.3
Quasi-elastic Laser Scattering (QELS)

Quasi-elastic Laser Scattering (QELS) provides a versatile tool for monitoring *in situ* the frequency of thermally induced capillary waves at the liquid/liquid junction [26, 27]. Unlike the previous techniques based on an absorption/fluorescence signal arising from molecular probes at the interface, QELS is not sensitive to specific molecules but to the interfacial tension of the molecular junction. A schematic representation of an experimental arrangement for QELS studies under potentiostatic control is illustrated in Fig. 4.11 [28]. The incident laser beam normal to the interface is quasi-elastically scattered by the capillary wave at a characteristic angle (ϖ) determined by

$$K \tan \varpi = k \tag{19}$$

where K and k are the wavenumbers of the incident beam and the capillary wave, respectively. In order to determine ϖ, a diffraction grating featuring a grating constant d is placed along the optical path. It follows that [26, 29]

$$n\lambda = d \sin \varpi \tag{20}$$

where n is the order of the diffraction spot and λ is the wavelength of the incident beam. The Fourier transform of the optical beat of a high order spot detected by a photomultiplier provides a power spectrum in which the maximum frequency (f_0) corresponds to [30]

$$f_0 = \frac{1}{2\pi} \sqrt{\frac{\gamma k^3}{\rho^{\mathrm{w}} + \rho^{\mathrm{o}}}} \tag{21}$$

where γ, ρ^{w} and ρ^{o} are the interfacial tension and the densities of water and the organic phase, respectively.

Su et al. have recently employed QELS to study the potential-induced assembly of Au nanoparticles at the water/DCE interface [31]. The water-soluble Au nanoparticles featured an average diameter of 1.5 ± 0.4 nm and were capped by mercaptosuccinic acid (MSA). The capacitance-voltage curves in Fig. 4.12a show an increase in the differential capacitance (C_{dl}) at negative $\Delta_{\mathrm{o}}^{\mathrm{w}}\phi$ with increasing concentration of the nanoparticles in the aqueous phase. As illustrated in

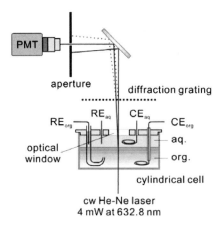

Fig. 4.11 Schematic representation of a QELS set up for measurements at liquid/liquid interfaces under potentiostatic control. Reprinted with permission from Ref. [28]. Copyright (2003) American Chemical Society.

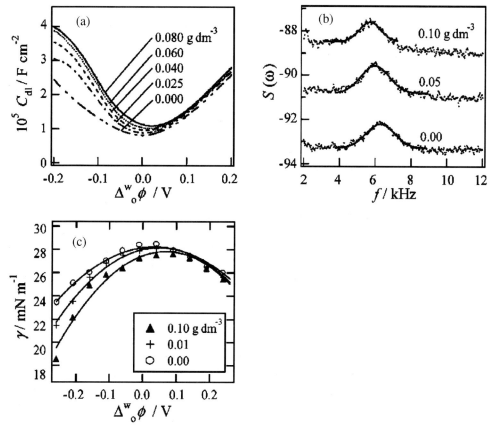

Fig. 4.12 Differential capacitance (C_{dl}) at the water/DCE interface as a function of the applied potential in the presence of water-soluble Au nanoparticles (a). The Au particles stabilized by negatively charged groups are reversibly assembled at the interface upon negative polarization of the water phase with respect to the organic phase. The evolution of the power spectra of the third-order spot of QELS at –0.160 V and various concentrations of Au particles is shown in (b). The increasing number density of particles at the interface manifests itself by a decrease in the frequency of the capillary waves. Electrocapillary curves constructed from the concentration dependence of the capillary wave frequency are displayed in (c). Reprinted with permission from Ref. [31]. Copyright (2004) American Chemical Society.

Fig. 4.12 b, the power spectra of the third-order spot of QELS measured at negative potentials consistently shift to lower frequencies with increasing concentration of the particles, reflecting the voltage-induced assembly of the negatively charged particles at the water/DCE interface. Estimation of the interfacial tension from Eq. (21) allowed the construction of electrocapillary curves as exemplified in Fig. 4.12 c. The increment of the C_{dl} and the decrease of γ at negative potential do indicate that the assembly of nanoparticles can be reversibly controlled by tuning the Galvani potential difference. By comparing the excess charge from the capacitance data and the number density of particles from the concentration dependence of γ, the effective charge per particle at the interface was estimated. This technique has also been employed to study the specific adsorption and aggregation of ionic porphyrin at the polarizable water/DCE interface [28].

More recently, Samec and coworkers investigated the line shape of the fluctuation spectrum at the polarizable water/DCE interface in the presence of the phospholipid DL-a-dipalmitoyl-phosphatidylcholine (DPPC) [32]. The line shape of experimental power spectra similar to those exemplified in Fig. 4.12 b was analyzed in terms of the mean vertical displacement of the interface generated by capillary waves. The experimental results in the presence and absence of DPPC at a wide potential range appear consistent with the description of the liquid/liquid boundary as "molecularly sharp". However, it is not entirely clear from this analysis how sensitive the spectrum line shape is to the molecular organization at the liquid/liquid boundary. As discussed in Section 4.3.2, vibrational sum frequency generation studies of the neat water/DCE interface provide a rather different conclusion.

4.2.4
Other Linear Spectroscopic Studies at the Neat Liquid/Liquid Interface

A variety of spectroscopic methods have been adapted to study the structure and organization of molecular probes at the neat liquid/liquid interface. In this section, some of the approaches compatible to electrochemical control of the Galvani potential difference will be briefly reviewed. It is important to realize that the surface specificity of most of the linear spectroscopic methods relies on the TIR condition. Therefore, changes in the reflection geometry arising from changes in the shape of the interface (meniscus) can lead to severe interference in signal detection. This problem may be particularly significant at potentials close to the edges of the ideally polarizable window, where drastic changes in the interfacial tension may occur. In addition, ion transfer reactions can also lead to important changes in the density at the liquid/liquid boundary, which could induce anomalous beam deflections.

Fujiwara and Watarai recently employed *Total Internal Reflection Resonance Raman Microspectroscopy* to study the adsorption of a water-soluble metalloporphyrin at the toluene/water interface [33]. The experimental set-up schematically shown in Fig. 4.13 a, b features a hemispherical cell which enables the scattering Raman signal to be collected in the normal direction to the interface. The TIR Raman spectra in Fig. 4.13 c shows a band at 394 cm^{-1} corresponding to the symmetric breathing mode of the $Mn(TMPyP)^{5+}$ ring adsorbed at the interface. The spectra were collected at two different concentrations of dihexadecyl phosphate (DHP) in the organic phase. The signal enhancement at higher concentrations of DHP was interpreted in terms of an increase of the interfacial concentration of $Mn(TMPyP)^{5+}$ induced by the adsorption of DHP at the liquid/liq-

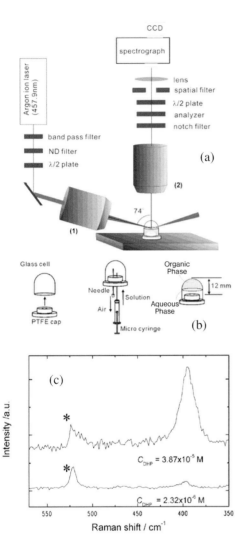

Fig. 4.13 Schematic representation of the optical set-up (a) and hemispherical cell (b) for TIR Raman microspectroscopy. TIR Raman spectra of Mn(TMPyP)$^{5+}$ at the toluene/water interface with two different concentrations of dihexadecyl phosphate (DHP) in the organic phase are shown in (c). The band at 394 cm^{-1} corresponds to the symmetric breathing mode of the porphyrin ring. The spectral features were normalized by the toluene band at 520 cm^{-1}. The spectra were collected with s-polarization. Reprinted with permission from Ref. [33]. Copyright (2003) American Chemical Society.

uid boundary. These studies also provided information on the mean tilt angle of the porphyrin ring with respect to the liquid/liquid plane.

Under a similar configuration, Yamamoto et al. also studied the *Surface-Enhanced Raman Scattering* (TIR-SERS) of oleate ions adsorbed on 5 nm Ag colloids assembled at the toluene/water interface [34]. Through careful comparisons of the SERS spectra of the oleate-stabilized Ag colloids in the bulk water phase and at the liquid/liquid interface, the authors proposed that oleate molecules exhibit two distinctive adsorption modes on the particle surface. At the region of the particles in contact with the aqueous side of the interface, the oleate molecules are adsorbed via the ethylene group, exposing the carboxyl group to the aqueous phase. On the other hand, the oleate is preferentially adsorbed by the carboxyl groups on the particle surface in contact with the organic phase. Furthermore, bands associated with toluene appear to be enhanced at the interface by the local electromagnetic field induced by the particle surface.

Moriya et al. have systematically studied the reflectance spectra of porphyrins adsorbed at the dodecane/water interface over a wide range of angles of illumination [35, 36]. The prism cell design illustrated in Fig. 4.14a allowed the recording of transmission spectra as well as *attenuated total internal reflection* (ATR) and *external reflection* (ER) spectra by adjusting the height of the cell with respect to the light beam. The second design shown in Fig. 4.14b involves a cylindrical cell placed on the center of a goniometer, featuring optical fibres, focusing lenses, and a polarizer for spectroscopic measurements in ER, ATR,

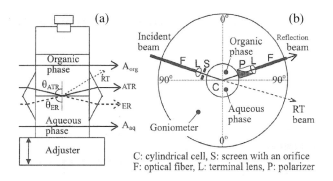

C: cylindrical cell, S: screen with an orifice
F: optical fiber, L: terminal lens, P: polarizer

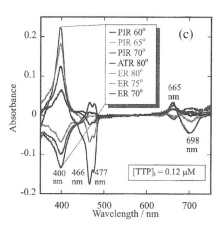

Fig. 4.14 Schematic illustration of a prism cell (a) and variable-angle cell (b) for reflectance spectroscopy at the liquid/liquid interfaces. These optical arrangements allow reflectance measurements with variable angle of illumination extending from attenuated TIR (ATR) and partial internal reflection (PIR) to external reflection (ER). The angle-dependent *p*-polarized reflection spectra of *p*-tolyl-porphyrin at the dodecane/aqueous interface are displayed in (c). Reprinted with permission from Ref. [36]. Copyright (2004) The Japan Society for Analytical Chemistry.

and *partial internal reflection* (PIR) geometries. A set of reflectance spectra obtained in the presence of *meso*-tetra-*p*-tolyl-porphyrin (TTP) at different angles and geometries employing *p*-polarized light are exemplified in Fig. 4.14 c. Depending on the sign and magnitude of the reflectance spectra, the authors obtained information on the interfacial density and average molecular orientation of the porphyrin species. Fujiwara and Watarai also employed polarized ER spectroscopy to study the orientation of MnTPPCl at the toluene/aqueous sodium hydrogen sulfate interface [37]. These interesting approaches reveal that surface-specific information can also be obtained from linear spectroscopy beyond the TIR geometry.

One of the most powerful approaches to studying not only orientation but also excited-state lifetimes and rotational dynamics of probes adsorbed at the liquid/liquid boundary is based on *time-resolved fluorescence/absorbance* in TIR. Wirth and Burbage pioneered the use of time-resolved fluorescence decay anisotropy to study molecular reorientation dynamics at the water/hydrocarbon interface [38]. In this work, the fluorescence decay of acridine orange was recorded as a function of the polarization of the excitation beam in TIR. The authors interpreted the out-of-plane reorientation dynamics in terms of the molecular roughness of the liquid/liquid boundary, while the in-plane responses were associated with interfacial friction. Bessho et al. analyzed pico-second-resolved fluorescence decay profiles of 1-naphthalenesulfonate at the heptane/water interface in terms of two distinctive solvation structures found at different distances from the interface [39]. A long-lifetime signal was found at regions within 1 nm of the interface, revealing either a local low polarity or high viscosity with respect to the bulk aqueous phase. More recently, Ishizaka et al. combined time-resolved fluorescence and dynamic fluorescence anisotropy to study interfacial molecular recognition at the water/CCl_4 junction [40]. It was observed that the fluorescence decay of riboflavin (RF) in water showed a double-exponential profile upon addition of [1,3,5]triazine-2,4,6-triamine (DTT) in the CCl_4 phase. In addition, the rotational dynamics exhibited two responses at 160–220 ps and 650–750 ps in the presence of both species. The faster response is comparable to the free RF signal, while the slower one was associated with an RF-DTT complex formed by triple hydrogen bonding at the interface. The same group has reported the interfacial-induced hybridization of single-stranded DNA (ssDNA) at the water/CCl_4 interface in the presence of octadecylamine by time-resolved fluorescence in TIR [41]. The fluorescence probe employed for the DNA structure was ethidium bromide, which shows that the ssDNA does not hybridize in the bulk of the aqueous phase but only at the liquid/liquid boundary.

Watarai and coworkers employed time-resolved spectroscopy to monitor the excited state lifetime of porphyrin species adsorbed at the toluene/water interface [37, 42]. Illumination with a white light probe and monochromatic pulse excitation in TIR revealed that the porphyrin H_2TMPyP^{4+} exhibits a triplet state lifetime significantly shorter at the interface than in the bulk aqueous phase [42]. On the other hand, Doung et al. reported under similar experimental conditions (only pulse excitation in TIR) that the lifetime of H_2TMPyP^{4+} only appears to be affected upon addition of the redox quencher tetracyanoquinodimethane to the organic phase [43]. However, it should be recalled that the interfacial behavior of TMPyP^{4+} derivatives are rather complex, as demonstrated by PMF studies at the water/DCE interface (see Figs. 4.8 and 4.10). Consequently, although TIR can effectively enhance the surface specificity of the spectroscopic signals, contributions of molecules located within the region of the evanescent wave can mask responses from species assembled at the molecular boundary.

4.3
Non-linear Spectroscopy

Since the pioneering works by Shen, Eizenthal, Corn, and Richmond, *Second Harmonic Generation* (SHG) and *Sum Frequency Generation* (SFG) have provided some of the most detailed studies of the structure of the liquid/liquid interface at the molecular level [44–49]. As discussed in the previous section, one of the major issues in *in-situ* spectroscopy at liquid/liquid interfaces is surface specificity. In linear spectroscopy, surface specificity has been enhanced by working in TIR as well as examining the effects of light polarization and of the Galvani potential difference. Under the so-called electric dipole approximation, non-linear optical responses are forbidden in centrosymmetric media, consequently these signals only arise from molecules assembled at the liquid/liquid boundary. In this section, I shall focus the discussion on recent non-linear optical studies performed under electrochemical control of the Galvani potential difference. Further information on non-linear optics can also be found in Chapter 5 of this book.

4.3.1
Second Harmonic Generation

Second harmonic generation (SHG) is a non-linear optical process in which two incident photons of frequency ω are converted into a single photon with frequency 2ω. The intensity of the SHG signal $I(2\omega)$ involves contribution from the resonant and non-resonant components of the non-linear susceptibility $\chi^{(2)}$,

$$I(2\omega) \propto \left| \chi_N^{(2)} + \chi_R^{(2)} \right|^2 (I(\omega))^2 \tag{22}$$

where the subscripts N and R stand for non-resonant and resonant components, respectively. As the latter contribution is usually overwhelmingly larger than the former, the SHG responses are extremely sensitive to the electronic structure of the probe. The relationship between $\chi_R^{(2)}$ and the molecular organization of the probe at the interface is given by

$$\chi_R^{(2)} = N \sum_{k,e} \frac{(\mu_{g,k}\mu_{k,e}\mu_{e,g})}{(\omega_{gk} - \omega - i\Gamma)(\omega_{eg} - 2\omega + i\Gamma)} \tag{23}$$

where N is the number of molecules contributing to the SHG responses, $\mu_{i,j}$ is the transition matrix element between two defined states, and Γ is the transition line width. The subscripts "g", "k" and "e" indicate the ground state, the intermediate virtual state, and the first excited state, respectively. Eqs. (22) and (23) indicate that the intensity of the SHG signal is proportional to the square of the intensity of the fundamental (incident) beam and also on the square of the number density of the molecular probe assembled at the liquid/liquid boundary. The upper term in Eq. (23) represents an orientational averaged distribution of the transition elements in the laboratory frame of reference. Consequently, evaluation of $\chi_R^{(2)}$ provides information on the molecular orientation of the probes. The maximum of the SHG response is obtained as 2ω approaches ω_{eg}, providing access to the effective electronic transitions of the probe at the interface.

Higgins and Corn published the first application of SHG to externally polarized liquid/liquid interfaces [50]. In this work, the surface coverage of 2-(n-octadecylamino)naphthalene-6-sulfonate (ONS) at the water/DCE interface was studied as a function of the applied Galvani potential difference. As illustrated in Fig. 4.15, the interfacial tension strongly decreases and the intensity of the SHG signal increases as the Galvani potential difference is shifted positively. At positive potentials, the ONS is assembled at the liquid/liquid interface by "anchoring" the sulfonate group in the aqueous phase. The potential-dependent adsorption followed a Frumkin isotherm with a linear dependence of the Gibbs-free energy of adsorption on the applied potential. The authors recognized the potential use of this technique to probe the potential distribution across the interface. However, despite efforts by this and other groups for over 10 years [51, 52], this issue remains a tremendous challenge.

More recently, Nagatani et al. have studied the adsorption of water-soluble meso-substituted porphyrins at the polarizable water/DCE interface [28, 53]. Figure 4.16 contrasts the absorption spectra associated with S_0–S_2 transition (Soret band) of the porphyrins H_2TMPyP^{4+} and $ZnTMPyP^{4+}$ in aqueous and DCE so-

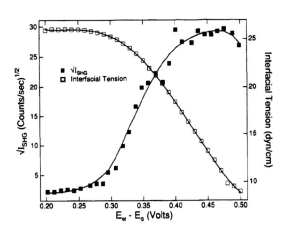

Fig. 4.15 Interfacial tension and square root of SHG intensity as a function of the applied potential at the water/DCE interface in the presence of 2-(n-octadecylamino)-naphthalene-6-sulfonate (ONS). Reprinted with permission from Ref. [50]. Copyright (1993) American Chemical Society.

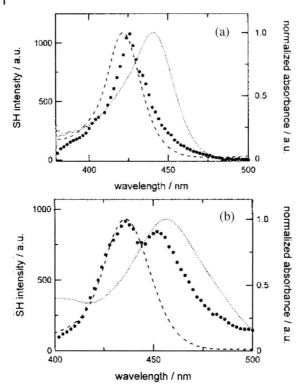

Fig. 4.16 SHG spectra of the water-soluble porphyrin H$_2$TMPyP^{4+} (a) and ZnTMPyP^{4+} (b) at the neat water/DCE interface. The SHG spectra are contrasted with the corresponding absorption spectra in aqueous (dashed line) and organic (dotted line) phases. Reprinted with permission from Ref. [53]. Copyright (2002) American Chemical Society.

lutions and the corresponding SHG spectra [53]. Both porphyrins exhibit 20-nm positive solvatochromic shifts, i.e. the soret band is shifted to longer wavelength with decreasing solvent polarity. The SHG spectrum of H$_2$TMPyP^{4+} showed a maximum at 426 nm, which is 5-nm red-shifted with respect to the maximum of the Soret band in the aqueous phase. The relative intensity of the SHG spectrum is somewhat remarkable considering the high molecular symmetry of the probe, which translates into low values of the first hyperpolarizability [54, 55]. The responses illustrated in Fig. 4.16 show that the liquid/liquid environment does enhance the non-linear optical activity of these molecules. As discussed later on, the red shift of the SHG spectrum with respect to the absorption spectra in aqueous phase can be interpreted in terms of the local polarity probed by the molecule at the interface. In contrast, the SHG spectra of ZnTMPyP^{4+} exhibits two well-defined maxima at 436 and 452 nm (see Fig. 4.16b). The positions of the two SHG maxima are remarkably similar to those of the absorption maxima in water (436 nm) and DCE (456 nm). The relative intensity of the two peaks is affected by the applied Galvani potential difference, as illustrated in Fig. 4.17. As the Galvani potential difference was swept to potentials more positive than the formal ion transfer potential, the transfer of the pophyrin from water to DCE induces a sharp decrease of the SHG intensity. The fact that the SHG signals at 436 and 532 nm show a different potential dependence indicates that this signal arises from two "different" species.

Although the authors suggested that the red-shifted SHG signal of ZnTMPyP^{4+} is related to the formation of J-aggregates at the interface [53], recent results by Steel et al. hinted at the possibility that solvatochromic effects due to the interfacial polarity provide a more plausible explanation [56–58]. The approach employed by Steel and coworkers involved systematic studies of the SHG spectra of the solvatochromic probe p-nitroanisole linked to a sulfonate group by a linear alkyl chain. As shown in Fig. 4.18, the maximum of the SHG spectra at the water/cyclohexane interface shifts toward the absorption maximum in the organic phase as the length of the alkyl chain increases [56]. This important result indicates that the sulfonate group is "anchored" at the aqueous side of the liquid/liquid boundary, and the p-nitroanisole group probes regions of different polarity at the interface, depending on the length of the alkyl chain. Transposing this behavior to that observed for the porphyrin molecules in Fig. 4.16, it can be proposed that the ZnTMPyP^{4+} is adsorbed in two regions of the interface with distinctive polarities. These results can be correlated with the

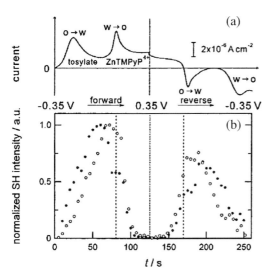

Fig. 4.17 Time-dependent current (a) and SHG intensity (b) during a potential cycle at 5 mV s^{-1} of the water/DCE interface in the presence of ZnTMPyP^{4+}. The full and hollow circles correspond to the SHG intensities at 436 and 452 nm, respectively. A sharp decrease in the SHG signal is observed in the potential range where the ZnTMPyP is transferred from water to DCE. Reprinted with permission from Ref. [53]. Copyright (2002) American Chemical Society.

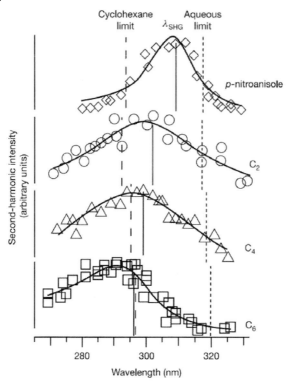

Fig. 4.18 Resonance-enhanced SHG spectra of p-nitroanisole and "molecular rulers" C_2, C_4 and C_6 at the water/cyclohexane interface. The number of carbons of the molecular ruler relates to the number of $-CH_2-$ groups separating the p-nitroanisole moiety from a sulfonate group. Reprinted with permission from Ref. [56]. © Nature Publishing Group 2003.

interpretation of the PMF studies on the same system (see Section 4.2.2.2). The phase shift of the potential-dependent fluorescence revealed the presence of two distinctive adsorption planes at the aqueous and the organic side of the interface [15, 23]. According to the SHG spectra, the interfacial behavior of H_2TMPyP^{4+} and $ZnTMPyP^{4+}$ appear rather different, and it is yet to be clarified what role is played by the metal center on the solvation properties of the porphyrin at the liquid/liquid boundary.

Another interesting example of the effect of the Galvani potential difference on the organization of porphyrins is given by the water-soluble *meso*-tetrakis(4-carboxyphenyl)porphyrinato zinc(II) ($ZnTPPC^{4-}$) at the water/DCE interface [28]. The SHG spectra at various potentials illustrated in Fig. 4.19 reveal the co-existence of various species, depending on the coverage of the porphyrin. At positive potentials, the coverage of the porphyrin is rather low, and the SHG maximum approaches the value of the absorption spectra in the aqueous phase. As the potential is shifted to more negative values, the coverage increases, and

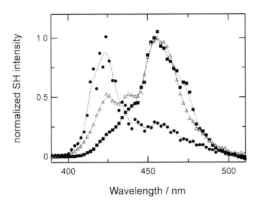

Fig. 4.19 Resonant SHG spectra of the porphyrin ZnTPPC^{4-} at the polarized water/DCE interface at -0.15 V (full circles), -0.20 V (hollow circles) and -0.25 V (full squares). Reprinted with permission from Ref. [28]. Copyright (2003) American Chemical Society.

the spectra exhibit several maxima at longer wavelengths, suggesting the possibility of interfacial aggregation. Previous studies based on photoelectrochemical techniques revealed that ZnTPPC is very likely to undergo lateral interactions through cooperative hydrogen bonding [59]. The results displayed in Fig. 4.19 demonstrate that not only the interfacial coverage but also the extent of the aggregation can be effectively controlled by the applied potential.

4.3.2
Vibrational Sum Frequency Generation

Vibrational Sum Frequency Generation (SFG) has provided some of the most detailed pictures of the structure of water at liquid/liquid and liquid/air interfaces [60–64]. In this technique, the overlap of two laser beams at the interfacial region, one visible and one tuneable in the IR region, generates a third beam with a frequency equal to the sum of the two incident beams. The intensity of the generated sum-frequency beam ($I(\omega_{SF})$) is proportional to the intensities of the visible ($I(\omega_{vis})$) and IR beams ($I(\omega_{IR})$)

$$I(\omega_{SF}) \propto \left| \chi_N^{(2)} + \sum_v \chi_v^{(2)} \right|^2 I(\omega_{vis}) I(\omega_{IR}) \tag{24}$$

where $\chi_N^{(2)}$ and $\chi_v^{(2)}$ correspond to the non-resonant and resonant non-linear susceptibilities, respectively. As in the case of SHG, the resonant component provides the most important contribution to the SFG signal, and can be expressed as

$$\chi_v^{(2)} \propto \frac{A_K M_{IJ}}{(\omega_v - \omega_{IR} - i\Gamma_v)} \tag{25}$$

where a_K is the IR transition moment, M_{IJ} is the Raman transition polarizability, ω_v is the vibrational transition frequency, and Γ_v is the transition line width. Tuning the IR frequency ω_{IR} of the incident laser to ω_v induces an enhancement of the SFG spectrum. The intensity of the SFG signal is substantially enhanced by having both incident beams in TIR configuration [65]. Furthermore, measurements of the SFG intensity as a function of the polarization of the two incident beams allow estimations of the orientation of the vibrational modes.

The work by Richmond and coworkers has led to the identification of various vibrational modes associated with water molecules at neat liquid/liquid interfaces. Figure 4.20 contrasts the SFG spectra obtained at the CCl_4/water, hexane/water and vapor/water interfaces [66, 67]. Three regions can be identified in these spectra: (a) 3100–3400 cm^{-1} associated with strongly hydrogen-bonded water, (b) 3400–3700 cm^{-1} associated with weakly hydrogen-bonded water, and (c) 3700 cm^{-1} representing the free OH bond of water molecules that straddle the interface. The relative intensity of the bands in each region is affected by the nature of the molecular interface. At the vapor/water interface, interfacial water molecules exhibit vibrational responses characteristic of tetrahedrally coordinated structures involved in strong hydrogen bonds. On the other hand, the shift of the SFG responses toward higher frequencies at the liquid/liquid interfaces indicates that water molecules are weakly hydrogen bonded. This blue shift of the SFG signal can be rationalized not only in terms of the weakening of individual hydrogen bonds between water molecules, but also as a decrease in the coordination number of the water molecules interacting with the organic phase. On the other hand, the stretching of the free OH bond observed around 3700 cm^{-1} corresponds to the OH protruding into the organic phase. The red shift of this band at the CCl_4/water interface with respect to the vapor/water interface suggests the presence of attractive interaction between the OH and the organic molecules. Detailed analysis of the SFG spectra at different pH values

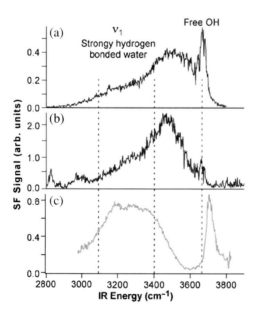

Fig. 4.20 Vibrational SFG spectra of the neat CCl_4/water (A), hexane/water (B) and vapor/water (C) interfaces, indicating the difference in hydrogen bonding at the various interfaces. Reprinted with permission from Ref. [66]. Copyright (2001) American Association for the Advancement of Science.

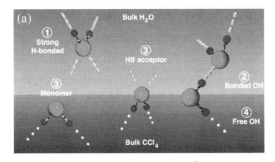

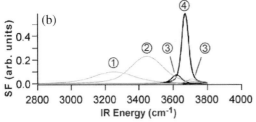

Fig. 4.21 Schematic representation of the water molecules contributing to the vibrational SFG responses at the CCl_4/water interface (A). Dashed lines highlight H_2O-H_2O interactions, while dotted lines correspond to H_2O-CCl_4 interactions. The peak position of the various OH modes identified from the SFG spectra are illustrated in (B). Reprinted with permission from Ref. [66]. Copyright (2001) American Association for the Advancement of Science.

and employing HDO enabled the resolution of a series of vibrational modes within the region of weakly hydrogen-bonded water. Figure 4.21 summarizes the various interfacial water species and orientations obtained from these studies, providing a good illustration of the tremendous sensitivity of this technique.

Although SFG studies are yet to be carried out at liquid/liquid interfaces under electrochemical control, valuable information has recently been gathered for electrochemically relevant junctions. For instance, Walker et al. have reported a systematic comparison of the SFG features of the neat CCl_4/water and DCE/water interfaces [68]. Figure 4.22 shows that the intensity of the free OH band observed at the CCl_4/water interface progressively decreases with increasing mole fraction of DCE in the organic phase. At the neat DCE/water interface, the SFG spectrum is rather featureless, with a weak broad band extending from 3000 to 3700 cm^{-1}. The authors concluded that although there is measurable interfacial thickness, as obtained from interfacial tension measurements (as well as neutron scattering [69]), the absence of the free OH band is indicative of a mixed-phase interfacial region featuring randomly oriented water molecules.

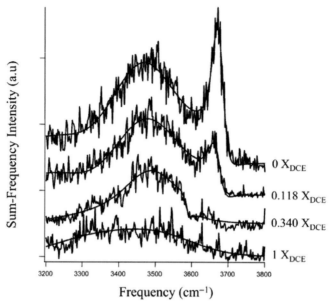

Fig. 4.22 Vibrational SFG spectra of the (CCl$_4$+DCE)/water interface with various mole fractions of DCE in the organic phase. The sharp spectral features observed at the CCl$_4$/water interface are significantly diminished with pure DCE as the organic phase. Reprinted with permission from Ref. [68]. Copyright (2004) American Chemical Society.

4.4
Summary and Outlook

The implementation of *in-situ* spectroscopic techniques to study electrochemical processes at liquid/liquid interfaces has provided key information on the dynamics of adsorption, transfer of ionic species, molecular organization, and local solvation structure of ionic probes. As summarized in Table 4.1, the field has evolved from kinetic studies of ion transfer reactions by fluorescence and reflectance measurements in TIR to more sophisticated approaches based on time-resolved and potential-modulated fluorescence anisotropy. The impact of non-linear optical techniques such as SHG and SFG has been tremendous, although the bulk of these studies have been carried out at neat interfaces. *In-situ* spectroscopic studies employing probes such as water-soluble metalloporphyrins demonstrate that the Galvani potential difference can have a profound effect on the coverage, orientation, and lateral interactions at the interfacial region. Consequently, adapting electrochemical control to spectroscopic techniques can significantly enhance the surface sensitivity of linear spectroscopic approaches. Furthermore, the interfacial environment probed by the non-linear optical techniques can be effectively modulated by the changes in the applied voltage.

Table 4.1 Summary of spectroscopic techniques for studying the liquid/liquid interfaces.

	Spectroscopic technique		Liquid/liquid junction	Ref.
Linear spectroscopy	Reflection – TIR, ATR, ER	Non-Polarizable Interface	water/dodecane	35, 36
	Fluorescence		water/alkane	38, 39
	Raman		water/toluene	3
	Transient Grating		methanol/decalin	70, 71
	Near-Field Optical Microscopy		p-xylene/ethylene glycol	71
	Ellipsometry		water/alkane	72
	Brewster Angle Microscopy		water/hexane	89
	X-Ray and Neutron Reflectivity		water/alkane	77, 80
	Differential Cyclic Voltabsorptiometry	Polarizable Interface	water/DCE	19
	Differential Cyclic Voltfluorometry		water/DCE	12
	Potential-Modulated Reflectance		water/DCE	15
	Potential-Modulated Fluorescence		water/DCE	20
	Quasi-Elastic Laser Scattering		water/DCE	31, 32
Non-Linear Spectroscopy	Second Harmonic Generation		polarizable water/DCE	50
	Sum Frequency Generation		water/CCl$_4$	66

Novel approaches have been reported in the last couple of years which can potentially provide valuable information on the structure and properties of these interfaces. Brodard and Vauthey have implemented *transient grating* studies with evanescent wave probing for investigating the speed of sound at liquid/liquid interfaces [70, 71]. At the methanol/decalin interface, this method was sensitive enough to probe the speed of sound as a function of the penetration depth of the evanescent wave. Deckert and coworkers combined Raman spectroscopy and Near-Field Optical Microscopy in order to study solvent density profiles at the p-xylene/ethylene glycol interface [72]. This extremely challenging approach can potentially provide information on the characteristic thickness of the interface, as well as lateral organization of molecular probes. It could be interesting to employ these techniques at liquid/liquid interfaces featuring self-assembling nanoparticles. In the case of transient grating measurements, it might be expected that large numbers of dense particles could affect the propagation of acoustic waves at the interface. On the other hand, Ag nanoparticles can enhance the Raman signal of interfacial species and the sensitivity of the Near-Field Optical detection.

A recent report by Lei and Bain highlights the use of *ellipsometry* to study surfactant-induced surface freezing at the water/alkane interface [73]. These fascinating experiments concluded that the frozen layer is of molecular thickness and contains chains of the surfactant that are oriented near the surface normal. Sawada and co-workers are currently developing a *total-internal-reflection ultrafast transient lens* (TIR-UTL) and *second harmonic generation coherent vibrational spectroscopy* (SHG-VCS) to study ultrafast dynamic processes at liquid interfaces [74]. TIR-UTL probes refractive index changes with sub-picosecond resolution, allowing the investigation of processes such as temperature increase and energy transfer from solute to solvent molecules. On the other hand, SHG-VCS is expected to deliver information on low-frequency vibrational modes connected to corrective motion of molecules at interfaces.

Other exciting developments involve *X-ray and neutron reflectivity* for investigating solvent density profiles and interfacial roughening [75–82]. Schlossman pointed out that the width of the interface measured by these techniques has two contributions: (a) the intrinsic width determined by the solvent density profiles and (b) the statistical width arising from the averaging capillary wave fluctuations [78]. X-ray reflectivity profiles of the neat water/alkane junction as a function of the *n*-alkane carbon number appear to show that the intrinsic width is determined by gyration radius for shorter alkanes and by the bulk correlation length for the longer alkanes [77]. The interfacial roughness of the water/hexadecane interface as estimated from neutron reflectivity studies was close to 15 Å [80], which is more than twice the value calculated from X-ray measurements [77]. The reasons behind these discrepancies remain unknown; however, the overall picture emerging from both approaches provides a highly valuable link between experiments and powerful computer simulations [83–87].

It is beyond doubt that our understanding of fundamental aspects of structure and dynamics at the liquid/liquid interfaces has significantly increased with the implementation of this wide range of *in-situ* spectroscopic techniques. However, numerous questions remain to be fully answered. Bain recently summarized some of these, including (a) factors determining whether an interface is molecularly "sharp" or "diffuse", (b) how interfacial properties can be predicted from the bulk properties of the electrolyte, (c) the influence of electrolyte on the structure of the interface, and (d) a more accurate description of the potential distribution across the interface [88]. The combination of powerful theoretical modeling and spectroscopic techniques is set to produce major advances in these issues in the next few years. Furthermore, virtually every new approach opens up a new and exciting question of fundamental relevance to the structure and properties of water in contact with soft condensed matter. Electrochemistry plays a central role in this field as it provides accurate means to control the local electrical potential and the overall energy of the system.

Acknowledgments

I would like to acknowledge the financial support by the Swiss National Science Foundation through the project PP002-68708. I am also grateful to Prof. Geraldine Richmond for providing an advanced copy of Ref. 64

Symbols

Roman Symbols

A	Surface area of the liquid/liquid junction
A_K	IR transition moment
A_{TIR}	Absorbance in TIR
C_{dl}	Differential capacitance
c_i^o	Concentration of species i in the organic phase
c^w	Bulk concentration in the aqueous phase
D	Diffusion coefficient
D^w	Diffusion coefficient in the aqueous phase
d	Grating constant
F	Faraday constant
F^o	Fluorescence intensity
ΔF_{re}	Real component of the AC fluorescence
ΔF_{im}	Imaginary component of the AC fluorescence
f_0	Maximum frequency
I	Evanescent field
I_0	(a) Intensity of the incident light
	(b) Photon flux of the excitation beam
$I(\omega_{IR})$	Intensity of the IR beam
$I(\omega_{SF})$	Intensity of the generated sum-frequency beam
$I(\omega_{vis})$	Intensity of the visible beam
i_F	Faradaic current
K	Wavenumber of the incident beam
k	Wavenumber of the capillary wave
k_b	Backward electron transfer rate constant
k_f	Forward electron transfer rate constant
k_{tr}	Apparent ion transfer rate constant
M_{IJ}	Raman transition polarizability
N	(a) Order of the diffraction spot
	(b) Number of molecules contributing to the SHG responses
n_{DCE}	Refractive index of 1,2-dichloroethane
n_i	Refractive index of the incident phase
n_r	Refractive index of the refracted phase
n_w	Refractive index of water
R_{ct}	Resistance of charge transfer
t	Time

z	Distance from the interface in the phase of lower refractive index
z_i	Charge of the transferring ion i

Greek Symbols

Γ	Transition line width
γ	Interfacial tension
ε^o	Absorption coefficient in the organic phase
ζ	Penetration depth of light
θ_c	Critical angle of incidence
θ_i	Angle of incident light, often with respect to the interface normal
θ_r	Angle of refracted light, often with respect to the interface normal
λ	Wavelength of the incident light
$\mu_{i,j}$	Transition matrix element between two defined states
ρ^o	Density of the organic phase
ρ^w	Density of water
σ	Warburg parameter
Φ^o	Fluorescence quantum yield in the organic phase
$\Delta_o^w \phi$	Galvani potential difference
$\Delta_o^w \phi^{o'}$	Formal ion-transfer potential
$\chi_N^{(2)}$	Non-resonant component of the non-linear susceptibility
$\chi_R^{(2)}$	Resonant component of the non-linear susceptibility
ω	(a) Frequency of potential modulation
	(b) Frequency of two incident photons
ω_{IR}	IR frequency of the incident light
ω_v	Vibrational transition frequency
ϖ	Characteristic angle

Abbreviations

ATR	Attenuated total internal reflection
DCE	1,2-Dichloroethane
DHP	Dihexadecyl phosphate
DPPC	Phospholipid DL-a-dipalmitoyl-phosphatidylcholine
DTT	[1,3,5]Triazine-2,4,6-triamine
EB	Eosin B
ER	External reflection
MSA	Mercaptosuccinic acid
ONS	2-(n-Octadecylamino)naphthalene-6-sulfonate
PIR	Partial internal reflection
PMF	Potential modulated fluorescence
PMR	Potential modulated reflectance
QELS	Quasi-elastic laser scattering

RF	Riboflavin
ssDNA	Single-stranded DNA
SERS	Surface-enhanced Raman scattering
SFG	Sum frequency generation
SHG	Second harmonic generation
SHG-VCS	Second-harmonic-generation coherent vibrational spectroscopy
TCNQ	Tetracyanoquinodimethane
$TCNQ^-$	Radical anion tetracyanoquinodimethane
TIR	Total internal reflection
TIR-UTL	Total-internal-reflection ultrafast transient lens
TTP	*meso*-Tetra-*p*-tolyl-porphyrin
$ZnTMPyP^{4+}$	*meso*-Tetrakis(*N*-methypyridyl)porphyrinato zinc(II)
$ZnTPPC^{4-}$	*meso*-Tetrakis(4-carboxyphenyl)porphyrinato zinc(II)

References

1 A. G. Volkov (ed.), *Liquid Interfaces in Chemical, Biological and Pharmaceutical Applications*, Marcel Dekker, Inc., New York, **2001**.

2 A. G. Volkov (ed.), *Interfacial Catalysis*, Marcel Dekker, Inc., New York, **2003**.

3 F. Reymond, D. Fermín, H. J. Lee, H. H. Girault, *Electrochim. Acta* **2000**, *45*, 2647–2662.

4 Z. Samec, T. Kakiuchi, *Adv. Electrochem. Sci. Eng* **1995**, *4*, 297–361.

5 H. H. Girault, *Mod. Aspects Electrochem.* **1993**, *25*, 1–62.

6 Z. Samec, *Chem. Rev.* **1988**, *88*, 617–632.

7 P. D. Beattie, A. Delay, H. H. Girault, *J. Electroanal. Chem.* **1995**, *380*, 167–175.

8 P. D. Beattie, R. G. Willington, H. H. Girault, *J. Electroanal. Chem.* **1995**, *396*, 317–323.

9 A. L. Barker, J. V. Macpherson, C. J. Slevin, P. R. Unwin, *J. Phys. Chem. B* **1998**, *102*, 1586–1598.

10 A. L. Barker, P. R. Unwin, S. Amemiya, J. F. Zhou, A. J. Bard, *J. Phys. Chem. B* **1999**, *103*, 7260–7269.

11 T. Kakiuchi, Y. Takasu, M. Senda, *Anal. Chem.* **1992**, *64*, 3096–3100.

12 T. Kakiuchi, Y. Takasu, *Anal. Chem.* **1994**, *66*, 1853–1859.

13 T. Kakiuchi, Y. Takasu, *J. Electroanal. Chem.* **1994**, *365*, 293–297.

14 T. Kakiuchi, K. Ono, Y. Takasu, J. Bourson, B. Valeur, *Anal. Chem.* **1998**, *70*, 4152–4156.

15 H. Nagatani, R. A. Iglesias, D. J. Fermín, P. F. Brevet, H. H. Girault, *J. Phys. Chem. B* **2000**, *104*, 6869–6876.

16 Z. F. Ding, R. G. Wellington, P. F. Brevet, H. H. Girault, *J. Phys. Chem.* **1996**, *100*, 10658–10663.

17 Z. F. Ding, R. G. Wellington, P. F. Brevet, H. H. Girault, *J. Electroanal. Chem.* **1997**, *420*, 35–41.

18 Z. F. Ding, F. Reymond, P. Baumgartner, D. J. Fermín, P. F. Brevet, P. A. Carrupt, H. H. Girault, *Electrochim. Acta* **1998**, *44*, 3–13.

19 Z. F. Ding, P. F. Brevet, H. H. Girault, *Chem. Commun.* **1997**, 2059–2060.

20 Z. F. Ding, D. J. Fermín, P. F. Brevet, H. H. Girault, *J. Electroanal. Chem.* **1998**, *458*, 139–148.

21 L. Tomaszewski, Z. F. Ding, D. J. Fermín, H. M. Cacote, C. M. Pereira, F. Silva, H. H. Girault, *J. Electroanal. Chem.* **1998**, *453*, 171–177.

22 D. J. Fermín, Z. Ding, P. F. Brevet, H. H. Girault, *J. Electroanal. Chem.* **1998**, *447*, 125–133.

23 H. Nagatani, D. J. Fermín, H. H. Girault, *J. Phys. Chem. B* **2001**, *105*, 9463–9473.

24 N. Nishi, K. Izawa, M. Yamamoto, T. Kakiuchi, *J. Phys. Chem. B* **2001**, *105*, 8162–8169.

25 T. Kakiuchi, *J. Electroanal. Chem.* **2001**, *496*, 137–142.

26 S. Takahashi, A. Harata, T. Kitamori, T. Suwada, *Anal. Sci.* **1994**, *10*, 305–308.

27 H. Yui, Y. Ikezoe, T. Sawada, *Anal. Sci.* **2004**, *20*, 1501–1507.

28 H. Nagatani, Z. Samec, P. F. Brevet, D. J. Fermín, H. H. Girault, *J. Phys. Chem. B* **2003**, *107*, 786–790.

29 A. Trojanek, P. Krtil, Z. Samec, *Electrochem. Commun.* **2001**, *3*, 613–618.

30 W. Lamb, D. M. Wood, N. W. Ashcroft, *Phys. Rev. B* **1980**, *21*, 2248.

31 B. Su, J. P. Abid, D. J. Fermín, H. H. Girault, H. Hoffmannova, P. Krtil, Z. Samec, *J. Am. Chem. Soc.* **2004**, *126*, 915–919.

32 Z. Samec, A. Trojanek, P. Krtil, *Faraday Discuss.* **2005**, *129*, 301–313.

33 K. Fujiwara, H. Watarai, *Langmuir* **2003**, *19*, 2658–2664.

34 S. Yamamoto, K. Fujiwara, H. Watarai, *Anal. Sci.* **2004**, *20*, 1347–1352.

35 Y. Moriya, T. Hasegawa, K. Hayashi, M. Maruyama, S. Nakata, N. Ogawa, *Anal. Bioanal. Chem.* **2003**, *376*, 374–378.

36 Y. Moriya, S. Nakata, H. Morimoto, N. Ogawa, *Anal. Sci.* **2004**, *20*, 1533–1536.

37 K. Fujiwara, H. Watarai, *Bull. Chem. Soc. Jpn.* **2001**, *74*, 1885–1890.

38 M. J. Wirth, J. D. Burbage, *J. Phys. Chem.* **1992**, *96*, 9022–9025.

39 K. Bessho, T. Uchida, A. Yamauchi, T. Shioya, N. Teramae, *Chem. Phys. Lett.* **1997**, *264*, 381–386.

40 S. Ishizaka, S. Kinoshita, Y. Nishijima, N. Kitamura, *Anal. Chem.* **2003**, *75*, 6035–6042.

41 S. Ishizaka, Y. Ueda, N. Kitamura, *Anal. Chem.* **2004**, *76*, 5075–5079.

42 S. Tsukahara, H. Watarai, *Chem. Lett.* **1999**, 89–90.

43 H. D. Duong, P. F. Brevet, H. H. Girault, *J. Photochem. Photobiol. A* **1998**, *117*, 27–33.

44 S. G. Grubb, M. W. Kim, T. Rasing, Y. R. Shen, *Langmuir* **1998**, *4*, 452.

45 H. F. Wang, E. Borguet, K. B. Eizenthal, *J. Phys. Chem. B* **1998**, *102*, 4927–4932.

46 R. R. Naujok, H. J. Paul, R. M. Corn, *J. Phys. Chem.* **1996**, *100*, 10497–10507.

47 K. B. Eizenthal, *Chem. Rev.* **1996**, *96*, 1343–1360.

48 A. V. Benderskii, K. B. Eizenthal, *J. Phys. Chem. A* **2002**, *106*, 7482–7490.

49 R. M. Corn, D. A. Higgins, *Chem. Rev.* **1996**, *96*, 9688.

50 D. A. Higgins, R. M. Corn, *J. Phys. Chem.* **1993**, *97*, 489–493.

51 J. C. Conboy, G. L. Richmond, *J. Phys. Chem. B* **1997**, *101*, 983–990.

52 A. Piron, P. F. Brevet, H. H. Girault, *J. Electroanal. Chem.* **2000**, *483*, 29–36.

53 H. Nagatani, A. Piron, P. F. Brevet, D. J. Fermín, H. H. Girault, *Langmuir* **2002**, *18*, 6647–6652.

54 A. Sen, P. C. Ray, P. K. Das, V. Krishnan, *J. Phys. Chem.* **1996**, *100*, 19611–19613.

55 K. S. Suslick, C. T. Chen, G. R. Meredith, L. T. Cheng, *J. Am. Chem. Soc.* **1992**, *114*, 6928–6930.

56 W. H. Steel, R. A. Walker, *Nature* **2003**, *424*, 296–299.

57 W. H. Steel, R. A. Walker, *J. Am. Chem. Soc.* **2003**, *125*, 1132–1133.

58 W. H. Steel, Y. Y. Lau, C. L. Beildeck, R. A. Walker, *J. Phys. Chem. B* **2004**, *108*, 13370–13378.

59 H. Jensen, J. J. Kakkassery, H. Nagatani, D. J. Fermín, H. H. Girault, *J. Am. Chem. Soc.* **2000**, *122*, 10943–10948.

60 M. C. Messmer, J. C. Conboy, G. L. Richmond, *J. Am. Chem. Soc.* **1995**, *117*, 8039–8040.

61 G. L. Richmond, *Anal. Chem.* **1997**, *69*, A536–A543.

62 G. L. Richmond, R. A. Walker, B. L. Smiley, *Spectroscopy* **1999**, *14*, 18–+.

63 G. L. Richmond, *Chem. Rev.* **2002**, *102*, 2693–2724.

64 M. A. Leich, G. L. Richmond, *Faraday Discuss.* **2005**, *129*, 1–21.

65 J. C. Conboy, M. C. Messmer, G. L. Richmond, *J. Phys. Chem.* **1996**, *100*, 7617–7622.

66 L. F. Scatena, M. G. Brown, G. L. Richmond, *Science* **2001**, *292*, 908–912.

67 L. F. Scatena, G. L. Richmond, *J. Phys. Chem. B* **2001**, *105*, 11240–11250.

68 D. S. Walker, M. Brown, C. L. McFearin, G. L. Richmond, *J. Phys. Chem. B* **2004**, *108*, 2111–2114.

69 J. Strutwolf, A. L. Barker, M. Gonsalves, D. J. Caruana, P. R. Unwin, D. E. Williams, J. R. P. Webster, *J. Electronanal. Chem.* **2000**, *483*, 163–173.

70 P. Brodard, E. Vauthey, *Rev. Sci. Instrum.* **2003**, *74*, 725–728.

71 P. Brodard, E. Vauthey, *J. Phys. Chem. B* **2005**, *109*, 4668–4678.

72 M. De Serio, A. N. Bader, M. Heule, R. Zenobi, V. Deckert, *Chem. Phys. Lett.* **2003**, *380*, 47–53.

73 Q. Lei, C. D. Bain, *Phys. Rev. Lett.* **2004**, *92*, 176103-1–176103-4.

74 H. Yui, Y. Hirose, T. Sawada, *Anal. Sci.* **2004**, *20*, 1493–1499.

75 G. Luo, S. Malkova, S. V. Pingali, D. G. Schultz, B. Lin, M. Meron, T. J. Graber, J. Gebhardt, P. Vanysek, M. L. Schlossman, *Faraday Discuss.* **2005**, *129*, 23–34.

76 D. M. Mitrinovic, Z. J. Zhang, S. M. Williams, Z. Q. Huang, M. L. Schlossman, *J. Phys. Chem. B* **1999**, *103*, 1779–1782.

77 D. M. Mitrinovic, A. M. Tikhonov, M. Li, Z. Q. Huang, M. L. Schlossman, *Phys. Rev. Lett.* **2000**, *85*, 582–585.

78 M. L. Schlossman, *Curr. Opin. Colloid Interface Sci.* **2002**, *7*, 235–243.

79 A. M. Tikhonov, D. M. Mitrinovic, M. Li, Z. Q. Huang, M. L. Schlossman, *J. Phys. Chem. B* **2000**, *104*, 6336–6339.

80 J. Bowers, A. Zarbakhsh, J. R. P. Webster, L. R. Hutchings, R. W. Richards, *Langmuir* **2001**, *17*, 2548.

81 J. Strutwolf, A. L. Barker, M. Gonsalves, D. J. Caruana, P. R. Unwin, D. E. Williams, J. R. P. Webster, *J. Electroanal. Chem.* **2000**, *483*, 163–173.

82 A. Zarbakhsh, J. Bowers, J. R. P. Webster, *Meas. Sci. Technol.* **1999**, *10*, 738–743.

83 I. Benjamin, *Chem. Phys. Lett.* **2004**, *393*, 453–456.

84 K. J. Schweighofer, I. Benjamin, *J. Electroanal. Chem.* **1995**, *391*, 1–10.

85 I. Benjamin, *J. Phys. Chem.* **1991**, *95*, 6675–6683.

86 P. A. Fernandes, M. N. D. S. Cordeiro, J. A. N. F. Gomes, *J. Phys. Chem. B* **1999**, *103*, 6290–6299.

87 P. Jedlovszky, A. Kereszturi, G. Horvai, *Faraday Discuss.* **2005**, *125*, 35–46.

88 C. D. Bain, *Faraday Discuss.* **2005**, *129*, 89.

89 S. Uredat, G. H. Findenegg, *Langmuir* **1999**, *15*, 1108–1114.

5
Sum Frequency Generation Studies
of the Electrified Solid/Liquid Interface

Steven Baldelli and Andrew A. Gewirth

5.1
Introduction

In-situ vibrational spectroscopy has long been used to study the electrified solid/liquid interface. By using the information given by peak position, width, and lifetime, vibrational spectroscopy can provide the chemical identity of the adsorbate, an estimation of surface coverage, and the orientation and even dynamics of molecules at the electrode. Three different types of vibrational spectroscopy are relevant to the solid/liquid interface. The first two of these, Raman and infrared spectroscopy, are thoroughly discussed in this book. A third technique successfully used to probe the liquid/solid electrochemical interface is vibrational sum frequency generation (SFG). SFG was developed as a surface probe some 20 years ago [1], and its use was extended to the electrochemical interface by Tadjeddine over a decade ago [2]. Several reviews examining the use of SFG in non-electrochemical environments exist [3–11]. Tadjeddine wrote two reviews on the application of SFG to electrochemical problems [12, 13]. This chapter updates the Tadjeddine work and focuses on the promise and problems of state-of-the-art electrochemical SFG.

5.1.1
Theoretical Background

Sum frequency generation is a second-order non-linear optical technique that has unique advantages for probing the vibrational spectrum of molecules adsorbed at a surface. The vibrational SFG process occurs when two laser beams, one in the visible spectral region and one in the infrared spectral region, are incident on the sample so that a third beam at the sum frequency of the incident beams is emitted, as shown in (1).

$$\omega_{SFG} = \omega_{IR} + \omega_{vis} \tag{1}$$

Advances in Electrochemical Science and Engineering Vol. 9.
Edited by Richard C. Alkire, Dieter M. Kolb, Jacek Lipkowski and Philip N. Ross
Copyright © 2006 WILEY-VCH Verlag GmbH & Co. KGaA, Weinheim
ISBN: 3-527-31317-6

When high-intensity light is incident on a sample, the response is no longer a linear function of the input field, and the induced polarizability, **P**, is described as a series expansion in terms of the electric fields, **E**, present [(2)].

$$\mathbf{P} = \chi^{(1)}\mathbf{E} + \chi^{(2)}\mathbf{EE} + \chi^{(3)}\mathbf{EEE}\ldots \tag{2}$$

$\chi^{(n)}$ is the nth order macroscopic susceptibility. $\chi^{(1)}$ is the usual first-order susceptibility that is involved in linear optical measurements. $\chi^{(2)}$ is the second-order response that is important for SFG and is a rank 3 tensor. Under the electric dipole approximation, $\chi^{(2)}$ is equal to zero in centrosymmetric media. The origin of this behavior is easily derived. The inversion operation (i) takes $\mathbf{E}(x, y, z) \xrightarrow{i} -\mathbf{E}(-x, -y, -z)$ and $\mathbf{P}(x, y, z) \xrightarrow{i} -\mathbf{P}(-x, -y, -z)$ in a centrosymmetric environment. The second-order term in (2) is $\mathbf{P}^{(2)} = \chi^{(2)}\mathbf{EE}$ and can only be valid under inversion (in a centrosymmetric environment) when $\chi^{(2)} = 0$. This fact means that SFG can arise only in non-centrosymmetric media. One intrinsically non-centrosymmetric region is found at the interface between two media. In particular, SFG is a surface-selective optical probe because the interface is inherently non-centrosymmetric. For electrochemical studies, this interfacial sensitivity is critical because the bulk electrolyte (solvent) and the surface often contain the same molecules; thus $\chi^{(2)}$ spectroscopy is a useful approach to distinguish bulk from surface phenomenon. A main benefit of SFG for applications in electrochemistry is that the SFG signal does not require background subtraction to obtain surface sensitivity.

5.1.2
SFG Intensities

The intensity of the sum frequency signal is given by the square of the induced second-order polarization, $\mathbf{P}^{(2)}$, as shown below [14–18].

$$I_{\text{SF}} = |\mathbf{P}^{(2)}|^2 \tag{3}$$

where $\quad \mathbf{P}_{\text{I}}^{(2)} \propto \sum_{\text{J,K}} \chi_{\text{IJK}}^{(2)} \mathbf{E}_{\text{J}}(\text{vis})\mathbf{E}_{\text{K}}(\text{IR})$

IJK refers to the laboratory Cartesian axes. **E** is the input electric field of the light beams at the surface and is modified by Fresnel's factors (see below). The macroscopic susceptibility, $\chi_{\text{IJK}}^{(2)}$, has a contribution that varies with incident IR wavelength and is referred to as $\chi_{\text{res}}^{(2)}$ and a wavelength-invariant component $\chi_{\text{NR}}^{(2)}$ as shown in Eq. (4).

$$\chi_{\text{IJK}}^{(2)} = \chi_{\text{res}}^{(2)} + \chi_{\text{NR}}^{(2)} \tag{4}$$

5.1.3
Resonant Term

The information about molecular adsorbates at the interface is contained in the second-order susceptibility, $\chi_{res}^{(2)}$. In particular, $\chi_{res\,ijk}^{(2)} = N\langle\beta_{ikj}\rangle$, where N is the number of oscillators and β is the molecular hyperpolarizability given as

$$\beta_{abc} = \frac{\langle g|a_{ab}|v\rangle\langle v|\mu_c|g\rangle}{\omega_{IR} - \omega_q + i\Gamma} . \qquad (5)$$

Here a_{ab} is the Raman transition moment, μ_c is the infrared transition moment, g and v refer to ground and excited vibrational states, ω_{IR} is the input infrared frequency, ω_q is the resonance frequency of the adsorbate, and Γ is a damping factor [8, 14–17]. Thus, the SFG intensity is related to the product of an (anti Stokes) Raman transition and an infrared transition. The SFG intensity is enhanced when the input infrared wavelength coincides with a vibrational mode of the adsorbate and the result of an SFG spectrum corresponds to the vibrational levels of the molecule. This situation is shown schematically in Fig. 5.1. From (5), non-zero SFG intensity will occur only for transitions that are both Raman and IR allowed. This situation occurs only for molecules lacking inversion symmetry [19].

Since it is the orientational average, $\langle\,\rangle$, of the molecular hyperpolarizability that gives SFG intensity, this average clearly cannot go to zero. This means that SFG is sensitive to the presence of order at the interface, and surface molecules must have a net polar orientation to be observed in SFG. For multiple oscillators, q (vibrational modes)

$$\chi_{res}^{(2)} = \sum_q N\langle\beta_q\rangle = \sum_q \frac{A}{\omega_{IR} - \omega_q + i\Gamma} \qquad (6)$$

where A defines the strength of the oscillator, related to the product of the infrared and Raman transition moments.

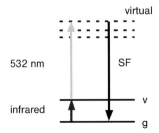

Fig. 5.1 Schematic diagram of the SFG process.

5.1.4
Non-resonant Term

A major difference between SFG and other surface vibrational spectroscopy techniques is the presence of a non-resonant background, because, in part, of the metal substrate. This background is usually treated as independent of the frequency and characterized as a constant (χ_{NR}), although this treatment is not always possible. In electrochemical systems, χ_{NR} is not usually independent of the applied potential. This is because of potential-dependent changes in the electronic state

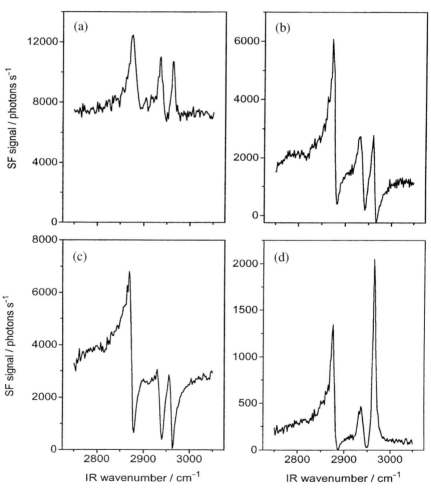

Fig. 5.2 SF spectra showing the dependence of the spectra on excitation frequency of (a) octadecanethiol (ODT) on Au with λp = 532 nm, (b) ODT on Au with λp (pump wavelength) = 1064 nm, (c) ODT on Ag with λp = 532 nm, (d) ODT on Ag with λp = 1064 nm [20]. Reprinted from Ref. [20] with permission from Elsevier.

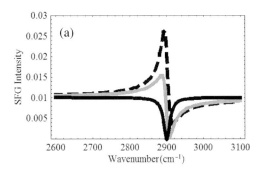 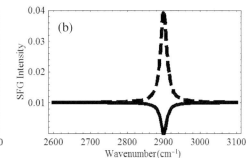

Fig. 5.3 Line shapes that occur in SFG spectroscopy of electrochemical interfaces with metal electrodes. Simulated curves according to Eq. (7) for a resonance at $\omega_{IR}=2900$ cm^{-1}, $\Gamma=10$, and $A_{res}=1$ (see text), where $\theta=\text{Arg}[\chi_{res}]=-\pi/2$ and each metal is given the typical amplitude and phase for experiment in co-propagating direction. **(A)** (---) non-resonant amplitude $(A_{NR})=0.1$, $\phi=0°$; (——), $(A_{NR})=0.1$, $\phi = 45°$; (——) $(A_{NR})=0.1$, $\phi=90°$ **(B)** (——), $(A_{NR})=0.1$, $\phi=90°$, and (- -) for gold with monolayer with orientation of monolayer flipped 180° with respect to surface normal (Au-inverted).

of the metal as well as potential-induced changes in adsorbate coverage and to double layer charging [13]. Further, χ_{NR} is related to the identity of the metal. χ_{NR} is influenced by electronic transitions in the substrate metal. The NR background is therefore dependent on the wavelength of the visible pump beam. The magnitude and phase of the NR background will depend on the electronic states present at the surface. Consequently, the frequency independence of χ_{NR} is only approximate. The frequency dependence of χ_{NR} was demonstrated by Bain et al. for alkanethiols on gold and silver surfaces [20] (Fig. 5.2). These properties of χ_{NR} will complicate the SFG line shape, as demonstrated in Fig. 5.3, but is useful as an internal phase reference for the SFG analysis [3, 21]. The importance of this will be demonstrated in the case studies below.

One method developed to minimize the contribution of χ_{NR} to the nonlinear spectrum is the Difference Frequency Generation (DFG) approach [13]. Since χ_{NR} is dependent on the electronic states of the metal, shifting the output frequency to lower energy avoids exciting the resonance modes in the metal. Although DFG signal is susceptible to contributions of fluorescence from the sample, it has been shown to help in the analysis of electrochemical interfaces [22].

5.1.5
Phase Interference

The final SFG intensity is the square of the sum of the non-linear susceptibilities and the input electric fields, as shown in (3). In general, the second-order polarizability is a complex quantity, as shown in (7)

$$\mathbf{P}^{(2)} = (|\chi_{res}^{(2)}|e^{i\theta} + |\chi_{NR}^{(2)}|e^{i\phi})\mathbf{E}_{vis}\mathbf{E}_{IR} \,. \tag{7}$$

Here, the exponents, θ and ϕ, describe the phase of the resonant term and the non-resonant term, respectively. The phase difference appears in the cross-product of the two terms [20, 23]. Therefore, they can interfere either constructively or destructively with each other. The phase of the non-linear signal is related to the direction of the oscillating dipole [24]. For example, the phase factor, $e^{i(\theta-\phi)}$, is +1 or −1 for up or down orientations. Often the resonant and non-resonant susceptibilities will interfere and give rise to a more complicated line shape in the spectrum. As an example, Fig. 5.3 A shows the variation in peak shape for a resonant peak with phase $\theta = -\pi/2$ (on resonance), as the non-resonant phase varies from ϕ to $\varphi = \pi/2$.

An additional complication arises because most spectra involve several resonance peaks, that will result in multiple interferences and sometimes complicated-looking spectra that require careful analysis to interpret properly (see Fig. 5.21).

5.1.6
Orientation Information in SFG

Since the susceptibility $\chi^{(2)}$ is a rank 3 tensor attached to the lab-frame Cartesian coordinate system, elements of the tensor are accessed using different polarization combinations of light. Consequently, SFG is very sensitive to the orientation of molecules on the surface. The orientation of functional groups at the surface can be estimated by performing polarization-dependent SFG spectroscopy. χ is equal to the orientation average, $\langle \rangle$, of the molecular hyperpolarizability, β. The hyperpolarizability is also a $3\times3\times3$ tensor that describes the response of the molecule to the external fields incident on the surface, as given in (3). Experimental parameters such as laser beam characteristics are most easily described in the lab coordinate system (xyz), while the transition dipoles and bond geometry are better described in a molecular based coordinate system (abc). Thus, several coordinate transformations must be performed to deduce susceptibility properties of the system from experimental changes in polarization. The measured elements of the χ tensor are then related to orientationally averaged values of the β tensor, with the result that molecular orientations can be obtained. However, the elements of the hyperpolarizability tensor are generally unknown, but their values are necessary in order to discern the orientation of functional groups. These tensor elements can be estimated by bond dipole models additively [25–27], *ab initio* calculations [28–30], or by measuring Raman depolarization ratios [31–33]. Additionally, symmetry relations depending on the point group symmetry of the functional group can reduce the 27-element β tensor to a smaller set.

For surfaces that are isotropic in the x–y plane there are four independent macroscopic susceptibilities that contribute to the SFG signal: $\chi_{yyz}=\chi_{xxz}$, χ_{zzz}, $\chi_{xzx}=\chi_{yzy}$ and $\chi_{zxx}=\chi_{zyy}$. Each of these susceptibilities can be accessed through different combinations of input and output polarizations. These are referred as s- and p-polarized (i.e. electric vector perpendicular or parallel to the plane of incidence, respectively). I is the field intensity for each beam at the surface, where the subscript indices indicate the polarization of the sum frequency, visible, and infrared beams, respectively. (8) shows how these different polarization combinations access different elements of the susceptibility tensor for the isotropic surface example [14–17, 33–35].

$$I_{ssp} \propto |\chi_{yyz}|^2 I(vis) I(IR) \tag{8}$$

$$I_{ppp} \propto |\chi_{zzz} + \chi_{yyz} + \chi_{xzx} + \chi_{zxx}|^2 I(vis) I(IR)$$

$$I_{sps} \propto |\chi_{xzx}|^2 I(vis) I(IR)$$

$$I_{pss} \propto |\chi_{zxx}|^2 I(vis) I(IR)$$

5.1.7
Phase Matching

Since the SFG process is coherent, the emitted photons are phase matched with the incident beams in a manner determined by the wavelength of the input light and the geometry of the set-up. The usual situation in SFG spectroscopy is a system with the three beams in the same plane, as shown in Fig. 5.4. The emitted photon direction is given by [36]:

$$\theta_{SFG} = \sin^{-1}(n_{IR}\omega_{IR}\sin\theta_{IR} + n_{vis}\omega_{vis}\sin\theta_{vis})/n_{SFG}\omega_{SFG} \tag{9}$$

where θ is the angle from the surface normal and n is the refractive index of the medium in which the SFG signal is present [37]. This directionality allows for an efficient collection of the SFG signal.

5.1.8
Surface Optics

Fresnel factors also influence the SFG signal by altering the optical field the molecules experience at the surface [38–41]. These Fresnel factors appear in **E** of (3) and (7). The optical fields, **E**, are frequently calculated with modified Fres-

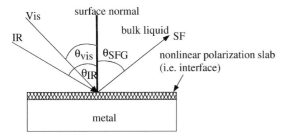

Fig. 5.4 Typical beam geometry for SFG in reflection.

nel's equations [40, 42–44]. One consequence of the electrochemical interface is that the solvent will limit the angle of incidence relative to the surface normal because of refraction by the solvent. Further, the solvent will modify the optical properties such that the surface is less reflective (compared to the vacuum), and causes the x and y field components of the visible radiation to be larger than in the gas/solid interface. Figure 5.5 provides a comparison of the surface intensities for 532-nm light on a Pt surface in water (Fig. 5.5 A) and air (Fig. 5.5 B). The figure shows that the ratio between the intensity of the x- or y-polarized light and the z-polarized light is greater in liquid than it is in air. Notice that in Fig. 5.5 C the intensities for the x- and y-polarization components are near zero. The larger x- and y-components of the visible light field allow one to access polarization information important for orientation analysis using SFG.

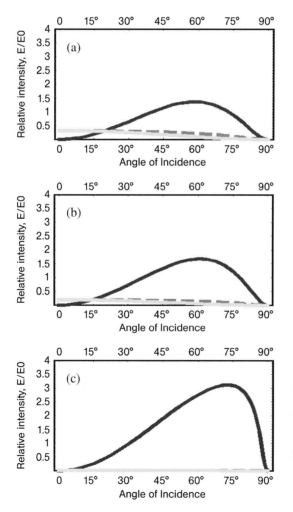

Fig. 5.5 Surface intensities for (- -)x, (—)y, and (—)z components of the electric field. Pt/H$_2$O (**A**) and Pt/air (**B**) interface. Note relative enhancement of z-component for the liquid/air interface. $\lambda = 532$ nm. (**C**) For Pt/H$_2$O interface at $\lambda = 3300$ nm (3000 cm^{-1}).

Since the intensity of IR light for s-polarization is very small at the metal surface, only vibrations with a transition dipole component along the surface normal are detected. This necessitates using p-polarized light for the IR beam. The transition dipole vector of the molecule and electric field vector of the IR light must have a non-zero projection on each other; the consequence is that molecules with their transition dipole parallel to the metal are not seen in the SFG spectrum. This effect is referred to as the IR dipole surface selection rule [45, 46].

5.1.9
Data Analysis Reference

Although the SFG signal inherently arises from to the interface only, it is affected by the bulk solvent through adsorption of infrared light. The adsorbed infrared will result in a decreased SFG signal and could be falsely interpreted as an SFG feature. This situation is avoided by careful experimental design and referencing of the SFG signal. Typically, the thin-layer electrochemical cell avoids most of the problem and working in spectral windows of the solvent is also beneficial, although not sufficient. To assure that the SFG signal is free of infrared absorption artifacts, it is useful to collect the infrared light reflected from the surface and use it to correct the acquired SFG spectrum. In practice, very sensitive infrared detectors or mixing the IR with visible in a nonlinear crystal to generate new SFG photons [47] provides a reference to adjust the spectra for infrared energy fluctuations over the spectral region. These methods correct for solvent absorption and laser fluctuations.

Assuming the other optical elements to be negligible (not always true), the beam path (l) through the liquid is equal to $2d/\cos\theta$, where θ is the angle from normal at the liquid/solid interface and d is the thickness of the solvent layer, as shown in Fig. 5.6. Thus, the beam pathlength to the surface is $d/\cos\theta$, and the absorbance is $A=\log[I_0/I]=\varepsilon cl$. I and I_0 are the incident and transmitted in-

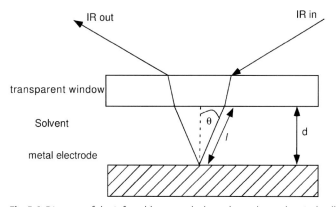

Fig. 5.6 Diagram of the infrared beam path through an electrochemical cell.

tensities. ε and c are absorption coefficient and concentration, respectively. According to Beer's law, $A/2$ of the infrared light will be adsorbed before it can be used to generate SFG. A measurement of the relative infrared intensity before and after the sample cell can be used to correct the SFG spectra for absorption due to solvent in the thin layer. The thin layer is typically on the order of 10 μm, which is a long enough path length to absorb nearly all the incident IR light for solvent molecules with a large ε values (i.e. water).

SFG spectra, once corrected and normalized, are curve fit to a chosen line shape function. Typically the curve is a Lorentzian line shape function as given in (6), although other spectral functions have been used as well [48].

5.1.10
Experimental Designs

Because SFG is a non-linear spectroscopy, the incident laser beams must have high intensity, which necessitates the use of pulsed lasers with pulse widths in the nanosecond to femtosecond range [3, 5, 8, 14–16]. Typical power densities used at the surface are 500 MW cm^{-2}.

There are several methods available to generate IR light of sufficient intensity for SFG, including Optical Parametric Generators (OPG)/Optical Parametric Amplifiers (OPA) [49], Raman shifters [50], and Free-Electron Lasers [51]. However, two laser-based systems are most commonly used. The two basic laser systems used in SFG experiments are referred to as scanning and broadband. The first typically use picosecond Nd:YAG lasers as the source of both the visible light and the narrow-band tunable infrared. IR light is generated using schemes based on non-linear crystals. Individual IR frequencies are generated by scanning to the wavelength of interest.

A common scanning configuration is shown in Fig. 5.7. This system is based on a commercial pump laser and parametric generator, while the remainder of the table is customized for flexibility in experimental design. A complete commercial system is available from the Lithuanian company, Eskpla, or from the Belgian company Euroscan. Since for a fixed input energy the efficiency of the nonlinear signal increases as pulse width decreases, picosecond pulses are ideal for scanning systems, since the nonlinear process is relatively efficient while the bandwidth is less than most vibrational resonances, <7 cm^{-1}. The scanning SFG system involves computer-controlled motors to vary the angle of nonlinear crystals in the optical parametric generator/amplifier (OPG/OPA). SFG experiments are performed in the continuous scanning (scan rate 1 cm^{-1} s^{-1})/data collecting mode, or stepping to individual wavelengths and collecting data for a set time period (i.e. 100 shot/point). For example, the spectra presented in Fig. 5.21 are the average of 5 scans each at 1 cm^{-1} s^{-1}.

The broadband configuration pioneered by Richter et al. [52] and van der Ham et al. [53] makes use of broadband IR pulses generated by OPAs pumped by Ti:sapphire lasers. These typically collect the entire SFG spectrum over a ca. 200–300 cm^{-1} window at one shot as the SFG is dispersed onto a CCD. The

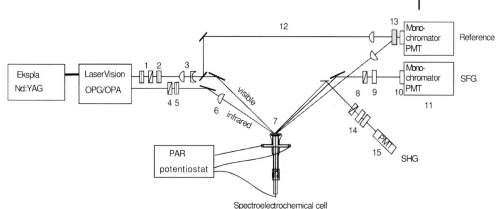

Fig. 5.7 SFG spectrometer at the University of Houston. A picosecond-pulsed Nd:YAG laser (Eskpla) at 1064 nm is split into two parts. One part is frequency doubled to 532 nm. The other part is used to pump a KTP/AgGaSe2 OPG/OPA system (LaserVision) that generates infrared light (IR). Both beams are directed to the liquid sample. (1) Attenuator (half-wave plates/polarizer). (2) $\lambda/2$ plate (polarization control). (3) Collimating reducing telescope. (4, 5) Polarizer/half-wave plate (polarization control). (6) Long focal length BaF$_2$ lens (500 mm). (7) Sample. (8) Analysis polarizer. (9) Kaiser Notch Plus filter. (10) 515 nm short pass filter. (11) 0.25 m monochromator with PMT. (12) Reference arm. (13) Single-crystal quartz (nonlinear signal reference). (14) 532 nm filters. (15) PMT.

window can be moved to obtain as much of the spectrum as desired. The advantage of the broadband technique lies in the increased signal-to-noise ratio of the collected SFG spectrum. This increase originates in the multiplex advantage attendant on the broadband process and the higher repetition rate of Ti:sapphire lasers (ca. 1 kHz) relative to Nd:YAG lasers ($\sim$20 Hz). However, the scanning system may have better spectral resolution relative to the broadband measurement (<5 cm^{-1} compared to >10 cm^{-1}). The scanning system is also significantly less expensive than the broadband instrument.

Both fs and ps lasers can damage metal surfaces through ablation and heating, therefore the SFG as a function of the input intensity should be checked for each new system to assure no damage/heating is occurring. One method is to plot the SFG signal as a function of input intensity for both the IR and visible separately; a linear plot should result.

5.1.11
Spectroscopy Cell

Electrochemical SFG is typically conducted in thin-layer spectroscopy cells similar to those used in infrared spectroscopy. A large angle of incidence will enhance the intensity of the light fields at the surface and therefore increase the signal, as shown in Fig. 5.5 for visible light. The light beams enter and exit the cell through a flat window, prism, or hemispherical window. The windows

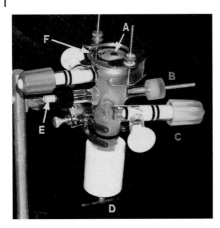

Fig. 5.8 SFG cell for electrochemistry and spectroscopy. (**A**) Pt working electrode, (**B**) reference electrode, (**C**) vacuum stopcock, (**D**) connection for working electrode, (**E**) counter electrode, (**F**) CaF$_2$ window.

must be able to transmit the infrared, visible, and SF light. Further, they need to be stable to the electrolyte (and solvent). In practice this limits the window material to CaF$_2$, infrared quartz, BaF$_2$, or diamond. Materials that are birefringent or have nonlinear activity should be avoided to simplify the experiment and analysis. The electrode is pressed against the window to minimize the path length of the IR light through the solution. The intensity of light will be reduced at the surface (see above). Pressing the window to the electrode will trap a layer of electrolyte about 1–10 μm thick. A thin-layer cell used for SFG is shown in Fig. 5.8.

5.2
Applications of SFG to Electrochemistry

To date, a number of electrochemical systems have been examined using *in-situ* SFG spectroscopy. A list of these is given in Table 5.1.

Vibrational spectroscopy studies examining adsorption and reactivity on single-crystal electrodes is important in developing structure-function relationships in electrochemistry. Often different crystallographic planes exhibit unique adsorption or reactivity [54, 55]. This phenomenon can be observed in vibrational spectroscopy including SFG. Most research thus far on single-crystal electrodes is limited to Au and Pt, because of their ease of preparation, with some recent work on Cu and Ag. The work on polycrystalline platinum is motivated toward understanding processes in electrocatalysis. Substantial effort has provided methods of preparing well-defined single-crystal surfaces of Au [56, 57], Pt [58], Ag, Cu, and other metals for use in electrochemical systems [59].

Table 5.1 Summary of electrochemical systems studied with SFG

Topic	Surface	System	Vibrational mode	Reference
Single crystal/ catalysis	Pt(poly) Pt(111) Pt(110) Pt(100)	CO	C≡O	51, 54–63
	Pt(110), Pt(111), Pt(100)	H	Pt-H	64–66
	Pt/Ru	CO	C≡O	67
Adsorption from solution	Pt(poly)	CN⁻, CO	C≡N, C≡O	2
	Au(110), Au(100), Au(111)	CN⁻	C≡N	22, 68
	Pt(110), Pt(111)	CN⁻	C≡N	69–71
	Pt(111)	CN⁻, CO	C≡N, C≡O	72
	Pt, Au, Ag	CN⁻	C≡N	73, 74
	Ag	OCN⁻	C≡N	75
	Au, Ag	SCN⁻	C≡N	76
	Au(111)	4-cyanopyridine	C≡N, ring modes	77–80
	Au(111)	pyridine	Ring modes	81, 82
	Glassy carbon	phenylalanine	C–H	83
Dynamics	Au, Pt, Ag	CN⁻	C≡N	84
	Cu	CN⁻	C≡N	85
	Pt(111)	CO	C≡O	86, 87
Solvent structure	Pt(111)	Acetonitrile	C–H, C≡N	21
		Water, acetonitrile	O–D, C–H, C≡N	88
	Ag(100)	Water	O–H	89
	Pt	1-butyl-3-methyl imidazolium tetrafluoroborate [BMIM][BF₄] (ionic liquid)	C–H	47
	Au	water	O–H	90
	Pt(110) Pt(111)	Water/HClO₄	NR, Cl–O	91, 92
Monolayers and corrosion	Au	16-mercaptohexadecanoic acid (SAM)	C–H	93
	Au(111) Pt(111) Ag(111)	CH₃CₙH₂ₙ₊₁SH (n=9, 10, 18) (SAM)	C–H	94
	Cu(111) Cu(100)	benzotriazole	Ring modes	95
	Au(111)	CH₃CₙH₂ₙ₊₁SH (n=6, 12, 18) (SAM)	C–H	96

5.2.1

CO Adsorption

Adsorbed CO has been studied on a number of electrodes using many *in-situ* techniques. The CO vibration exhibits a large infrared cross-section which is located in a spectral window for the commonly used water solvent. Additionally, the CO vibrational spectrum is influenced by the adsorption site and its geometry on the surface. Finally, CO is a poisoning intermediate in the oxidation reaction of many organic molecules, and the studies of CO may help to understand fuel cell processes.

Tadjeddine and coworkers published the first SFG spectra examining CO adsorbed onto Pt in the electrochemical environment. In a series of papers start-

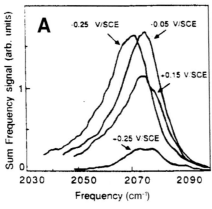

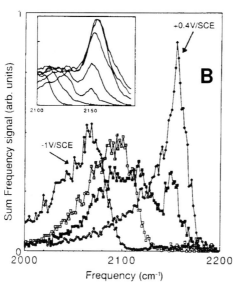

Fig. 5.9 (A) SFG spectra of CO adsorbed on a platinum electrode in 0.1 M HClO$_4$ aqueous electrolyte. **(B)** SFG spectra of CN$^-$ adsorbed on a platinum electrode in 0.1 M NaClO$_4$ and 0.025 M KCN. Starting from an electrode cleaned at −1.25 V/SCE, going toward positive potentials. Filled diamonds, −0.2 V/SCE; open diamonds +0.4 V/SCE. The insert shows the growth of the band at 2150 cm^{-1} in spectra taken from −0.8 V/SCE to +0.4 V/SCE every 200 mV. The dotted line is for SCN$^-$ adsorption at −1 V/SCE in the same conditions. Reprinted from Guyot-Sionnest et al., Chem. Phys. Lett. Vol. 172, No 5, 1990, pp 341–345, with permission from Elsevier.

ing in the early 1990s, they investigated CO generated by methanol decomposition on various faces of Pt [2, 51, 60–62]. This work leveraged the well-understood changes in CO stretching frequency with change in adsorption site. Specifically, in FTIR work dating from the 1980s, Ito et al. showed that CO exists on atop (2000–2080 cm^{-1}), bridge (1800–1860 cm^{-1}) and/or three-fold (1700–1790 cm^{-1}) sites on Pt, depending on the electrode potential.

The spectrum of CO on Pt(poly) is shown in Fig. 5.9 [2]. This spectrum is significant as it demonstrated the possibility of obtaining SFG from electrochemical interfaces with very good S/N ratio. Further, the results are comparable to those obtained using infrared spectroscopy. In the same report, Tadjeddine and coworkers also presented potential-dependent SFG spectra of CN$^-$ on a Pt electrode. Since the SFG process is coherent, the phase of each resonance is present in the spectra. As noted earlier, phase information can be used to determine molecular orientation at the surface. For example, the phase information can determine whether CN$^-$ is surface coordinated through the nitrogen or the carbon atom [24].

Tadjeddine and coworkers showed that CO formed by methanol decomposition from a solution containing 0.01 M HClO$_4$ and 1 M CH$_3$OH occupied both bridge and atop sites. The bridge and atop site coordination was apparent in the SFG spectra as peaks at 1977 and 2053 cm^{-1}, respectively. Further, they showed that CO is present on the surface at potentials as low as 0.05 V vs NHE. This result, significant for reasons related to catalysis, was made possible because complicating reference spectra are not needed using SFG, as discussed in the introduction. Further SFG studies of CO generated by methanol decomposition on single-crystal Pt electrodes [63–65] demonstrate potential and crystal orientation dependence, similar to that found in studies of direct CO adsorption on single-crystal Pt. Figure 5.10 presents the surface site and the peak position observed in SFG vibrational spectroscopy which are consistent with the FTIR results [66].

Additional detailed information about the state of adsorbed CO on Pt electrode surfaces was obtained from more recent SFG measurements.

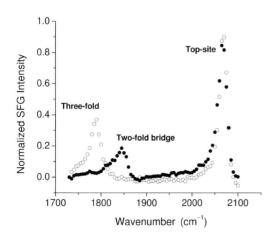

Fig. 5.10 Vibrational (SFG) spectrum of CO/Pt(111) with CO in solution. Peak position and site assignment ○=0.0 mV, ●= 400 mV vs Pd/H$_2$ reference electrode. ssp polarization [103].

5.2.1.1 Polarization Studies

Studies of the energy of the CO stretches on Pt electrodes are complemented by polarization measurements. Using Equations 5 and 6, the relative magnitude of polarizability tensor elements of CO can be determined by analysis of the polarization-dependent SFG spectrum shown in Fig. 5.11. SFG experiments determined that the polarizability of CO adsorbed at a three-fold site (1780 cm^{-1}) is larger than CO at a top site (2070 cm^{-1}) and that both of these polarizabilities are greater than that of the free CO molecule [67]. This conclusion is based on the appearance of the CO peaks in both ssp- and ppp-polarized spectra (Fig. 5.11). In the SFG results, the increased polarizability found for adsorbed CO arises from electron donation from the metal into the π^* orbital of CO.

The manifestation of increased polarizability for surface-coordinated CO is consistent with the Blyholder model [68] for CO adsorption. In the Blyholder model, the 5σ orbital [the highest occupied molecular orbital (HOMO)] and the 2π orbital [the lowest unoccupied molecular orbital (LUMO)] of CO determine the surface chemical bond. Electrons are donated from the 5σ orbital to the substrate and electrons from the substrate are back-donated to the 2π orbital of CO. The magnitude of π back-donation governs the bond strength of CO to the metal surface.

5.2.1.2 Potential Dependence

Sudden changes in the CO structure as a function of electrode potential have been detected *in situ* on Pt electrodes. Stripping voltammetry of CO on Pt(111) indicates a small amount ($\sim$10–15%) of the adsorbed CO is oxidized at about 300 mV negative of the main oxidation wave. Since this "prewave" feature occurs without oxidization of the entire CO layer, it was suspected that the prewave was associated with a change in the surface CO structure. Measurements using SFG found that CO on Pt(111) reorients on the surface, going from a perpendicular to a more parallel orientation relative to the metal surface upon oxi-

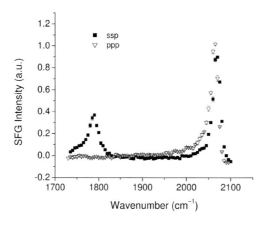

Fig. 5.11 SFG spectra of CO/Pt(111) at 50 mV NHE without CO in solution. ■ = ssp, ▽ = ppp.

dation of ~10–15% of the monolayer [67]. Others observed similar results with a different interpretation [69], including a possible observation of interfacial water [70]. The difference in the final observations and interpretation is likely due to the different system preparation and time scale for the SFG measurement. The CO does not exhibit an ordered structure in the pre-wave region as determined by *in-situ* X-ray diffraction [71, 72].

5.2.1.3 CO on Alloys

Other studies involving CO focused on the behavior of this molecule on surfaces more relevant for fuel cells. Somorjai and Ross examined CO on Pt-Ru bulk alloys. They showed that CO adsorbed at atop sites exhibiting only one CO stretching frequency. Additionally, the CO stretch exhibits a frequency shift that is linear with increasing Pt content in the alloy, as shown in Fig. 5.12 A and B. Mixtures of ^{13}CO and ^{12}CO co-adsorbed on the alloy surface exhibited a similar peak shift compared to the pure Pt, as was found in equivalent FTIR measurements [73–76]. The CO exhibited one peak in the SFG spectra for each alloy,

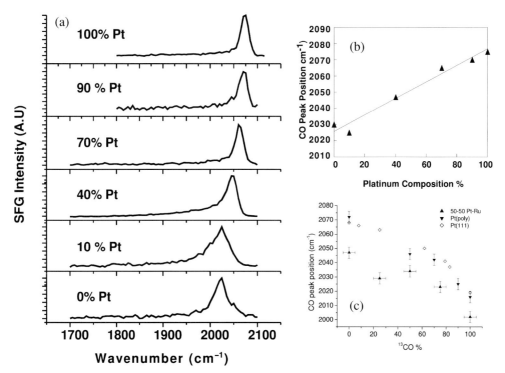

Fig. 5.12 (a) SFG spectra of CO on Pt/Ru alloys as a function of bulk Pt composition. **(b)** Plot of peak position vs bulk Pt composition. **(c)** CO peak position as a function of ^{13}CO isotope composition for Pt and Pt/Ru alloy.

and the $\Delta\nu_{CO}$ is the same on Pt as on the alloy samples. These results indicate that the electronic structure of the alloy is almost a linear function of metal composition and that CO adsorbed on Pt and Ru is dipole coupled.

CO on Pt(111)/Ru surface alloys has also been recently studied by SFG [77]. In contrast to the bulk alloy, CO adsorbed on the surface alloy exhibits two peaks: one at 2073 cm^{-1} and one at 2000 cm^{-1}, which are interpreted as originating from two distinct sites for CO adsorption, namely the atop sites of atoms of Pt and Ru, respectively. Thus the CO molecules on this surface are distinct and vibrationally uncoupled, an effect likely due to the large domains of Ru islands in the surface alloy, in contrast to the bulk alloy material.

5.2.1.4 Solvent Effects

CO adsorbed on a Pt electrode in 0.1 M LiSO$_3$CF$_3$/acetonitrile electrolyte was studied by SFG. In this system, the peak associated with atop site CO exhibited a Stark shift of ~ 20 cm^{-1} V^{-1}, as monitored with SFG. This study demonstrated the ability of SFG to obtain information from a surface immersed in the non-aqueous acetonitrile solvent [78].

5.2.2
Adsorption of upd and opd H

Adsorption of H$^+$ and evolution of H$_2$ are some of the most fundamentally important electrochemical reactions. SFG spectra examining H on Pt(111), (110) and (100) were obtained in both the under-potential and over-potential region [79, 80]. Terminally bonded Pt-H exhibits a peak near 2000 cm^{-1}, and this resonance was observed on all three low Miller index faces of Pt. The peak increased in intensity as the potential approached and exceeded the potential of the hydrogen evolution reaction (HER). At potentials close to the potential of the HER, a peak at 1770 cm^{-1} appeared in the spectra. This peak was assigned to a dihydride species and proposed as an intermediate in the HER [81]. This work extended greatly the insight from an earlier FTIR study, which observed only atop Pt-H [82]. Finally, SFG examining the role of adsorbed H on Pt in the presence of methanol at potentials in the H OPD region also revealed a peak at 1770 cm^{-1} which was related to the bridging Pt-H species [83].

5.2.3
CN on Pt and Au Electrodes

5.2.3.1 CN/Pt

Daum and coworkers used SFG to study the adsorption and structure of CN$^-$ on Pt(111) and Pt(110) electrodes [69, 84, 85]. The spectra, shown in Figs. 5.13 and 5.14, indicate the presence of two adsorption sites for CN$^-$. A low frequency peak at ~ 2080 cm^{-1} is assigned to a CN stretch for CN$^-$ adsorbed on well-ordered Pt sites. This peak exhibits a potential-dependent shift of nearly 30 cm^{-1}

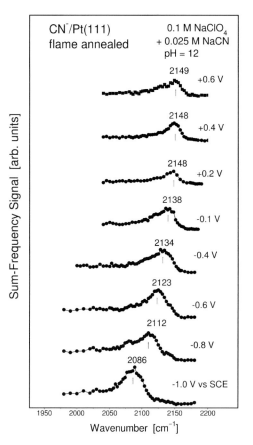

Fig. 5.13 SFG spectra of CN⁻ on Pt(111). The sample was cleaned by flame annealing and displayed a cyclic voltammogram as shown in Fig. 3a of Ref. [84]. Electrolyte: 0.1 M NaClO₄ + 0.025 M NaCN, pH = 12 (with kind permission of Springer Science and Business Media).

V^{-1}. The peak at $\sim$2145 cm^{-1} is nearly potential independent and assigned to a CN$^-$ at disordered Pt sites, which develop after a flame-annealed sample is subjected to oxidation-reduction cycles (see Fig. 5.15) [86].

The measurements performed by the Daum group are similar to those examining CN$^-$ on Pt(110) and Pt(111) performed by Tadjeddine et al. [87]. These latter investigators also concluded there are two forms of CN on Pt(110): one with C coordinated to Pt and the with N coordination to the metal. The C-N stretch for the Pt-C≡N species is assigned to the high-frequency 2150 cm^{-1} peak and the C-N stretch for Pt-N≡C is assigned to the lower frequency near 2060 cm^{-1}. These authors note that immersion potential is critical to determine the resulting form of CN on the surface, while others noted that the source of the CN is also important [85]. SFG results by Tadjeddine et al. were supported by density-functional calculations, which were very useful in interpreting the SFG results of CN on Ag [88] and gold electrodes [89].

Later SFG [90] and STM [91] experiments attributed the two peaks in the spectrum to CN$^-$ at different surface sites. FTIR experiments on Pt(111) and

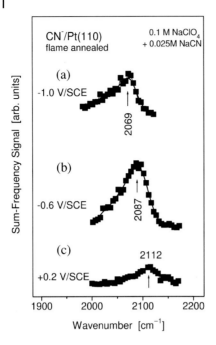

Fig. 5.14 SFG spectra of CN⁻ on Pt(110). The sample was cleaned by flame annealing and displayed a cyclic voltammogram as shown in Fig. 3b of Ref. [71]. Spectra (a–c) were recorded after cyanide adsorption starting at –1.0 V (with kind permission of Springer Science and Business Media).

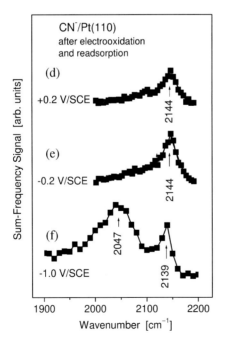

Fig. 5.15 Prior to spectrum (d), the electrode was oxidized at 0.72 V and cyanide was readsorbed at +0.2 V. Spectra (d–f) were then recorded starting at +0.2 V (with kind permission of Springer Science and Business Media).

Pt(100) also indicate that only top-site CN^- is present but that CN^- is decomposed to NO and CO_2 at high potential on Pt(100) [92]. Combined STM and FTIR studies by Weaver et al. indicate that CN^- binds at an atop site and forms an ordered $(2\sqrt{3}\times2\sqrt{3})R30°$ structure that changes to a 2×2 lattice above 0 V vs SCE [91].

5.2.3.2 **CN/Au**

Tadjeddine's group performed several SFG and DFG [13] measurements on the CN^-/Au(*ijk*) system and correlated the results with *ab-initio* calculations [12, 13, 22, 88, 89, 93]. A particularly noteworthy feature of this work is the use of DFG to minimize the non-resonant contribution to the SFG spectrum on Au, resulting in a more easily interpretable spectrum. CN^- is tightly bound to the Au surface atoms in an atop-site linear geometry characterized by a resonance between 2090 and 2130 cm^{-1} as the potential is changed from -1.2 to 0 V vs SCE. The intensity of the peak reaches a maximum as the potential is increased to more positive values in the order Au(110), Au(100) and Au(111). This result indicates that the packing density of CN^- on Au(*ijk*) also increases in this order.

5.2.4
OCN and SCN

The related systems of OCN^- and SCN^- on Ag and Au have been studied by Tadjeddine [94] and Bain [95], respectively. For SCN^- on Ag and Au, the thiocyanate bonds to the metal through the sulfur and does not show evidence for an N-bound species (Fig. 5.16). SFG would detect these by the presence of a CN stretching peak in the opposite direction since the phase of the $\chi^{(2)}_{Res}$ is also reversed (i.e. a peak to a dip and vice versa, see Fig. 5.3 B). On the basis of signal absence, the SFG was interpreted to suggest that the linear SCN^- ion is oriented parallel to the surface at potentials less than -0.4 V vs SCE. A recent study used SFG to study SCN^- adsorption on electrodeposited Co and Ni films. In this study the change in CN stretch band is used to justify a upd process of Co^{2+} induced by the anion [96]. For OCN^- on Ag, the ion is N-bound to the surface with two potential-dependent peaks in the $\sim 2150\ cm^{-1}$ and 2220 cm^{-1} regions. These corresponded to the bridging and terminal-bound OCN^-, respectively.

5.2.5
Pyridine and Related Derivatives

Adsorption of pyridine on Au has also been investigated by Tadjeddine et al. [97, 98]. Here, the advantage of the low-background signal from DFG is used to examine the potential-dependent structure of pyridine on Au electrodes. The ring vibrations at 1008 cm^{-1} and 1090 cm^{-1} provide two perpendicular modes (in-plane and out-of-plane) to report on the orientation of the molecule as a function of potential. The pyridine is bonded to the gold through lone pair elec-

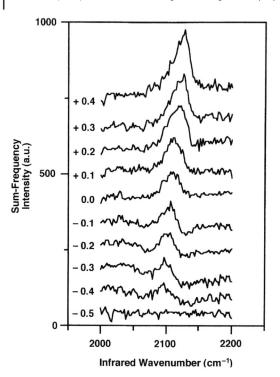

Fig. 5.16 Sum-frequency spectra of thiocyanate adsorbed on gold for various electrode potentials within the double-layer region. Potentials in volts vs SCE. Spectra have been offset vertically for clarity: the zero of the ordinate refers to the spectrum acquired at 0.0 V [95] (used with permission by The American Chemical Society).

trons on the nitrogen. The results suggest that pyridine is orientated mostly parallel to the surface but then reorients, becoming aligned with the surface normal and twisting about the C_2 axis or the ring as the surface charge goes from negative to positive.

There are two vibrational spectroscopic studies of 4-cyanopyridine in aqueous electrolyte on Au(111). FTIR measurements suggest that the molecule is coordinated through the nitrogen atom of the aromatic ring with potential-dependent orientation [99]. SFG experiments on the same system provide evidence that the molecular orientation and bonding changes from aromatic ring N coordination at negative potentials to coordination through π bonding from a molecule oriented with the ring parallel to the surface to CN coordination at positive potentials [100, 101]. Features in the CV were correlated to the orientation of the adsorbed molecule by monitoring the vibrations of both the ring modes and $C\equiv N$ bond. An additional focus was the use of SFG in combination with IR and CV to elucidate mechanisms of reorientation as well as hydrolysis of the 4-cyanopyridine molecule [102]. The Tadjeddine experiments benefited from ab-initio calculations to interpret the experimental results [103].

In what were the first SFG measurements obtained from a glassy carbon electrode, Somorjai and coworkers examined the adsorption of amino acids to the surface [104, 105]. Glassy carbon is a useful inert electrode used in many electrochemical applications and is difficult to interrogate using other vibrational spectroscopic methods. The results show that, at positive potentials and presumably positive surface charge, the phenylalanine is adsorbed to the electrode with the negatively charged carboxylate toward the electrode.

5.2.6
Dynamics of CO and CN Vibrational Relaxation

Since SFG is performed with pulsed lasers having pulse durations in the picosecond to femtosecond range, dynamic processes that occur on this time scale can be explored. In the electrochemical environment, Guyot-Sionnest et al. explored the vibrational relaxation of CN^- and CO on different metal electrodes and in different supporting electrolytes as a function of applied potential. In these experiments, a pump-probe experiment configuration is used where a pump infrared photon is directed to a surface-confined species to excite the vibration. The vibrational decay is probed using SFG, delayed relative to the input pump photon (Fig. 5.17). The time decay from the vibrational excited state is determined by plotting the SFG signal as a function of time normalized to the signal level at saturation (i.e. from maximum SFG signal).

For CO on Pt(111) in acetonitrile containing 0.1 M $NaClO_4$, the lifetime of the CO vibration increased from 1.5 to 2.1 ps with decreasing potential over a range of 2.6 V. These observations are consistent with the Blyholder model since, as the potential is adjusted to more negative values, electrons are expected to fill the $2\pi^*$ orbital on CO, leading to a less efficient decay process [106].

However, in the work of Tadjeddine et al. in aqueous and nonaqueous electrolyte (acetonitrile), a lifetime of 1.7 ps was observed with no effect of potential, even over a range of 2 V [107, 108]. The results indicate the CO vibrational lifetime is not dependent on electrode potential or solvent molecules or surface orientation [Pt(111), Pt(110)] [2, 108]. This discrepancy is small and could be due to an improved signal-to-noise ratio in the later results. Thus, CO vibrational relaxation is presumed to be dominated by the charge transfer-like mechanism because of the interaction between the CO $2\pi^*$ and the Fermi level of the metal. These measurements help in developing molecular models for bonding of molecules at electrode surfaces.

CN^- on Ag exhibits a vibrational lifetime ranging between several ps to up to 60 ps in D_2O solution (Fig. 5.17), a lifetime much longer than that found for the CO system. The lifetime of CN^- on both Pt and Au is potential dependent, with lifetime increasing as potential becomes more positive [90]. These results are discussed in terms of two dominant processes for the vibrational relaxation of an adsorbate on a metal surface: (a) the image-dipole and (b) the charge-transfer mechanisms. The image-dipole mechanism involves electromagnetic coupling of orbitals in CN to charge carriers in the metal. The charge-transfer mechanism involves fluctuation of charge between the metal and the adsorbate. If the vibrational lifetimes are long, the image-dipole mechanism is favored [90]. Interestingly, it is noted that the vibrational lifetime of CN^- correlates with

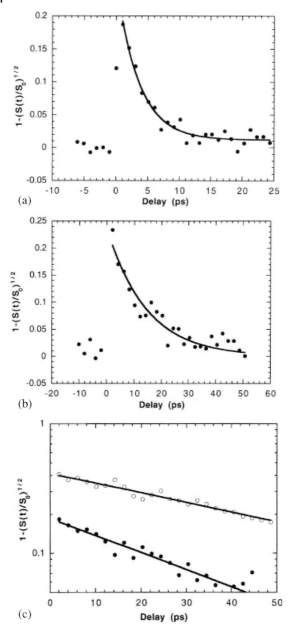

Fig. 5.17 (a) Saturation data for CN/Pt(111) (solid dots) at −0.8 V/SCE with a single exponential plus a constant fit (solid line) giving $T_1 = 3.6$ ps. (b) Saturation data for CN/Au (solid dots) at −0.8 V/SCE with a single exponential giving $T_1 = 14.5$ ps. (c) Saturation data for CN/Ag in H_2O (filled circles) and D_2O (open circles). The data are displayed on a log scale and single exponential fits yield lifetimes of 31 ps and 58 ps for H_2O and D_2O solvents, respectively [106]. (Used with permission of the American Institute of Physics).

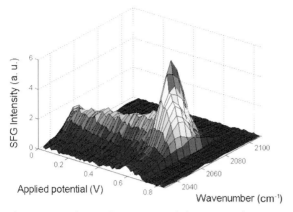

Fig. 5.18 CO vibrational spectra recorded every 22 s during a positive-going potential sweep at 0.5 mV s^{-1} [110]. (Used with permission by The American Chemical Society)

the metal interband transition, where the lifetime of the C≡N vibration is the shortest on Pt and the longest on Ag (Fig. 5.17). The interband transition energy also increases in this order, resulting in less mixing of the Fermi level with the $2\pi^*$ orbital of CN$^-$. With decreasing mixing, the energy transfer process becomes less efficient and the lifetime is therefore longer [109].

Dynamics with respect to applied potential have also been studied. Chou et al. [110] followed the oxidation of CO on Pt(111) using *in-situ* SFG collecting 1 spectrum every 22 s. A −100 mV shift in the CO stripping peak was observed in the CV for a scan rate of 0.5 mV s^{-1} and followed a simultaneous decrease in the SFG signal for top site CO. Also, a dramatic frequency shift and intensity increase is noted just before CO oxidation. The intensity and frequency shift are related to the Raman cross-section and coupling between CO dipoles while CO surface coverage changes, as shown in Fig. 5.18 [110].

5.2.7
Solvent Structure

Because SFG can discriminate molecules adsorbed at interfaces from those present in the bulk, it is an ideal technique to examine the organization of solvent molecules at the solid/liquid interface.

5.2.7.1 Nonaqueous Solvents
An area of active research using electrochemical SFG is in nonaqueous solvents systems such as DMSO, acetonitrile, and ionic liquids. An advantage of these systems over water is that the potential window can be much larger and the vibrational spectra are typically less complicated because these nonaqueous solvents do not hydrogen-bond to the same extent as water [111–116].

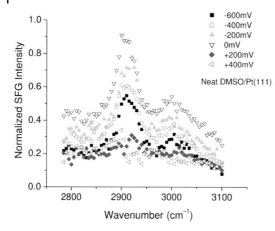

Fig. 5.19 SFG spectra of DMSO containing 0.1 M Li triflate at the Pt(111) surface. Polarization is ppp.

One solvent studied is DMSO on Pt(111). As shown in Fig. 5.19, peaks associated with the symmetric ($2900\ cm^{-1}$) and antisymmetric ($3000\ cm^{-1}$) CH stretches exhibit considerable potential dependence. An analysis of these spectra indicates that at potentials near $+200\ mV$ vs Ag (quasi reference electrode, QRE) the DMSO molecule must reorientate so that the preferred orientation of the CH-related dipoles is no longer normal to the electrode surface.

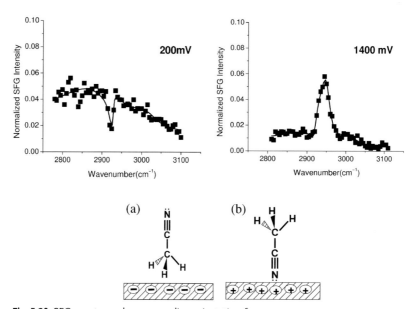

Fig. 5.20 SFG spectra and corresponding orientation for acetonitrile at Pt(111) electrode at two different potentials (vs Ag/AgCl) **(A)** $+200\ mV$, **(B)** $+1400\ mV$.

Another solvent studied with SFG spectroscopy is acetonitrile (CH_3CN) [21, 85, 117]. As was the case with DMSO, potential-dependent CH_3CN reorientation was observed [21]. These studies showed that acetonitrile orientation changes in response to the surface charge, with CH_3 pointing toward the metal at negative surface charge and away from the surface at positive surface charge. This reorientation is manifested in the potential-dependent phase change of the resonant CH_3 vibration at 2945 cm^{-1}, as shown in Fig. 5.20A and B. In these experiments there was also evidence for displacement of acetonitrile from the surface by trace water at negative potentials [117].

An interesting application of SFG is to study the structure of room-temperature ionic liquids at the electrode interface [47]. Ionic liquids are composed of

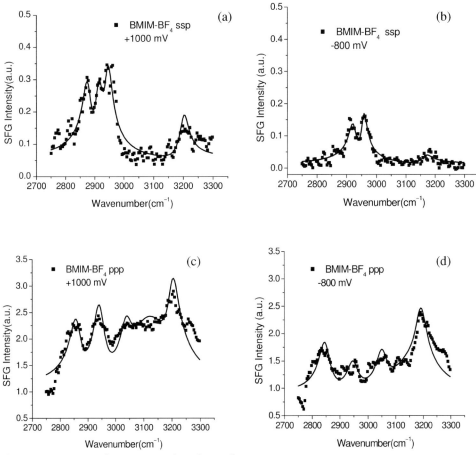

Fig. 5.21 SFG spectra of BMIM-BF$_4$ at the polycrystalline electrode interface. (**A**) ssp −800 mV, (**B**) −800 mV, (**C**) ssp +1000 mV, (**D**) ppp +1000 mV. Lines are a fit to Eq. (2) [47].

organic cations and inorganic anions that are liquid at room temperature. They have the advantage of being a solventless electrolyte.

The SFG spectra of [BMIM][BF$_4$] at ppp and ssp polarization on polycrystalline platinum are shown below in Fig. 5.21. The spectra are relatively complicated, as there are several resonances observed in the C-H stretching region as well as an intense non-resonant background. The peaks at 2850 and 2875 cm^{-1} are from the methylene and methyl groups, respectively. Resonances due to CH$_3$ directly attached to the ring are observed at 2970 cm^{-1}, and the symmetric stretch of the aromatic C-H at the C(4)-C(5) position is at 3170 cm^{-1}. The peak at 3140 cm^{-1} is called the interaction peak and is due to hydrogen-bonding interaction of the aromatic C-H groups with the anion [118], or possibly the electrode.

The orientation analysis is accomplished by the same method as that used by Wang [33, 34] and Hirose et al. [14–16]. Qualitatively, the peaks in the SFG spectra indicate that the imidazolium ring is oriented with the C$_2$ axis along the surface normal. Simulations of peak intensity versus tilt angle (of the C$_2$ axis) are plotted for several twist angles in Ref. [47]. At +1000 mV, the C$_2$ axis is tilted ~ 45° from the surface normal and increases to ~ 64° at −800 mV. The twist angle is restricted to near 0°; this is reasonable since the side chains, butyl and methyl, restrict the twist. Also, the H-C(4)C(5)-H antisymmetric stretch does not appear, so the orientation of the imidazolium plane is mostly a function of tilt angle θ. For [BMIM][PF$_6$], the orientation change is mostly the result of a twisting of the imidazolium ring (see Fig. 5.22) [47].

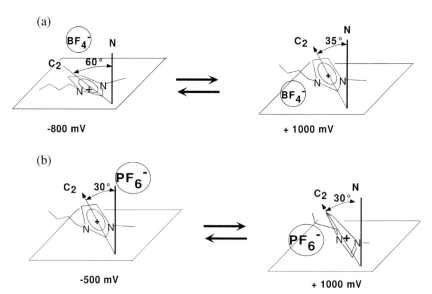

Fig. 5.22 Possible orientations of ionic liquid ions at the electrochemical interface [BMIM]$^+$ with (**A**) [BF$_4$]$^-$ and (**B**) [PF$_6$]$^-$.

These results suggest that the solvent (ions) organize themselves at the surface in order to effectively screen the charge on the metal electrode. When the metal is negatively charged, the cation will orient with the imidazolium ring more parallel to the surface, while when the surface is positively charged the ring is more tipped along the surface normal as shown in Fig. 5.22.

5.2.7.2 Aqueous Solvents

The structure of water at the electrode interface has long been a subject of substantial interest in electrochemistry. To examine the structure of water, Schultz and Gewirth used SFG to interrogate the OH stretching region for water with and without electrolyte on an Au(100) electrode [119]. SFG spectra obtained in the absence of electrolyte exhibit a single broad resonance in the OH stretch spectral region associated with weakly H-bonded water, which the authors correlated to disordered water clusters previously reported in UHV systems. In the presence of electrolyte, two new potential-dependent bands in the OH stretch region were observed, as shown in Fig. 5.23. One of these was associated with more strongly H-bonded water (band 2 in Fig. 5.23), while the other was attributed to a surface-coordinated species (band 1). These studies show that the H-bonding facility of water at the electrode surface is influenced by the presence

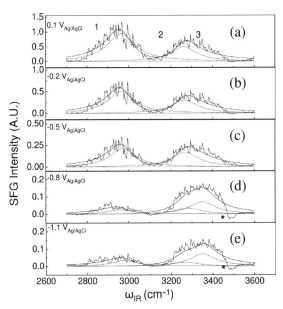

Fig. 5.23 SFG spectra obtained from an Ag(100) surface in 0.1 M KF electrolyte at the applied potentials indicated (a–e). The numbers indicate peak labels as described in the text. Open circles represent experimental data points: the solid lines are the fitting result and the dashed lines are the deconvolution.

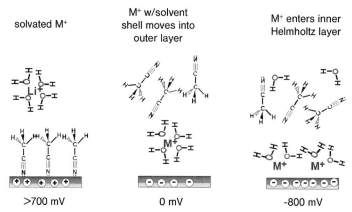

Fig. 5.24 Model of solvent structure at the Pt electrode interface as a function of applied voltage [117].

of electrolyte and that the electrolyte – or the consequences of an applied potential – also influences the orientation of the water at the interface.

Zheng et al. have noted the influence of potential on the NR background in the SFG signal in the 0.1 M HClO$_4$/Pt(110) system. However, because of the interference of solvent (water), noted in the experimental section, no resonant features due to water were observed [120]. These authors did observe a Pt-H peak at 2042 cm^{-1} at 0.12 V vs NHE, which disappears at more positive potentials, and a new feature appears at $\sim$1000 cm^{-1} associated with ClO$_4^-$. ClO$_4^-$ will specifically adsorb to the electrode with a partial solvation shell intact. The appearance of ions at the interface with an intact solvation shell was observed earlier on Pt(111) for acetonitrile/D$_2$O electrolyte with triflate ions in solution. Here, SFG is used to observe the migration of cations into the double layer as the potential is made negative; this migration is manifested as a decrease in the acetonitrile signal and an increase in that for D$_2$O. The appearance of a partial solvation shell as the cations become adsorbed to the Pt surface is evident by the O-D stretching vibration (Fig. 5.24) [117]. Recently, the SFG signal of the OH stretch of water has been observed on gold electrodes [121]. The signal follows the expected change as the water dipole reorients under the influence of surface charge.

These results constitute one of the most recent applications of SFG spectroscopy to obtain a molecular understanding of solvent configuration at the electrode surfaces. The relatively few *in-situ* experiments conducted so far should inspire additional efforts, which will undoubtedly result in an improved understanding of the interplay between solvent, electrolyte, and electrode.

5.2.8
Monolayers and Corrosion

In a very unusual application of SFG, Bittner et al. studied the change in orientation of alkanethiols on an Au electrode occurring during the electrodeposition of Cu [122]. Their results showed that while the original well-ordered monolayer is oriented with the alkyl chain at a tilt angle of 22° relative to surface normal prior to deposition, this angle changes to a value of about 83° from surface normal during the deposition process. The monolayer can regain some of its original orientation once the Cu is electrochemically removed.

Alkanethiols are a versatile platform for derivatizing the surface of metal electrodes. To understand aspects of this surface chemistry, SFG is used to study their properties *in situ* as a function of electrode potential [123]. *In-situ* SFG measurements revealed that the alkanethiol monolayer on Pt is much less dense than that on Au and Ag. SFG studies show that, because of the high packing density on these latter two surfaces, changes in potential do not cause any conformational change in the alkyl chain until the monolayers are oxidized [3]. On Pt, ion migration into and out of the double layer with changes in applied potential cause the chains to alter their conformation as a function of potential. In a novel application, Langer and coworkers used SFG to show that sparse arrays of carboxylate-terminated thiols adsorbed on Au reorient in response to surface charge. Either the carboxylate or the alkyl chain is exposed to the solution at negative or positive surface charge, respectively [124].

SFG can provide considerable information regarding the buried interface that is of central importance to corrosion inhibition processes. A model system in this respect is the monolayer of benzotriazole (BTA) that forms beneath a thick multilayer of the same molecule on Cu. Two SFG studies have examined this system thus far [125, 126]. In the study by Schultz et al., SFG showed that BTA forms a relatively well-ordered monolayer on Cu(100) between −0.7 and +0.2 V, while on Cu(111) this order is only present at high potential. Titration with Cl⁻ showed that the monolayer was destabilized at lower Cl⁻ concentrations than those needed to destabilize the polymeric and somewhat more inaccessible multilayer. Work performed by Romero et al. using 5-methylbenzotriazole on Cu(poly) show that the 5-methylbenzotriazole is stable on the surface with no orientation changes with potential [125]. Similarly to the system studied by Schultz, the degree of preferential ordering of BTA on Cu(111) seems to be less than that on the Cu(poly) surface.

5.3
Conclusion

The results presented above show the huge potential of SFG for obtaining information from systems in the electrochemical environment. To date, only a few systems have been examined, with the usual suspects – CO and other small molecules on Pt and Au – predominating. As instrumentation and expertise percolate into the electrochemical community, the number and diversity of systems examined will certainly increase. Because of its interfacial specificity, SFG is most appropriately applied to buried interface problems, and these occur in abundance in the electrochemical environment. For example, an area of considerable importance that is only now receiving attention using this *in-situ* spectroscopy relates to the structure of solvents above the interface and the interaction of the solvent with anions and cations that form the electrochemical double layer. Additional advantages accrue to SFG because of its ability to obtain detailed structural information at the interface and because of its intrinsic high time resolution. One can anticipate that SFG might provide important information concerning the dynamics of the interfacial charge transfer event, for example.

These advantages relating to SFG come with a corresponding cost regarding (a) instrumental complexity and (b) complications due to data analysis. For the former of these cost areas, the appearance of commercially available laser systems is a welcome improvement. For the latter, future attention will have to be directed at developing methodology to cleanly separate resonant and non-resonant components. The use of DFG and/or tunable visible sources may prove quite helpful in this regard. A detailed theoretical framework for SFG in the electrochemical environment will have to describe relative phase factors of different resonances with each other and will be necessary to obtain quantitative information from the measurement.

Fifteen years ago, STM and AFM were poised to move into the realm of electrochemistry. Their introduction – leveraged by the wide availability of instrumentation – certainly enlivened the field, such that it is difficult to find an electrochemical program today not influenced by these microscopies. While the vibrational spectroscopic information available from SFG is perhaps somewhat less unique than the structural information available from STM and AFM, this technique is ready today to continue to make important contributions to the study of the electrified solid/liquid interface.

Acknowledgments

We thank Zachary D. Schultz for reading this manuscript and for helpful comments. AAG is indebted to Prof. Abderrahmane Tadjeddine for introducing him to SFG and for his interest and friendship. AAG acknowledges the support of the U.S. Department of Energy, Division of Materials Sciences under Award No. DEFG02-91ER45439, through the Frederick Seitz Materials Research Laboratory at the University of Illinois at Urbana-Champaign. The Baldelli group acknowledges funding from the Welch Foundation and The Petroleum Research Fund. SB is grateful to Prof. Gabor Somorjai, Prof. Y. R. Shen and Dr. Phil Ross for starting him out into the SFG/EC field.

References

1 Shen, Y. R. *Nature* **1989**, *337*, 519.

2 Guyot-Sionnest, P.; Tadjeddine, A. *Chem. Phys. Lett.* **1990**, *172*, 341.

3 Bain, C. D. *J. Chem. Soc. Faraday Trans.* **1995**, *91*, 1281.

4 Bloembergen, N. *Appl. Phys. B* **1999**, *68*, 289.

5 Buck, M.; Himmelhaus, M. *J. Vac. Sci. Technol. A* **2001**, *19*, 2717.

6 Du, Q.; Freysz, E.; Shen, R. *Science* **1994**, *264*, 826.

7 Eisenthal, K. B. *Acc. Chem. Res.* **1993**, *26*, 636.

8 Huang, J. Y.; Shen, Y. R. In *Laser Spectroscopy and Photochemistry on Metal Surfaces*; Dai, H. L., Ho, W., Eds.; World Scientific, Singapore, 1995.

9 Miranda, P. B.; Shen, Y. R. *J. Phys. Chem. B* **1999**, *103*, 3292.

10 Richmond, G. L. *Analytical Chemistry* **1997**, 536A.

11 Shultz, M. J.; Baldelli, S.; Schnitzer, C.; Simonelli, D. *J. Phys. Chem. B* **2002**, *106*, 5313.

12 Tadjeddine, A.; LeRille, A.; Pluchery, O.; Vidal, F.; Zheng, W. Q.; Peremans, A. *Phys. Stat. Sol. A* **1999**, *175*, 89.

13 Tadjeddine, A.; LeRille, A. In *Interfacial Electrochemistry*; Weickowski, A., Ed.; Marcel Dekker, New York, 1999, p 957.

14 Hirose, C.; Akamatsu, N.; Domen, K. *Appl. Spectrosc.* **1992**, *46*, 1051.

15 Hirose, C.; Akamatsu, N.; Domen, K. *J. Chem. Phys.* **1992**, *96*, 997.

16 Hirose, C.; Yamamoto, H.; Akamatsu, N.; Domen, K. *J. Phys. Chem.* **1993**, *97*, 10064.

17 Yamamoto, H.; Akamatsu, N.; Wada, A.; Hirose, C. *J. Electron Spectrosc. Relat. Phenom.* **1993**, *64/65*, 507.

18 Akamatsu, N.; Domen, K.; Hirose, C. *J. Phys. Chem.* **1993**, *97*, 10070–10075.

19 Cotton, F. A. *Chemical Applications of Group Theory*, 3rd ed.; John Wiley and Sons: New York, 1990.

20 Potterton, E. A.; Bain, C. D. *J. Electroanal. Chem.* **1996**, *409*, 109.

21 Baldelli, S.; Mailhot, G.; Ross, P. N.; Shen, Y. R.; Somorjai, G. A. *J. Phys. Chem. B* **2001**, *105*, 654.

22 Tadjeddine, A.; LeRille, A. *Electrochim. Acta* **1999**, *45*, 601.

23 Ward, R. N.; Davies, P. B.; Bain, C. D. *J. Phys. Chem.* **1993**, *97*, 7141.

24 Superfine, R.; Huang, J. Y.; Shen, Y. R. *Chem. Phys. Lett.* **1990**, *172*, 303.

25 Gough, K. M.; Murphy, W. F.; Stroyer, T.; Svendsen, E. N. *J. Chem. Phys.* **1987**, *87*, 3341.

26 Gough, K. M. *J. Chem. Phys.* **1989**, *91*, 2424.

27 Gough, K. M.; Srivastava, H. K.; Belohorcova, K. *J. Chem. Phys.* **1993**, *98*, 9669.

28 Maroulis, G. *J. Chem. Phys.* **1998**, *108*, 5432.

29 Maroulis, G. *J. Chem. Phys.* **1991**, *94*, 1182.

30 Maroulis, G. *J. Chem. Phys.* **1994**, *101*, 4949.

31 Bell, G. R.; Bain, C. D.; Ward, R. N. *J. Chem. Soc.* **1996**, *92*, 515.

32 Fitchett, B. D.; Conboy, J. C. *J. Phys. Chem. B* **2004**, *108*, 20255.

33 Lu, R.; Gan, W.; Wu, B. H.; Chen, H.; Wang, H. F. *J. Phys. Chem. B* **2004**, *108*, 7297.

34 Lu, R.; Gan, W.; Wu, B. H.; Zhang, Z.; Guo, Y.; Wang, H. F. *J. Phys. Chem. B* **2005**, *In press*.

35 Wang, J.; Clarke, M. L.; Chen, Z. *Anal. Chem.* **2004**, *76*, 2159.

36 Muenchausen, R. E.; Keller, R. A.; Nogar, N. S. *Opt. Soc. Am. B* **1987**, *4*, 237.

37 Bloembergen, N.; Pershan, P. S. *Phys. Rev.* **1962**, *128*, 606.

38 Campion, A. *Annu. Rev. Phys. Chem.* **1985**, *36*, 549.

39 Braun, R.; Casson, B. D.; Bain, C. D.; Ham, E. W. M. v. d.; Vrehen, Q. H. F.; Eliel, E. R.; Briggs, A. M.; Davies, P. B. *J. Chem. Phys.* **1999**, *110*, 4634.

40 Zhuang, X.; Miranda, P. B.; Kim, D.; Shen, Y. R. *Phys. Rev. B.* **1999**, *59*, 12632.

41 Pettinger, B. In *Adsoption of Molecules at Metal Electrodes*; Lipkowski, J., Ross, P. N., Eds.; VCH Publishers Inc., New York, 1992; Vol. 1, p 414.

42 Feller, M. B.; Chen, W.; Shen, Y. R. *Phys. Rev. A* **1991**, *43*, 6778.

43 Roy, D. *Phys. Rev. B* **2000**, *61*, 13283.

44 Wei, X.; Hong, S. C.; Zhuang, X.; Goto, T.; Shen, Y. R. *Phys. Rev. E* **2000**, *62*, 5160.

45 Greenler, R. G. *J. Chem. Phys.* **1966**, *44*, 310.

46 Greenler, R. G. *J. Chem. Phys.* **1969**, *50*, 1963.

47 Rivera-Rubero, S.; Baldelli, S. *J. Phys. Chem. B* **2004**, *108*, 15133.

48 Bain, C. D.; Davies, P. B.; Ong, T. H.; Ward, R. N. *Langmuir* **1991**, *7*, 1563.

49 Krause, H. J.; Daum, W. *Appl. Phys. B* **1993**, *56*, 8.

50 Rabinowitz, P.; Perry, B.; Levinos, N. *IEEE J. Quant. Elec.* **1986**, *QE22*, 797.

51 Peremans, A.; Guyot-Sionnest, P.; Tadjeddine, A.; Glotin, F.; Ortega, J. M.; Prazeres, R. *Nuclear Instruments and Methods in Physics Research A* **1993**, *331*, ABS 28.

52 Richter, L. J.; Petralli-Mallow, T. P.; Stephenson, J. C. *Opt. Lett.* **1998**, *23*, 1594.

53 van der Ham, E. W.; Vrehen, Q. H.; Eliel, E. R. *Surf. Sci.* **1996**, *368*, 96.

54 Lipkowski, J.; Ross, P. N. *Electrocatalysis*, Wiley-VCH, New York, 1998.

55 Lipkowski, J.; Ross, P. N., Eds. *Adsorption of Molecules at Metal Electrode*; VCH, New York, 1992.

56 Chen, A.; Shi, Z.; Bizzotto, D.; Lipkowski, J.; Pettinger, B.; Bilger, C. *J. Electroanal. Chem.* **1999**, *467*, 342.

57 Wandlowski, T.; Ocko, B. M.; Magnussen, O. M.; Wu, S.; Lipkowski, J. *J. Electroanal. Chem.* **1996**, *409*, 155.

58 Clavilier, J.; Faure, R.; Guinet, G.; Durand, R. *J. Electroanal. Chem.* **1980**, *107*, 205.

59 Itaya, K. *Prog. Surf. Sci.* **1998**, *58*, 121.

60 Peremans, A.; Tadjeddine, A.; Suhren, M.; Prazeres, R.; Glotin, F.; Jaroszynski, D.; Ortega, J. M. *J. Electron Spectrosc. Relat. Phenom.* **1993**, *64/65*, 391.

61 Peremans, A.; Tadjeddine, A. *Chem. Phys. Lett.* **1994**, *220*, 481.

62 Peremans, A.; Tadjeddine, A.; Guyot-Sionnest, P.; Prazeres, R.; Glotin, F.; Jaroszynski, D.; Berset, J. M.; Ortega, J. M. *Nucl. Instrum. Methods Phys. Res. A* **1994**, *341*, 146.

63 Vidal, F.; Busson, B.; Six, C.; Pluchery, O.; Tadjeddine, A. *Surf. Sci.* **2002**, *502/503*, 485.

64 Vidal, F.; Busson, B.; Six, C.; Tadjeddine, A.; Dreesen, L.; Humbert, C.; Peremans, A.; Thiry, P. A. *J. Electroanal. Chem.* **2004**, *563*, 9.

65 Vidal, F.; Busson, B.; Tadjeddine, A. *Electrochim. Acta* **2004**, *49*, 3637.

66 Iwasita, T.; Nart, F. C. *Prog. Surf. Sci.* **1997**, *55*, 271.

67 Baldelli, S.; Markovic, N.; Ross, P. N.; Shen, Y. R.; Somorjai, G. A. *J. Phys. Chem. B* **1999**, *103*, 8920.

68 Blyholder, G. *J. Phys. Chem.* **1964**, *68*, 2772.

69 Dederichs, F.; Friedrich, K. A.; Daum, W. *J. Phys. Chem. B* **2000**, *104*, 6626.

70 Friedrich, K. A.; Daum, W.; Dederichs, F.; Akemann, W. *Z. Phys. Chem.* **2003**, *217*, 527.

71 Markovic, N. M.; Grgur, B. N.; Lucas, C. A.; Ross, P. N. *J. Phys. Chem. B* **1999**, *103*, 487.

72 Lucas, C. A.; Markovic, N. M.; Ross, P. N. *Surf. Sci.* **1999**, *425*, L381.

73 Korzeniewski, C. *Crit. Rev. Anal. Chem.* **1997**, *27*, 81.

74 Kim, C. S.; Korzeniewski, C.; Tornquist, W. J. *J. Chem. Phys.* **1994**, *100*, 628.

75 Severson, M. W.; Stuhlmann, C.; Villegas, I.; Weaver, M. J. *J. Chem. Phys.* **1995**, *103*, 9832.

76 Chang, S.; Leung, L. H.; Weaver, M. J. *J. Phys. Chem.* **1989**, *93*, 5341.

77 Lu, G. Q.; White, J. O.; Weickowski, A. *Surf. Sci.* **2004**, *564*, 131.

78 Hoffer, S.; Baldelli, S.; Chou, K.; Ross, P. N.; Somorjai, G. A. *J. Phys. Chem. B* **2002**, *106*, 6473.

79 Peremans, A.; Tadjeddine, A. *Phys. Rev. Lett.* **1994**, *73*, 3010.

80 Peremans, A.; Tadjeddine, A. *J. Chem. Phys.* **1995**, *103*, 7197.

81 Climent, V.; Rodes, A.; Orts, J. M.; Feliu, J. M.; Aldaz, A. *J. Electroanal. Chem.* **1999**, *467*, 20.

82 Ogasawara, H.; Ito, M. *Chem. Phys. Lett.* **1994**, *221*, 213.

83 Peremans, A.; Tadjeddine, A. *J. Electroanal. Chem.* **1995**, *395*, 313.

84 Daum, W.; Friedrich, K. A.; Klunker, C.; Knabben, D.; Stimming, U.; Ibach, H. *Appl. Phys. A* **1994**, *59*.

85 Dederichs, F.; Petukhova, A.; Daum, W. *J. Phys. Chem. B.* **2001**, *105*, 5210.

86 Friedrich, K. A.; Daum, W.; Klunker, C.; Knabben, D.; Stimming, U.; Ibach, H. *Surf. Sci.* **1995**, *335*, 315.

87 Tadjeddine, A.; Peremans, A.; LeRille, A.; Zheng, W. Q.; Tadjeddine, M.; Flament, J. P. *J. Chem. Soc. Faraday Trans.* **1996**, *92*, 3823.

88 Tadjeddine, M.; Flament, J. P.; Tadjeddine, A. *J. Electroanal. Chem.* **1996**, *408*, 237.

89 Tadjeddine, M.; Flament, J. P. *Chem. Phys.* **1999**, *240*, 39.

90 Matranga, C.; Guyot-Sionnest, *J. Chem. Phys.* **2000**, *112*, 7615.

91 Stuhlmann, C.; Villegas, I.; Weaver, M. J. *Chem. Phys. Lett.* **1994**, *219*, 319.

92 Huerta, F. J.; Morallon, E.; Vazquez, J. L.; Aldaz, A. *Surf. Sci.* **1998**, *396*, 400.

93 LeRille, A.; Tadjeddine, A.; Zhang, W. Q.; Peremans, A. *Chem. Phys. Lett.* **1997**, *271*, 95.

94 Bowmaker, G. A.; Leger, J. M.; LeRille, A.; Melendres, C. A.; Tadjeddine, A. *J. Chem. Soc. Faraday Trans.* **1998**, *94*, 1309.

95 Ong, T. H.; Davies, P. B.; Bain, C. D. *J. Phys. Chem.* **1993**, *97*, 12047.

96 Rambaud, C.; Cagnon, L.; Levy, J. P.; Tourillon, G. *J. Electrochem. Soc.* **2004**, *151*, E352.

97 Tadjeddine, A.; LeRille, A.; Pluchery, O.; Hebert, P.; Marin, T. *Nucl. Instrum. Methods A* **1999**, *429*, 481.

98 Hebert, P.; LeRille, A.; Zheng, W. Q.; Tadjeddine, A. *J. Electroanal. Chem.* **1998**, *447*, 5.

99 Chen, A. C.; Sun, S. G.; Yang, D. F.; Pettinger, B.; Lipkowski, J. *Can. J. Chem.* **1996**, *74*, 2321.

100 Pluchery, O.; Tadjeddine, A. *J. Electroanal. Chem.* **2001**, *500*, 379.

101 Pluchery, O.; Zheng, W. Q.; Marin, T.; Tadjeddine, A. *Phys. Stat. Sol. A* **1999**, *175*, 145.

102 Pluchery, O.; Climent, V.; Rodes, A.; Tadjeddine, A. *Electrochim. Acta* **2001**, *46*, 4319.

103 Tadjeddine, M.; Flament, J. P. *Chem. Phys.* **2001**, *265*, 27.

104 Kim, J.; Chou, K. C.; Somorjai, G. A. *J. Phys. Chem. B* **2002**, *106*, 9198.

105 Chou, K. C.; Kim, J.; Baldelli, S.; Somorjai, G. A. *J. Electroanal. Chem.* **2003**, *554-555*, 253.

106 Schmidt, M. E.; Guyot-Sionnest, P. *J. Chem. Phys.* **1996**, *104*, 2438.

107 Peremans, A.; Tadjeddine, A.; Guyot-Sionnest, P. *Chem. Phys. Lett.* **1995**, *247*, 243.

108 Peremans, A.; Tadjeddine, A.; Zheng, W. Q.; LeRille, A.; Guyot-Sionnest, P.; Thiry, P. A. *Surf. Sci.* **1996**, *368*, 384.

109 Matranga, C.; Wehrenberg, B. L.; Guyot-Sionnest, P. *J. Phys. Chem. B* **202**, *106*, 8172.

110 Chou, K. C.; Markovic, N. M.; Kim, J.; Ross, P. N.; Somorjai, G. A. *J. Phys. Chem. B* **2003**, *107*, 1840.

111 Roelfs, B.; Schroter, C.; Solomun, T. *Ber. Bunsenges. Phys. Chem.* **1997**, *101*, 1105.

112 Marinkovic, N. S.; Hecht, M.; Loring, J. S.; Fawcett, W. R. *Electrochim. Acta* **1996**, *41*, 641.

113 Fawcett, W. R.; Kloss, A. A.; Calvente, J. J.; Marinkovic, N. *Electrochim. Acta* **1998**, *44*, 881.

114 Faguy, P. W.; Fawcett, W. R.; Liu, G.; Motheo, A. J. *J. Electroanal. Chem.* **1992**, *339*, 339.

115 Marinkovic, N. S.; Calvente, J. J.; Kloss, A.; Kovacova, Z.; Fawcett, W. R. *J. Electroanal. Chem.* **1999**, *467*, 325.

116 Marinkovic, N. S.; Calvente, J. J.; Kovacova, Z.; Fawcett, W. R. *J. Electrochem. Soc.* **1996**, *143*, L171.

117 Baldelli, S.; Mailhot, G.; Ross, P. N.; Somorjai, G. A. *J. Am. Chem. Soc.* **2001**, *123*, 7697.

118 Dieter, K. M.; Dymek, C. J.; Heimer, N. E.; Rovang, J. W.; Wilkes, J. S. *J. Am. Chem. Soc.* **1988**, *110*, 2722.

119 Schultz, Z. D.; Gewirth, A. A. *J. Am. Chem. Soc.* **2005**, *127*, 15916.

120 Zheng, W. Q.; Pluchery, O.; Tadjeddine, A. *Surf. Sci.* **2002**, *502/503*, 490.

121 Nihonyanagi, S.; Ye, S.; Uosaki, K.; Dreesen, L.; Humbert, C.; Thiry, P. A.; Peremans, A. *Surf. Sci.* **2004**, *573*, 11.

122 Epple, M.; Bittner, A. M.; Kuhnke, K.; Kern, K.; Zheng, W. Q.; Tadjeddine, A. *Langmuir* **2002**, *18*, 773.

123 Hines, M. A.; Todd, J. A.; Guyot-Sionnest, P. *Langmuir* **1995**, *11*, 493.

124 Lahann, J.; Mitragotri, S.; Tran, T. N.; Kaido, H.; Sundaram, J.; Choi, I. S.; Hoffer, S.; Somorjai, G. A.; Langer, R. *Science* **2003**, *299*, 371.

125 Romero, C.; Baldelli, S. *In Preparation* **2004**.

126 Schultz, Z. D.; Biggin, M. E.; White, J. O.; Gewirth, A. A. *Anal. Chem.* **2004**, *76*, 604.

6
IR Spectroscopy of the Semiconductor/Solution Interface

Jean-Noël Chazalviel and François Ozanam

6.1
Introduction

The development of electrochemistry as a science has long been based on electrical measurements, that is the investigation of relationships between the two key parameters current and potential and also possibly time or frequency (typically, voltammetric or impedance techniques) (see e.g. Ref. [1]). Despite the great power of these techniques, it soon became clear that key pieces of information, such as the chemical nature of interface species, would be obtained only through the use of independent, for example spectroscopic, methods. However, whereas a wealth of fancy techniques had been developed for surface science, such as electron or ion spectroscopies, most of these tools were clearly inappropriate for the routine *in-situ* investigation of the electrode/electrolyte interface, the presence of the liquid representing a considerable hindrance. Significant development of techniques compatible with the liquid environment has taken place since the 1980s [2–14]. Among these, the optical techniques (spectroscopic ellipsometry [4, 5], Raman [6] and infrared (IR) spectroscopies [7–11], and second-harmonic (see e.g. Ref. [12]) and sum-frequency generation [13, 14]) have brought a wealth of new information. IR spectroscopy probably represents the most affordable of these techniques in terms of budget. Various reviews on its use have appeared [7–11], including the special cases of near-IR spectroscopy [10] and the IR spectroscopy of semiconducting electrodes [11]. A more detailed review of the latter will be presented here. We will first discuss the specificity of the technique at semiconductor electrodes, then review the various kinds of information that it may bring.

Advances in Electrochemical Science and Engineering Vol. 9.
Edited by Richard C. Alkire, Dieter M. Kolb, Jacek Lipkowski and Philip N. Ross
Copyright © 2006 WILEY-VCH Verlag GmbH & Co. KGaA, Weinheim
ISBN: 3-527-31317-6

ergy side, by lattice absorption (if one excepts oxides, this becomes significant only below 10^3 cm^{-1}). Free-carrier absorption also occurs at low energies. However, only for highly doped semiconductors ($>10^{17}$ cm^{-3}) does it appear above 10^3 cm^{-1}. At a semiconductor/electrolyte interface, this wide transparency range allows the IR beam to be sent to the interface through the electrode. Given the limitations due to free-carrier absorption, this geometry will be usable for doped semiconductors with resistivities down to the range 0.01–0.1 Ωcm. In this geometry, n_1 represents the semiconductor and n_2 the electrolyte, which in a first approximation can both be taken as real. One generally has $n_1 > n_2$, and for an incidence angle φ_1 above a rather small critical value $\sin^{-1}(n_2/n_1)$, the condition $n_1 \sin\varphi_1 = n_2 \sin\varphi_2$ can no longer be fulfilled: the beam is fully reflected into medium 1, and the refracted beam in medium 2 is replaced by an evanescent wave with a characteristic decay length $\delta = \lambda / \left(2\pi \sqrt{n_1^2 \sin^2 \varphi_1 - n_2^2}\right)$ [15, 21]. The expressions given above for the electric field at the interface remain valid, provided one makes some formal changes ($\sin\varphi_2$ is now larger than 1, which means that φ_2 and $\cos\varphi_2$ are imaginary). The presence of any absorbing species at the interface will result in decreasing the amplitude of the reflected wave.

This internal reflection geometry (also termed ATR for Attenuated Total Reflection) is schematized in Fig. 6.2 b [21, 22]. It exhibits several advantages as compared to external reflection: (a) It allows for the use of a classical electrochemical cell: there is no need for a thin-electrolyte cell; hence, there are no limitations in the presence of Faradaic processes and/or gas evolution at the electrode surface (mass transport may even be controlled by using a circulation cell [23]), and it is possible to use high-frequency potential modulation. (b) Multiple reflections can be used, thereby enhancing IR sensitivity. (c) The IR beam penetrates into the electrolyte over a thickness δ and is sensitive to the absorption of an electrolyte layer of effective thickness $\delta/2$, which is much smaller than λ in typical conditions ($\lambda/25$ at a silicon/water interface, with $\varphi_1 = 45°$). This makes the technique relatively immune to bulk electrolyte absorption. (d) In contrast to the external-reflection geometry, here the amplitudes of the three electric-field components ($E_{//}$ in s-polarization, $E_{//}$ and $E_{2\perp}$ in p-polarization) turn out to be comparable, which makes the technique about equally sensitive to species of any orientation on the surface [24]. The only practical requirements are that the sample must be shaped as a prism, and the surfaces must be maintained reasonably flat and smooth. Semiconductors exhibiting some IR absorption can be investigated in the form of thin layers deposited on a non-absorbing substrate. The deposit may be realized by any suitable method, for example, vacuum evaporation or sputtering, wafer bonding [25], spin coating, or electropolymerization for the case of semiconducting polymers [26]. It may even consist of a powder [27]. Silicon, germanium, ZnSe, and more recently diamond have been among the most popular substrates for such studies [28]. The only requirements are that the index of the layer be sufficiently high and the contact sufficiently tight (maximum allowable empty space $< \delta$) as to allow penetration of the IR beam into the layer. Finally, a last general requirement is that the Ohmic contacts be taken in a region not visited by the IR beam, because the condition of total reflection is not fulfilled at a semiconductor/metal interface and the reflectivity there is indeed significantly lower than unity.

As a basis for quantitative analysis of IR data in an ATR geometry, the absorption of a thin layer of absorbing species has been calculated, assuming that this layer, of thickness d, can be described as a medium with a complex dielectric function $\varepsilon(\sigma) = \varepsilon' - i\varepsilon''$ [24]. The latter may even be taken as anisotropic with principal directions x, y, z (hence ε_x, ε_y, ε_z) where z is the direction normal to the interface and xz is the incidence plane. The relative changes in IR reflection, calculated to first order in d, are then [24]

s-polarization:
$$\left(\frac{\Delta R}{R}\right)_s = \frac{2\pi}{\lambda n_1 \cos \varphi_1} A_y \varepsilon_y'' d \tag{1}$$

p-polarization:
$$\left(\frac{\Delta R}{R}\right)_p = \frac{2\pi}{\lambda n_1 \cos \varphi_1} \left(A_x \varepsilon_x'' + A_z \frac{n_2^4}{\varepsilon_z'^2 + \varepsilon_z''^2} \varepsilon_z'' \right) d \tag{2}$$

where A_x, A_y and A_z represent the relative squared amplitudes of the interface electric field (corresponding to $E_{//,p}$, $E_{//,s}$, and $E_{2\perp,p}$, respectively) [24]:

$$A_x = \frac{4n_1^2 \cos^2 \varphi_1 (n_1^2 \sin^2 \varphi_1 - n_2^2)}{n_2^4 \cos^2 \varphi_1 + n_1^4 \sin^2 \varphi_1 - n_1^2 n_2^2} \tag{3}$$

$$A_y = \frac{4n_1^2 \cos^2 \varphi_1}{n_1^2 - n_2^2} \tag{4}$$

$$A_z = \frac{4n_1^4 \sin^2 \varphi_1 \cos^2 \varphi_1}{n_2^4 \cos^2 \varphi_1 + n_1^4 \sin^2 \varphi_1 - n_1^2 n_2^2} \tag{5}$$

For a silicon/water interface (values at 2000 cm^{-1}: $n_1 = 3.43$, $n_2 = 1.32$) and $\varphi_1 = 45°$, one has $A_x = 1.94$, $A_y = 2.35$, and $A_z = 2.76$, so that knowledge of the $\varepsilon(\sigma)$ function in principle allows d to be determined.

6.3
Practical Aspects at an Electrochemical Interface

Even though the ATR geometry exhibits a fair immunity to electrolyte absorption, the latter will still be orders of magnitude larger than the absorption of interface species, which are typically in a concentration of a monolayer ($\delta/2 \approx \lambda/25 \gg 1$ Å). In practice, IR absorption will then be dominated by the electrolyte (see Fig. 6.3). However, interface species are expected to be sensitive to a number of parameters, while the electrolyte bulk is not. Such parameters may be the time (spontaneous evolution of the interface under given electrochemical

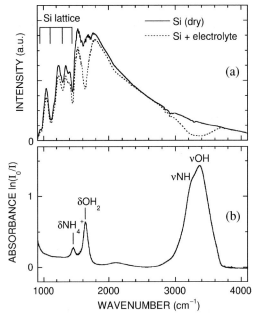

Fig. 6.3 Electrolyte absorption in ATR. Here a silicon electrode is in contact with 1 M NH₄Cl aqueous solution (beam path in Si is 15 mm; there are 10 reflections at the Si/electrolyte interface; s-polarized beam). (a) Raw spectra for the cell empty and filled with the electrolyte. Note the IR absorption of the silicon lattice. (b) Absorbance as obtained from the above two raw spectra. The absorption of silicon cancels out and only electrolyte absorption is seen. This absorption is much larger than that expected for interface species, which are typically in monolayer concentration.

conditions) or illumination of the electrode (a possible means for inducing electrochemical processes at semiconducting electrodes), but the most commonly used ones in practice will be the applied current or preferably the electrode potential. As is usual for surface IR spectroscopy, one then takes a reference spectrum under some well-defined experimental conditions and does not attempt to determine the "absolute" absorption of the interface, the aim being limited to measuring the change in IR absorption when some parameter (for example electrode potential) is changed.

6.3.1
How Potential can Affect IR Absorption

A change in electrode potential may affect the IR absorption of the interface region in various ways. It may change the concentration of some species (Faradaic processes, changes in surface composition, adsorption/desorption, and double-layer modulation) or change the IR absorption of fixed species (Stark effect or change in electron distribution of surface-attached species [29]). The variety of possible mechanisms may complicate data analysis. Especially the potential-in-duced variation of fixed species may yield spectral profiles distinct from the usual absorption shape (a change in vibrational frequency will yield a spectrum with derivative shape, and a change in bandwidth will result in an M- or W-shaped contribution). Nevertheless, these complications must be accepted, as the sensitivity to potential (or to some other parameter as cited above) is the necessary fingerprint of an interface-related species.

6.3.2
How to Isolate Potential-sensitive IR Absorption

Differential techniques The change in IR absorption associated with a change in potential can be determined through two main families of techniques. The most straightforward method consists in successively recording IR spectra for two distinct values of the potential, say V_1 and V_2, and taking the "difference" of the two spectra, or more precisely the change in absorbance $\ln[I_\sigma(V_1)/I_\sigma(V_2)]$, where $I_\sigma(V)$ is light intensity reaching the detector at wavenumber σ for potential set to V (equivalent to the relative change in reflectivity $\Delta R/R$ for small absorbances). This so-called differential technique is now routinely applied using standard Fourier-Transform IR (FTIR) spectrometers, which offer good sensitivity and almost perfect reproducibility of the wavenumber scale. It is sometimes referred to as SNIFTIRS (Subtractively Normalized Interfacial Fourier-Transform InfraRed Spectroscopy) [7, 30]. In practice, if the system is fully reversible, the $V_1 \rightarrow V_2$ sequence may be repeated several times and the data co-added in order to improve the signal-to-noise ratio. One of the two potentials may be taken as a reference and the second one may be varied, in order to explore the dependence as a function of potential. More complex sequences, such as voltammetric cycles, may also be used. Relative variations in reflectivity $\Delta R/R$ on the order of 10^{-4} may be measured with this technique (a figure that may be improved if multiple reflections are being used). However, this technique suffers from the inherent difficulty that it yields a small quantity by calculating the difference between two much larger ones. It is then very sensitive to long-term drift (of the source, the system, sample positioning, atmospheric absorption…) and to digitization noise. The problem of drift can be largely remedied if the system is sufficiently reversible to enable the above-mentioned cycling and co-addition technique to be used, but digitization noise generally sets the ultimate sensitivity limit of the method.

Modulation techniques For reversible systems, a more suitable kind of method consists in using a potential modulation coupled with a phase-sensitive detection (lock-in amplifier): the potential may be modulated, for example, as a square wave between V_1 and V_2 and the signal from the detector may be sent to a lock-in amplifier. The signal from the lock-in provides a function $\Delta I(\sigma)$ (change in intensity reaching the detector associated with the modulation), the average intensity $I(\sigma)$ being measured either simultaneously or sequentially. The signal sought is $\Delta R/R = \Delta I/I$. This modulation method, originally implemented on dispersive IR spectrometers and known as EMIRS (Electrochemically Modulated InfraRed Spectroscopy [7]) offers a sensitivity that is limited only by the ultimate performance of the detectors [31]. Unfortunately, this benefit cannot be straightforwardly combined with that of FTIR spectroscopy, because the convenient frequencies for a modulation (say 10 Hz to 10 kHz) fall right in the range used by common FTIR spectrometers for interferometer operation and data acquisition. This difficulty can be circumvented by using step-scan or slow-scan interferometers. The associated technique has been termed FTEMIRS (Fourier-Transform, Electrochemically Modulated InfraRed Spectroscopy) [32–34]. Conventional lock-in amplifiers [32, 34] as well as fast digital processing systems [33] have been used. The interesting feature of conventional lock-in amplifiers is their ability to work up to MHz frequencies, the only practical limitation being the response of the electrochemical cell. However, combining a conventional lock-in with a step-scan interferometer requires some care, as suitable circuitry or extra delays may be necessary in order to avoid saturation of the lock-in by the switching transients of the interferometer. To reach a given signal-to-noise ratio, the delays required may result in a large increase in the measurement time [33]. In view of these problems, if a conventional lock-in is used, the slow-scan approach may seem more appealing. Sensitivities on the order of $\Delta I/I = 10^{-6}$ at 1 kHz modulation frequency (10^{-7} per reflection for 10 reflections) have been reported using such an arrangement [32], a performance which deteriorates somewhat at lower frequencies because of the $1/f$ noise of the detectors. Unfortunately, only step-scan interferometers are commercially available. Their advantage is that they allow for a range of modulation frequencies in principle unlimited on the lower edge. However, the practical limitation there may be set by the stability of the electrochemical system itself.

An alternative technique The differential techniques are equivalent to a modulation at frequencies lower than $1/2T$, where T is the time required for recording one interferogram. On the other hand, the modulation techniques operate at frequencies higher than m/T, where m is the number of spectral elements (or the number of points in an interferogram, typically $\sim 10^3$). An alternative method has been proposed [35], equivalent to a modulation in the intermediate frequency range. The modulation frequency is such that $1/f = nT/m$, that is, a period is an integral multiple of the elementary time for acquiring a point of the interferogram. The principle of that method (schematized in Fig. 6.4 for $n=6$) rests on the successive recording of n interferograms, the potential modulation being applied with a variable phase shift with respect to the sampling scheme of the interferometer. The obtained $m \times n$ array of data points can be reordered in the form of n interferograms, each one corresponding to a given time in the modulation sequence. As an example, Fig. 6.4 indicates, by dots, how to select the data points to reconstruct the interferogram corresponding to the first sampling point at potential V_1. After Fourier transform, the differential spectrum can be obtained from the n obtained spectra (sum of the spectra at V_2 minus

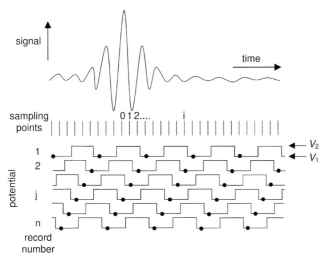

Fig. 6.4 Interferogram reconstruction. Modulation period is n times the sampling time (here for $n=6$). n interferograms (each of m points) are recorded, with different phases of the modulation with respect to the sampling grid. Through proper selection of the data points in the $n \times m$ array, the interferogram corresponding to a given phase can be reconstructed (see the black circles, providing the interferogram associated with the first sampling point after applying V_1).

sum of the spectra at V_1). Dynamic information down to the T/m time scale can even be obtained by considering the n spectra separately. This method looks rather appealing, as it offers flexibility for the choice of the modulation frequency while staying compatible with the use of a conventional fast-sweep interferometer. Unfortunately it has not really come to fruition, because it relies on delicate software, the sensitivity is limited by digitization noise, as for conventional SNIFTIRS, and long-time drifts of the system may affect the result in a complex manner.

To summarize, difference and modulation techniques can both be used. In principle, the modulation techniques offer higher sensitivity (by a factor of ~ 10) but at the cost of higher complexity of the experimental arrangement. This advantage is currently tending to diminish with the continued improvement in the performance of optical design and digital electronics [36]. Furthermore, it disappears completely with systems involving long time constants, so that today the differential techniques are by far the most widely used ones.

6.4
What can be Learnt from IR Spectroscopy at the Interface

We will here present some examples of the kind of information that can be extracted from IR spectroscopy at the semiconductor/solution interface.

6.4.1
Vibrational Absorption of Interfacial and Double-layer Species

Interface species The major motivation for developing *in-situ* IR spectroscopy at the electrode/electrolyte interface was the need to identify interfacial chemical species. Figs. 6.5 and 6.6 show two representative examples of the use of the technique for such applications.

In Fig. 6.5, the evolution of a Ge(100) surface was monitored when the potential was cycled in 1 M HClO$_4$ electrolyte [37]. A voltammogram was taken at a sweep rate of 10 mV s^{-1} and the IR spectra were recorded at the same time as the voltammogram, at intervals of 50 mV. Several potential cycles were made and the resulting spectra corresponding to each potential were accumulated in order to improve the signal-to-noise ratio. The absorbance spectra shown in Fig. 6.5 were obtained by taking as the reference spectrum that obtained at the positive potential bound. The most conspicuous feature is the appearance of two bands of opposite sign, a νGeH band at 1970 cm^{-1} and a νOH band around

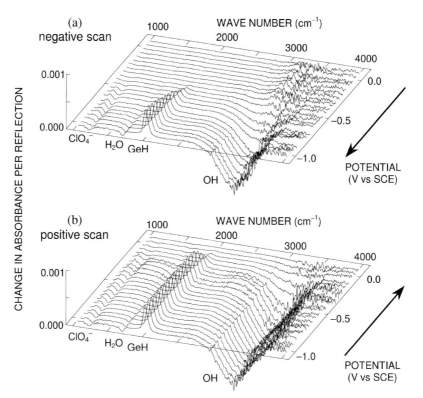

Fig. 6.5 n-Ge(100) in 1 M HClO$_4$ electrolyte. Spectra obtained during a voltammetric scan. Reference taken at the upper potential bound. Spectra taken (a) for decreasing potentials and (b) for increasing potentials. The two main bands demonstrate the quasi-reversible change of the interface between a GeH and a Ge-OH termination [37].

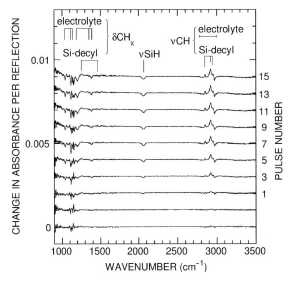

Fig. 6.6 p-Si(111) in 1 M decylmagnesium bromide in diethyl ether. Change of the surface from Si-H termination to Si-decyl termination on successively applying current pulses (300 μA cm^{-2}, duration 1.05^{n-1} ×100 ms, where n is pulse number, shown on the right). The reference spectrum is that of the initial surface. The two lowest spectra were taken before any pulse has been applied, which shows that the change in surface chemistry takes place on applying a current only [38].

3300 cm^{-1}, indicating the quasi-reversible change of the chemical state of the surface between a GeH termination at negative potentials and a GeOH termination at positive potentials. The change can be correlated with the current in the voltammogram, providing direct evidence for the origin of the voltammetric peaks [37].

Fig. 6.6 shows the evolution of a p-Si(111) surface when submitted to an anodic current in a Grignard electrolyte (1 M decylmagnesium bromide in diethyl ether) [38]. Here the evolution is irreversible, and no accumulation of successive cycles could be made. In practice, the anodic current was applied as a succession of galvanostatic pulses (300 μA cm^{-2}, pulse width $\Delta t_n = \Delta t_0 a^{n-1}$, where n is the pulse number, with $\Delta t_0 = 100$ ms and $a = 1.05$), and the spectra were recorded between the pulses. The initial surface had been prepared atomically flat and hydrogen-terminated by preparing a thin oxide by wet oxidation and etching it in NH$_4$F solution [39]. The spectra give evidence for the disappearance of the SiH bonds (negative band at 2070 cm^{-1}) and the simultaneous appearance of bands characteristic of the grafting of decyl groups covalently bonded to the Si surface (νCH$_2$ at 2855 and 2920 cm^{-1}, νCH$_3$ at 2970 cm^{-1}, δCH$_2$ at 1470 cm^{-1}) [38]. From an analysis of the advancement of the modification reaction as a function of the cumulated Faradaic charge, it was concluded that the reaction involves about two elementary charges per substituted SiH bond (a rather efficient process), and a detailed kinetic model was worked out.

A number of reports on similar applications can be found in the literature. The multiple-reflection geometry may bring the sensitivity to a level of a few percent of a monolayer (even for a weakly polar bond like GeH, see Fig. 6.5). In contrast to the case of external reflection, electrolyte absorption does not obscure full spectral windows (despite a lowered beam intensity and increased noise in the νOH absorption region for 10 reflections with a typical aqueous electrolyte, see Fig. 6.5). Examples of interesting results brought by this method are the hydrogenated character of the silicon surface in fluoride or alkaline electrolytes [40–45] and the absence of hydrogen at the surface during its dissolution in fluoroborate organic electrolytes [46], the role of SiOH species in the transition from porous-silicon formation to electropolishing [47], the formation of charged hydrated states at the surface of SiO_2 in contact with water [48], the formation of C-OH bonds on a diamond electrode after polarization to positive potentials [28], the formation of AsH groups at the GaAs surface at sufficiently negative potentials [49], and the presence of peroxide species at the surface of a TiO_2 electrode under UV illumination [27, 50].

Double-layer species Interestingly, in Fig. 6.5 as well as in Fig. 6.6, absorption bands from electrolyte species are seen to be present. Since these spectra are differential spectra, electrolyte absorption should in principle be cancelled out. The presence of electrolyte bands indicates that the change in surface chemistry or electrode potential does induce some change in electrolyte absorption. In the case of Fig. 6.6, the appearance of negative electrolyte contributions can readily be attributed to the appreciable thickness of the grafted decyl layer (~ 1 nm) and the loss of the absorption from the electrolyte being replaced by that layer [38]. Similar contributions are to be expected every time a new layer is created at the interface. Oxide layers are frequent examples of this behavior [23]. More interestingly, Fig. 6.5 exhibits a negative δOH_2 band, correlated with the negative νOH band, and an asymmetric M-shape contribution in the region of perchlorate absorption [37]. These features can be attributed to double-layer contributions: as a consequence of the hydrophilic character of the hydroxylated surface, water absorption appears reinforced (increased density of water near the interface) and perchlorate ions become strongly physisorbed (whence a change in shape of their absorption, associated with the asymmetric M-shape). Among similar double-layer contributions to IR absorption are the adsorption of acetonitrile molecules at Si [31], of acetonitrile, acetic acid [51] and CO_2 reduction intermediates [19] at CdTe, or of ferri-ferrocyanide ions at Si [52]. Also, the changes in ionic concentration in the diffuse double layer, though they represent rather weak contributions at a semiconducting electrode, may not be out of reach (for example at the Si/acetonitrile interface, see Ref. [31]).

The differential technique makes it easy to follow chemical changes in real time, inasmuch as their evolution is sufficiently slow (typically, time scales from seconds to hours are conveniently accessed), which may be useful for disentangling complex chemical or electrochemical mechanisms. Examples of phenomena investigated through this technique include the etching process of Si surfaces in NH_4F solution [53, 54], hydrogen exchange on an Si surface in water [55], the dissolution of an SiO_2 film on Si in dilute HF [56], the relaxation of surface composition at silicon anodes [57], the oscillations during anodic dissolution of silicon [58], and the oscillations during reduction of hydrogen peroxide at GaAs cathodes [59].

6.4.2
Vibrational Absorption of Species outside the Double-layer

Whether the geometry is that of external or internal reflection, the IR beam goes through one medium and penetrates into the other at least over a depth on the order of a fraction of the wavelength (δ in the ATR case). The IR absorption is then sensitive to possible changes in the two media, far from the double layer region.

Changes in electrolyte absorption   Changes in the electrolyte absorption outside the double layer commonly arise in the presence of redox reactions involving species in solution. Among reports using external-reflection geometry are the study of oxidation products of the ferrocyanide ion on n-GaAs [60] and the study of the decomposition of water and various organic solvents on TiO_2 [61–64]. In internal-reflection geometry, the successive reduction products of the heteropolyanions $SiW_{12}O_{40}^{4-}$ and $PW_{12}O_{40}^{3-}$ at a Ge electrode have been investigated [65], and the dissolution of silicon in fluoroborate electrolyte has been shown to produce BF_3 species [46]. In the ATR geometry, the region probed ($\delta/2 \approx \lambda/25$) is much shallower than the diffusion layer (typically 100 μm). In principle, IR absorption may then provide a quantitative determination of the local concentration of species near the electrode surface. With a cell design allowing controlled circulation of the electrolyte, this may be a useful tool for the study of reactions involving mass-transport effects [23]. In other cases (either fast potential modulation or short-lived intermediates), the changes in concentration may extend over a depth ($\sim \sqrt{D/\omega}$ or $\sqrt{D\tau}$, where D is diffusion constant, $\omega/2\pi$ modulation frequency and τ lifetime) comparable to the probed depth $\delta/2$. In such cases, changing the IR incidence angle may be used as a means of changing δ, which provides a tool to determine concentration profiles near the electrode surface.

In protic solvents, an important parameter that may be affected by mass-transport limitations is the local pH in the vicinity of the electrode. Infrared spectroscopy in the ATR geometry is a convenient tool for monitoring any changes of this parameter if some weak acid has been added to the electrolyte as a pH indicator. In practice, most weak acids, for example, inorganic oxoacids or carboxylic acids, can play this role, as the IR absorption of these acids and that of their conjugate anion always differ substantially (see Fig. 6.7). For pH values in the vicinity of the pK_a of the indicator, the method may be quite sensitive (sensitivity to a pH change down to 10^{-4} pH units has been reported [66]). Conversely, by choosing an indicator with multiple pK_a values, the method may be applied to the measurement of large pH variations (Fig. 6.7b) [67].

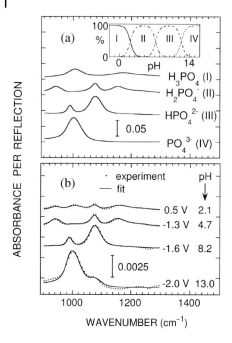

Fig. 6.7 Determination of the local pH at a GaAs/electrolyte interface. Electrolyte is 1 M KCl+0.025 M phosphoric acid, as an IR indicator. (a) IR spectra of the four phosphate species, taken in 1 M solution. The inset shows their respective concentrations as a function of pH. (b) IR spectra (referred to pure water) at the GaAs electrode, as a function of potential. Fitting these spectra as linear combinations of those in (a) leads to a determination of the pH at the interface. As the potential is made more negative, local pH is seen to increase from 2 to 13, a change associated with hydrogen evolution [67].

Another special case of interest is that where the IR absorption of an electrolyte species is changed because of its interaction with an IR-non-active species generated at the electrode. Such is the case when highly charged cations are generated upon electrode dissolution. For example, anodic dissolution of GaAs in H_2SO_4 electrolyte generates Ga^{3+} ions. The complexation of these cations

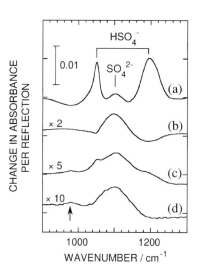

Fig. 6.8 Example of indirect detection of IR-non-active species, here Ga^{3+} ions produced by dissolution of GaAs in 0.5 M H_2SO_4+0.1 M H_2O_2 solution. Absorption changes are induced by (a) addition of 0.5 M H_2SO_4 to pure water; (b) a pH change from 0.3 to 0.5 at fixed sulphate concentration; (c) addition of 10 mM $Ga_2(SO_4)_3$ to 0.5 M H_2SO_4 solution; (d) 1 min GaAs dissolution at open-circuit potential. Note the small peak at 980 cm^{-1}, attributed to the breathing mode of SO_4^{2-}, which becomes weakly IR active because of complexation of Ga^{3+} by the sulphate ions. The spectrum in (d) can be regarded as a combination of (b) and (c) (addition of Ga^{3+} together with a pH increase) [68].

with electrolyte anions leads to a significant increase in SO_4^{2-} absorption at 1100 cm^{-1}, together with the appearance of a weak mode at 980 cm^{-1} (Fig. 6.8). The latter mode, associated with the breathing mode of the SO_4^{2-} ion, is normally IR-non-active, but it becomes active in the complex because the tetrahedral symmetry of the anion is broken. It is interesting to note that Ga^{3+} ions can be detected in this way even though IR is blind to the naked Ga^{3+} ions [11, 68]. The method can be made quantitative by calibration of the IR signals in solution.

Changes in electrode absorption   Changes in the IR absorbance of the electrode may take place in the presence of an intercalation reaction. Typical examples of such behavior are electrochromic materials (typically WO_3) [69, 70] or semiconducting polymers [26]. In either case, ion incorporation into the material leads to a change of its conducting and optical properties, including those in the IR. Especially, in the latter case, vibrational spectroscopy appears as a useful tool to characterize the various oxidation states of the polymer [71–74]. Similar experiments have been performed on fullerene films [75]. Though less common in more conventional semiconductors such as Si and Ge, the intercalation of hydrogen has been reported to occur under electrochemical conditions, and the resulting SiH and GeH species have been identified by their IR absorption, which is distinct from that of SiH and GeH at the surface [76]. Finally, in the context of semiconducting polymers, it is worth mentioning that electropolymerization reactions have often been followed *in situ* by ATR/FTIR spectroscopy by using a semiconducting ATR prism as the substrate [77, 78].

6.4.3
Electronic Absorption

At the semiconductor/electrolyte interface, the widespread use of IR spectroscopy for vibrational absorption has somewhat overshadowed the interest in studying electronic absorption. However, the latter may also be a valuable tool for interface characterization. At photon energies below the energy of the semiconductor bandgap, electronic absorption in a semiconductor electrode may proceed from free carriers or electronic levels lying within the forbidden gap, especially interface states. Changes in these absorptions will arise when the populations of these electron reservoirs are varied, which may occur through changes in the chemistry (for example, doping through ion intercalation and creation or removal of chemically induced interface states) or just through reversible physical effects (photocreation of carriers, potential-induced band bending, …).

Free-carrier absorption   IR absorption of free carriers in the semiconductor bands usually exhibits a Drude shape $\Delta I/I \propto \sigma^{-a}$, where $a \approx 2$ [20]. It may also be affected by interband transitions (for example, for electrons, transitions from the bottom of the conduction band to a higher band), which makes the absorption shape more complex than the simple Drude form. However, for usual semiconductors, free-electron and free-hole absorption have been investigated in detail, and the corresponding absorption shapes can be found in textbooks [20]. For organic semiconductors, the charge carriers are polarons or solitons rather than "free" carriers, and their optical absorption is usually dominated by several broad bands, generally extending down to the IR region [79]. In the latter case, changes in polaron absorption associated with oxidation or reduction of the material can be observed – a fingerprint of the change in its bulk electrical properties [72–75].

For more common inorganic semiconductors, no significant chemical modifications generally occur in the bulk properties. Changes in bulk free-carrier concentration may be induced by bandgap illumination and have indeed been studied using IR spectroscopy in a solid-state context [80–82]. At an electrode, such effects have been investigated by microwave reflectivity [83], but seldom by IR absorption [80]. More often, changes in free-carrier concentration occur near the semiconductor/electrolyte interface in the so-called space-charge region, which can be represented by a band bending (Fig. 6.9) (see e.g. Ref. [84]). If the surface is turned into accumulation (Fig. 6.9 a), the local concentration of free carriers may become very high, leading to strong free-carrier absorption [85, 86]. In more common cases, the semiconductor surface remains under depletion conditions, that is, there is a layer below the surface devoid of free carriers (Fig. 6.9 b). An increase in the thickness of this layer results in a decrease in the free-carrier absorption for the IR beam crossing the sample, and vice versa. Such changes may result from a modification in surface chemistry (for example, creation of interface states leading to a surface charge and a change in band bending [87]), but are routinely encountered when varying the electrode potential.

Interestingly, for a small potential modulation, the electrical behavior of the interface can be represented by the space-charge capacitance $C = \Delta\rho_S / \Delta V$, where

(a) *Accumulation*   (b) *Depletion*

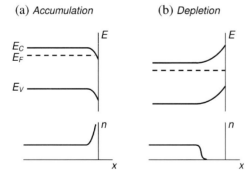

Fig. 6.9 Space-charge region at a semiconductor surface, here represented for the case of an n-type semiconductor. Energy band profiles and electron concentration $n(x)$. E_C, E_V, E_F respectively stand for conduction band energy, valence band energy, and Fermi energy. Case of an accumulation layer (a) and a depletion layer (b). The spatial extent of the depletion layer is actually much larger than that of the accumulation layer.

ΔV is the potential modulation and $\Delta\rho_S$ is the associated variation in electrode charge. Since $\Delta\rho_S = e\Delta n_S$, where Δn_S is the change in free-carrier concentration integrated over depth, the change in free-carrier absorption associated with the modulation of potential brings a piece of information equivalent to the space-charge capacitance. The resulting data can be plotted in the Mott-Schottky form $1/C^2$ vs V, leading to a linear plot that allows for a determination of the flat-band potential by extrapolation to zero ordinate (Fig. 6.10) [31, 86, 88]. The method exhibits several advantages over that based on classical electrochemical impedance measurements:

1. The determination of the space-charge capacitance from free-carrier IR absorption is free from interference with the interface-state capacitance, which plagues the use of common capacitance measurements. The method in this respect is equivalent to what can be obtained from microwave reflectivity measurements (Drude absorption is nothing but ac Joule dissipation at an IR frequency).
2. Although the contribution of electronic IR absorption can sometimes be difficult to extract from the spectrum because of its smooth shape, and can require a high sensitivity to be detected, its analysis may be especially useful for electrodes for which electrical measurements are difficult. Its use for elec-

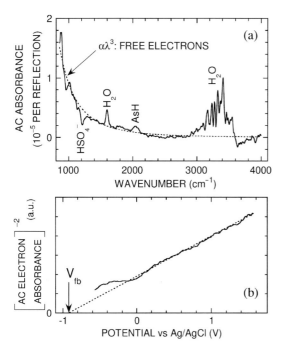

Fig. 6.10 (a) Changes in free-carrier concentration detected by IR absorption at an n-GaAs/0.5 M H_2SO_4 electrolyte interface. The baseline can be fitted as a power law, characteristic of free-carrier absorption. (b) Application of these measurements to the determination of the flat-band potential (Mott-Schottky plot) [88].

trodes made of small particles has been proposed [88]. One should yet keep in mind that the spectral shape may be affected by the size of the particles (probable decrease of IR absorption at low wavenumbers, in contrast to normal Drude behavior, if the mean free path of the carriers becomes limited by particle size).

3. Finally, when both types of carrier (electrons and holes) contribute to the interface capacitance, it may be possible to separate the two contributions if their spectral shapes are sufficiently different. This has indeed proved to be possible for germanium [11, 37].

Interface-state absorption   In the presence of interface states, new IR absorptions appear, associated with the optical transitions between the interface states and the bands. Such absorptions exhibit a low-energy threshold (Fig. 6.11) and can therefore be clearly distinguished from free-carrier absorption [31, 88]. When the IR absorption associated with a small potential modulation is studied, this feature is a decisive advantage over the electrochemical impedance method (where the interface-state and space-charge contributions can be disentangled only through an analysis of the frequency response) and over the microwave reflectivity method, which is blind to interface states. In the case of interface states energy-distributed through the semiconductor gap, the relevant interface-state energy at a given potential is that of the surface Fermi level, separating the occupied interface states from the empty ones [89]. IR absorption then provides a tool for determining the energy of the interface states and their filling as a function of electrode potential (Fig. 6.12).

Inasmuch as electronic absorption and vibrational absorption are both weak, the two effects are simply additive, and, for example, if vibrational absorption is of major interest, it can be extracted from the spectra by simply subtracting the electronic absorption background. However, in some cases, the existence of cross effects has been reported [80, 90, 91]. For example, when semiconductor doping is sufficiently high, modulation of the space charge layer may significantly affect the local refractive index and in turn the IR electric field in the evanescent-wave region, making electrolyte absorption reappear in the potential-modulated spectra [91]. Also, interference of electronic absorption with vibrational absorption may lead to changes in the vibrational absorption lineshapes,

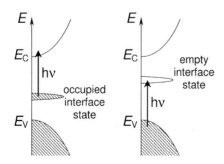

Fig. 6.11 Principle of interface-state absorption. The IR absorption associated with the two kinds of optical transitions (occupied interface states to conduction band and valence band to empty interface states) can be varied when the population of the interface states is changed, for example when changing the electrode potential.

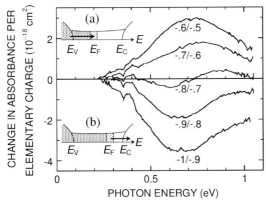

Fig. 6.12 An example of interface-state absorption: Interface states induced by deposition of Au atoms on the surface of n-Si, here investigated at the Si/ acetonitrile + 0.1 M $(C_4H_9)_4N^+ClO_4^-$ interface. The various spectra have been recorded under a square-wave modulation of the potential between the two values indicated below the curves. They represent the change in absorbance when the potential is changed from the lower to the higher value. The change in shape is consistent with a broad distribution of interface states through the gap. It is associated with the increasing energy of the surface Fermi level (increased occupancy of the interface states) as the potential is made more negative. The change in sign can be understood from insets (a) and (b), where the area shaded in light grey represents the states whose occupancy is affected by the potential modulation. It should be borne in mind that the transitions occurring at a lower threshold dominate the IR absorption. At more negative potentials (E_F closer to E_C), they are those from the occupied interface states to the conduction band, and at more positive potentials (E_F closer to E_V) they are those from the valence band to the empty interface states [89].

a well-known effect in other contexts [92, 93], as $\varepsilon'(\sigma)$ and not only $\varepsilon''(\sigma)$ may come and play a direct role in the vibrational absorption [80, 90].

6.5
Effect of Light Polarization in ATR Geometry

As shown in Section 6.2, when using an internal-reflection geometry, the direction of the electric field of the IR beam at a planar interface crucially depends on beam polarization. As a result, IR absorption depends on the polarization of the beam. Polarization in the IR range is conveniently fixed by grid polarizers, which are commercially available at moderate prices and can be easily inserted into the beam.

6.5.1
Selection Rules for a Polarized IR Beam

If the IR beam is s-polarized, the electric field at the interface is parallel to the interface plane; hence, the beam undergoes absorption only by species with a dynamic dipole parallel to the interface [see ε_y'' in Eq. (1)]. On the other hand, if the beam is p-polarized, the x and z components of the electric field are non-vanishing; hence, it undergoes absorption by species with dynamic dipoles either parallel (ε_x'') or perpendicular to the interface [ε_z'' in Eq. (2)]. Since the three coefficients A_x, A_y, A_z as given by Eqs. (3–5) exhibit comparable values for typical values of the refractive indexes and an incidence angle of 45°, practical consequences are as follows:

1. An interface species with dynamic dipole perpendicular to the interface leads to absorption only for a p-polarized beam.
2. An interface species with dynamic dipole parallel to the interface, and randomly oriented in the surface plane, leads to about the same absorption for s- and p-polarized beams (A_y compared to A_x).
3. An interface species with fully random orientation leads to twice as much absorption for a p-polarized beam as for an s-polarized beam (A_x+A_z compared to A_y).

However, a look at Eq. (2) shows that the $A_z\varepsilon_z''$ term enters the expression with an extra multiplying factor $n_2^4/(\varepsilon_z'^2+\varepsilon_z''^2)$, as compared to $A_x\varepsilon_x''$ or to $A_y\varepsilon_y''$ in Eq. (1). The presence of this extra factor makes the latter result actually depend on ε_z', the real part of ε_z. The quantity ε_z' represents the effective optical dielectric constant of the thin layer modeling the interface species. It is expected to be governed by electronic contributions and will usually be much larger than the imaginary part ε_z''. However, it is not generally clear what its value will be. For the case of a molecular group located on the electrolyte side of the interface, one will have $\varepsilon_z' \approx \varepsilon_z'' \approx n_2^2$, and the above statement will hold true. However, for a species buried in the semiconductor, one has rather $\varepsilon_z \approx n_1^2$, and the contribution of the A_z term in Eq. (2) is decreased by over an order of magnitude, leading to comparable absorption for s- and p-polarized beams. The physical origin of this effect is the continuity relation for the perpendicular electric field $E_\perp$ at the interface, and the much lower value of $E_\perp$ in the semiconductor, which has a dielectric constant much higher than that of the electrolyte.

6.5.2
Case of Strongly Polar Species: LO-TO Splitting

A material exhibiting vibrational IR absorption at frequency $\omega_0/2\pi$ can be described by a dielectric function of the type [24]

$$\varepsilon(\omega) = \varepsilon_\infty - \frac{Ne^{*2}/\varepsilon_0\mu}{\omega^2 - \omega_0^2 - i\gamma\omega} \tag{6}$$

where ε_0 is vacuum permittivity and ε_∞ the high-frequency dielectric constant (associated with electronic contributions); e^*, μ and γ are respectively the effective charge (dynamic dipole), the reduced mass, and a damping factor associated with the vibration mode under consideration, and N is the concentration of vibrators per unit volume in the material. Here, for the sake of simplicity, we have assumed the material to be isotropic (anisotropy could be introduced through distinct values of the parameters for ε_x, ε_y, and ε_z). If we now consider Eq. (2), which describes the absorption in p-polarization, we note that the x-term appears similar to the y-term in Eq. (1), but the z-term involves a more complex function of ε. Put in another way (and dropping the x, y, and z subscripts), the x-term and y-term [in Eq. (1)] involve $\varepsilon''=-\text{Im}(\varepsilon)$, whereas the z-term involves $\varepsilon''/(\varepsilon'^2+\varepsilon''^2)=\text{Im}(1/\varepsilon)$. This intuitively occurs because in the oblique incidence geometry the parallel components of the electric field E are imposed at the interface (and energy per unit volume must be written $\varepsilon\varepsilon_0 E^2/2$, where ε is in the numerator), whereas for the perpendicular component it is the electric induction $D=\varepsilon\varepsilon_0 E$ that is imposed (and energy per unit volume must then be written in the form $D^2/2\varepsilon\varepsilon_0$, where ε is in the denominator). The underlying physics here is similar to that for Joule dissipation in a (variable) resistor R: if a voltage V is imposed across the resistor, dissipation writes V^2/R, and is then proportional to $1/R$; if a current I is imposed, RI^2 must be used instead, and the dissipation becomes proportional to R.

For weakly polar (e^* small) or dilute (N small) IR-active species, the vibrational contribution to ε is much smaller than 1 for every ω, so that $\varepsilon'^2+\varepsilon''^2 \approx \varepsilon_\infty^2$. The two functions $\text{Im}(1/\varepsilon)$ and $-\text{Im}(\varepsilon)$ are then proportional to each other, and the absorption shapes associated with the three components of the interface electric field are similar.

In the case of strongly polar or concentrated species (which is typically the case for an oxide), the vibrational contribution to $\varepsilon(\omega)$ may become larger than ε_∞ in the region of the resonance. The shapes of the functions $-\text{Im}[\varepsilon(\omega)]$ and $\text{Im}[1/\varepsilon(\omega)]$ then become different, the former exhibiting its maximum for $\omega=\omega_0$, whereas for the latter the maximum turns out to be shifted to $\omega_0'=(\omega_0^2+Ne^{*2}/\varepsilon_0\varepsilon_\infty\mu)^{1/2}$. If one considers the phonon modes in the infinite 3D material, the two modes ω_0 and ω_0' appear as the zero-wavevector limit of the transverse-optical (TO) and longitudinal-optical (LO) phonon branches, and for that reason are generally termed "TO mode" and "LO mode" [94].

It is noteworthy that these two "modes" actually correspond to the same molecular vibration. This occurs, though the material is isotropic, just because it has been put in the form of a thin layer. The origin of the effect can be intuitively explained as follows (Fig. 6.13a): in a homogeneous isotropic material, the synchronous vibration of the polar species produces a dielectric polarization $P=Ne^*\delta z$ in the medium (where δz represents the instantaneous value of the normal coordinate associated with the vibration mode under consideration). If the sample is shaped in the form of a thin layer and the direction of the vibration is perpendicular to the layer, a depolarizing field $-P/\varepsilon_\infty\varepsilon_0$ will appear, acting as a restoring force $-e^* P/\varepsilon_\infty\varepsilon_0$ on the vibrators (Fig. 6.13a). This amounts to increasing the usual

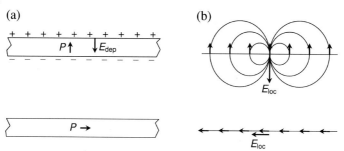

Fig. 6.13 Intuitive origin of the LO-TO splitting. (a) When the vibrators inside a planar layer (viewed as a homogeneous medium) oscillate perpendicular to the layer, the depolarizing field acts as an extra restoring force, increasing the oscillation frequency. This effect is absent when the vibrators oscillate parallel to the layer. (b) A similar effect is present for discrete vibrators in a plane. Interaction between the vibrators leads to a local field opposite to the dipole moments (increasing the oscillation frequency) when the vibration is perpendicular to the surface. In contrast, when it is parallel to the surface, the local field is in the direction of the dipole (lowering the oscillation frequency).

stiffness $\mu\omega_0^2$ by a quantity $e^* P/\varepsilon_\infty\varepsilon_0\delta z = Ne^{*2}/\varepsilon_\infty\varepsilon_0$, whence the expression of ω_0'. No such field will appear if the direction of the vibration is parallel to the layer. The same reasoning can be applied to the case of optical phonons in a 3D sample (LO and TO, respectively). The only difference here is that the map of P follows a sine wave profile instead of being uniform inside a layer of finite thickness. The results turn out to be identical, whence the terms LO and TO.

A similar situation may occur for discrete species adsorbed on the surface. Let us consider an assembly of synchronous vibrators, homogeneously distributed on a planar surface. If the vibration is directed perpendicular to the surface plane, the electric field experienced by one of these dipoles and caused by the other ones is obtained by summing over the neighbors located at distances r_i ($i=1, 2, \ldots$) from the vibrator under consideration. The sum, which is written $(-e^*\delta z/4\pi\varepsilon_\infty\varepsilon_0)\Sigma(1/r_i^3)$, acts as a restoring force on the other vibrators, thereby increasing their stiffness and their vibration frequency (Fig. 6.13b). In contrast, if the vibration is directed in the surface plane, upon averaging over all the possible r_i orientations in that plane, the electric field becomes $(e^*\delta z/8\pi\varepsilon_\infty\varepsilon_0)\Sigma(1/r_i^3)$, which tends to soften the vibration frequency as compared to the isolated vibrator. Despite some differences between this case and that of the continuous layer, they are both relevant to the same effect of dipolar interactions between vibrators, here in the limiting situation of an atomically thin (sub)monolayer. An intermediate case (oxide in the form of islands) has been discussed in the literature [95].

The case of discrete interacting vibrators on a surface, which is often encountered in adsorption studies [24, 96], is further complicated by the need to take into account image-potential effects. Up to now it does not seem to have been met at semiconductor/electrolyte interfaces. The former case of a film composed of a highly polar material is of more common relevance, as oxide films

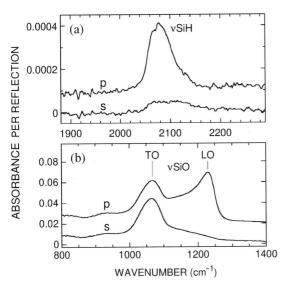

Fig. 6.14 Typical effects associated with polarization of the IR beam in ATR geometry. (a) SiH species on an Si(111) surface (here obtained by oxide dissolution in an alkaline NH₄F solution). The oriented dipoles, mostly normal to the surface, contribute to IR absorption in p-polarization only. (b) Thin (5 nm) anodic SiO₂ layer on the surface of silicon in dilute fluoride electrolyte. The absorption shape is strongly different for s- and p-polarization. This is because of the LO-TO splitting, the LO mode being observed in p-polarization only [23].

are a typical case where the LO-TO splitting may be quite significant. As an example, Fig. 6.14 shows s- and p-spectra taken at the Si/electrolyte interface under electropolishing conditions in the presence of an oxide film of thickness ~ 10 nm [23]. As expected, the TO mode appears with about the same magnitude for s- and p-polarizations $(A_x \sim A_y)$, whereas the LO mode is observed for p-polarization only. The LO-TO splitting is seen to be as large as 170 cm^{-1}. This value, governed by the quantity $Ne^{*2}/\varepsilon_\infty\varepsilon_0\mu$, can be used as an indication of the degree of perfection or purity of the oxide. In the present case, it must be compared with the value 189 cm^{-1}, reported for dense amorphous silica [97, 98]. A possible contribution to this discrepancy may arise from a lower value of the concentration of Si-O vibrators N in the anodic oxide due to the presence of voids or impurities. Another cause for a decreased value of the splitting may be a deviation from the ideal planar-layer geometry: maximum splitting is obtained for the geometry of a perfectly planar parallel layer, and the splitting would disappear for a spherical sample or a discontinuous layer consisting of weakly interacting spheroidal particles (in the latter case, because of dielectric-image effects, an inverted splitting might even appear, the perpendicular mode becoming softer than the parallel mode). In practice, roughness is then expected to decrease the value of the splitting and to broaden the TO and LO bands. Here, the weakness of the observed discrepancy (170 vs 189 cm^{-1} and very distinct bands) indicates that the chemical and structural quality of the anodic oxide film is fairly good.

6.5.3
Polarization Modulation

The strong effects of polarization for interface species might suggest using polarization modulation (which yields the difference in transmitted light between s- and p-polarization) as a tool to select interface absorption without the need for using potential or another parameter as an extra handle to the interface. Unfortunately, the map of the IR electric field is dependent on polarization in a region on the order of a wavelength from the interface. For the special case of ATR geometry, the penetration depth into the electrolyte is independent of polarization, but the amplitude of the electric field is different for s- and p-polarization ($A_x+A_z=2A_y$ for $45°$ incidence!) so that a polarization-modulation spectrum is dominated by electrolyte absorption. In other words, the method is not sufficiently selective to the interface. For reaching interface selectivity, it has to be combined with the change of a more relevant parameter, such as the potential [99]. This makes the method of little practical interest. Furthermore, at a semiconductor electrode, independent measurement of the absorption spectrum for s- and p-polarization is much more informative than the p-s-difference.

6.6
Dynamic Information from a Modulation Technique

As discussed in Section 6.3, modulation techniques (such as a potential modulation associated with a lock-in detection) are intrinsically more sensitive than differential techniques, an advantage that however disappears when the system under investigation is unable to follow the modulation. In practice, the response of the system may be limited by instrumental aspects (time response of the potentiostat or the cell), but it may also be determined by mechanistic aspects of interest for the understanding of the system (for example, kinetics of adsorption/desorption phenomena or of Faradaic processes). In such cases, a detailed investigation of the response as a function of modulation frequency may be helpful to gain understanding on the system. The obtained data can be termed electro-optical impedance spectra, the name being inspired by that of the well-known electrochemical impedance spectroscopy (EIS) technique (see, e.g. Ref. [100]). However, conventional EIS aims at understanding complex reaction pathways just through the use of the potential-current response function, that is, only the current associated with the potential modulation is measured, and the elementary steps at the interface are detected only through the associated transfer of charge, leaving much room for interpretation. Here, the species involved in the charge transfer can in principle be identified from their IR spectrum; hence, much more unambiguous information about their chemical nature is obtained. The accessible frequency range using EMIRS or FTEMIRS is typically from 10 Hz to 10 kHz. At frequencies higher than a few kilohertz, the time response of the electrochemical cell and potentiostat become a problem. At frequencies lower than 10 Hz, the low-frequency noise of the detectors leads to decreased sensitivity. Also, the scanning speed of the interferometer (if using a slow-scan FTEMIRS arrangement, see Section 6.3) may become a problem. For frequencies below a few Hertz, dynamic information is best obtained by using a fast-scan FTIR interferometer in the differential mode, a frequency limit that may soon extend up to a few hundred Hz with improving optics design and data acquisition systems [36]. The data are then taken in the time domain (typically, recording successive spectra in response to a small potential perturbation) but can be turned to the frequency domain by Fourier transform. In any case, the possibility to work at low frequencies relies on the stability of the system under investigation.

Electro-optical impedance data can be analyzed either as a collection of impedance curves, each one corresponding to a given wave number, or as a collection of complex spectra (each one consisting of an in-phase "real" spectrum and an out-of-phase "imaginary" spectrum) taken at the different frequencies. This represents a large amount of data and acquisition time. Fortunately, a complex spectrum taken at a single modulation frequency already represents a significant piece of information, as each spectral line may have a distinct phase, characteristic of the response of the associated species. This kind of analysis has been made in the study of a redox reaction involving species in solution. As was to be expected, under mass-transport limitation, the electro-optical impedance associated with the redox species exhibits a Warburg behavior and can easily be distinguished from the capacitive component associated with changes of ionic concentrations in the double layer (Fig. 6.15) [101].

The method may be especially useful in the study of reaction mechanisms. For example, if a reaction involves an intermediate species with a lifetime τ, the spectrum of that species will decrease with increasing modulation frequency $\omega/2\pi$ with a dependence of the type $1/(1+i\omega\tau)$. The catalytic mechanism of hydrogen evolution at a semiconducting electrode modified with heteropolyanions has been investigated with this method, providing evidence for a reduced intermediate with a characteristic lifetime in the range of milliseconds [65]. A problem is the rather narrow range of lifetimes that may be accessed, as the lifetime of long-lived intermediate species (typically, for $\tau > 0.1$ s) makes a modulation technique difficult to apply, and the concentration of short-lived intermediate species (short, that is, $\tau < 0.1$ ms) may fall below the IR sensitivity limit. This problem may explain the limited use that has been made of the method up to now.

An easier application of the modulation technique to extracting dynamic information is for electronic absorption. At semiconducting electrodes, interface states often respond with time constants in the ms range. For instance, at an n-Si/acetonitrile electrolyte interface, it has been possible to correlate interface states with SiOH species, as their respective IR absorptions exhibit the same distinctive phase lag in response to a potential modulation [31]. Also, response

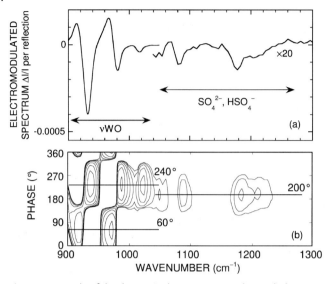

Fig. 6.15 Example of the distinction between electrolyte species through an analysis of their response to a modulation: n-Ge electrode in 1 M $H_2SO_4 + 10^{-2}$ M α-$H_4SiW_{12}O_{40}$ electrolyte. Potential is square-wave modulated at 125 Hz between –0.15 and –0.35 V vs SCE. The νWO bands are associated with the first reduction wave of the heteropolyanion $SiW_{12}O_{40}^{4-}$. The sulfate bands are associated with modulation of the double layer. (a) Spectrum for a phase setting of 40°. (b) Two-dimensional map showing isolevel lines of the spectrum as a function of wavenumber and phase setting (only positive range is shown). Note the distinct phases associated with the two kinds of species: the bands associated with the double-layer species appear maximal for a phase of 20° (or 20+180=200°), a phase lag due to the RC time constant of the cell, whereas those associated with species involved in the Faradaic process appear maximal for a phase setting of 60° (or 240°), an extra phase lag due to the process of mass transport in a diffusion layer [65].

to light modulation has been used for the study of photocarrier dynamics in a porous-Si film immersed in HF electrolyte. The photocarriers were created at the outer film surface by chopped UV-light illumination. The amplitude and phase of IR absorption of the photocarriers reaching the bulk-Si ATR substrate were found to exhibit a non-trivial dependence on film thickness, from which a determination of the lifetime and diffusion length of the photocarriers in porous Si was obtained [80].

6.7
Case of Rough or Complex Interfaces

Practical interfaces often exhibit complex behavior. As an example of such complex behavior, Fig. 6.16 shows a series of absorption spectra taken at a p-Si electrode in dilute fluoride electrolyte as a function of applied potential [47]. Refer-

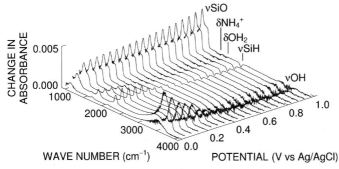

Fig. 6.16 An example of complex interface behavior: Si anodic dissolution in dilute fluoride medium (1 M NH₄Cl+0.025 M NH₄F+0.025 M HF, pH ≈ 3). The reference spectrum was taken at open circuit (ca. −0.3 V) and the spectra were recorded on increasing the potential. Formation of the oxide film (clearly present above 0.2 V, see SiO bands and decreased electrolyte absorption) is preceded by the appearance of SiOH species (broad band around 3300 cm⁻¹) and that of a porous, surface-hydrogenated, silicon layer (large SiH band and increased electrolyte absorption) [47].

ence potential was taken at open circuit. At high potential, the data show the presence of an oxide film (νSiO band at 1000–1200 cm⁻¹), the loss of the initial hydrogen coverage of the surface (negative band at 2100 cm⁻¹) and of the electrolyte displaced by the oxide layer (negative band at 3300 cm⁻¹). An interesting feature is the large positive band around 3200 cm⁻¹ in the intermediate potential range, a band attributed to SiOH species (intermediate species in the oxide formation process) [47]. More surprising is the behavior of the SiH band, which goes through a maximum together with other electrolyte bands (δOH₂ at 1650 cm⁻¹ and δNH₄⁺ at 1400 cm⁻¹) before reaching its asymptotic negative value. This surprising behavior is the fingerprint of the presence of a layer of porous silicon impregnated with electrolyte and with a large specific surface area covered with hydrogen. In the following we will try to discuss the issues related to such geometrically complex interfaces and the possibility of extracting quantitative information from such data.

6.7.1
Surface Roughness

Surface roughening is a common problem for optical measurements at electrodes. Especially in the presence of surface roughness, the rules regarding the electric-field map as described by Eqs. (1–5) and in Section 6.5 no longer hold. This problem persists, even for roughness on a scale much smaller than the wavelength. For example, for the case of a planar surface under an s-polarized IR beam, the IR electric field is parallel to the surface plane. In the presence of nanoscale roughness, because of the different dielectric functions of the electrode and the electrolyte, the field lines will bend and the local field along the interface will acquire a normal component, which will be all the more impor-

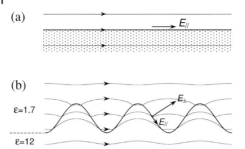

Fig. 6.17 At a planar interface and for s-polarization, the electric field is everywhere parallel to the surface plane (a). This does not hold any longer at a rough surface [here electric-field map at a sine wave profile, with $n_1 = 3.5$ ($\varepsilon = 12$) and $n_2 = 1.3$ ($\varepsilon = 1.7$)] (b).

tant as the local slopes associated with the roughness will be larger (see Fig. 6.17). As a result, the selection rules for polarization, as stated in Section 6.5 for a planar electrode, become unusable in the presence of surface roughness [37].

Another consequence of surface roughness is a loss in the specular reflection due to light scattering. This effect is wave number dependent and generally increases with increasing wave numbers, being at its maximum when the wavelength matches the scale of the roughness. It appears as a sloping background in the absorption spectra, the apparent absorption increasing toward higher wave numbers. This effect becomes detectable for an r.m.s. roughness down to the nanometer range. It has been used as an indication of surface roughening, for example during the chemical dissolution of GaAs [68] or the anodic dissolution of Si at very high potentials [102]. One might conceivably turn this method into a more quantitative tool for the analysis of surface roughness.

6.7.2
Composite Interface Films

The formation of porous films, as in the case of the data shown in Fig. 6.16, may be regarded as an extreme case of surface roughening. However, if the scale of the porosity is much smaller than the wavelength, diffusion losses are negligible, and the film can be described as a homogeneous medium with an effective dielectric function. Other examples of films that may be described as an effective medium are nanoparticle electrodes or discontinuous metal deposits. Here we will first outline the principle of effective-medium descriptions, then show the various kinds of effects that such composite films may cause.

Effective-medium theories As long as the inhomogeneities of the medium are on a scale a much smaller than the IR wavelength λ, the optical properties of the medium are described by an effective dielectric function, which can in principle be derived by solving Maxwell's equations in a piece of the material of a size $b \gg a$. By taking $b \ll \lambda$, Maxwell's equations reduce to those of electrostatics. One can then imagine a capacitor filled with the composite material and calculate the ratio of its capacitance to that of the empty capacitor [103]. It is clear that the effective dielectric function will turn out to be some average of the dielectric functions of the constituents, but the detail of the averaging depends on the morphology of the inhomogeneity: for example, a layered material consisting of an alternation of slices of equal thickness of materials with dielectric constants ε_1 and ε_2 will exhibit an effective dielectric constant $(\varepsilon_1 + \varepsilon_2)/2$ in the directions parallel to the layers (similar to capacitances in parallel), but $[(\varepsilon_1^{-1} + \varepsilon_2^{-1})/2]^{-1}$ in the direction perpendicular to the layers (capacitances in series). Unfortunately, taking into account the morphology in a realistic case is not a simple task. This has given rise to many effective-medium theories.

A most popular approach is that of Bruggeman [103, 104]. The simplest form of the theory addresses the case of an isotropic morphology. The medium is approximated by a dense packing of cells filled with different materials $i = 1, \ldots, n$ with respective dielectric functions ε_i and volume fractions f_i (with $\sum_i f_i = 1$). For a Maxwell field E in the sample, the field inside a cell of type i is given by $E_i = \varepsilon_{\text{eff}} E / [\varepsilon_{\text{eff}} + (\varepsilon_i - \varepsilon_{\text{eff}})/3]$, where the cell is assumed to be spherical (for space-filling cells, this is obviously an approximation, but it does express correctly the isotropy of the medium) and surrounded by the effective medium of dielectric function ε_{eff} (self-consistency). The factor $1/3$ arises from the depolarizing field and the spherical shape of the cell. Writing that E is the volume average of the E_I values, one gets the Bruggeman expression [103, 104]

$$\sum_i f_i \frac{\varepsilon_i - \varepsilon_{\text{eff}}}{\varepsilon_{\text{eff}} + (\varepsilon_i - \varepsilon_{\text{eff}})/3} = 0 \tag{7}$$

from which ε_{eff} can be calculated. The expression can be readily transposed to the case of an anisotropic morphology by replacing the spheres by ellipsoids, the $1/3$ factor being replaced by the relevant depolarizing factor A ($0 < A < 1$; for example, $A = 1$ for layers, and $A = 0$ for needles or layers with electric field parallel to the layer plane). Note that, in agreement with the above simple example of a layered material, this extension of Eq. (7) reduces to $\varepsilon_{\text{eff}} = \sum_i f_i \varepsilon_i$ for $A = 0$ and $1/\varepsilon_{\text{eff}} = \sum_i f_i / \varepsilon_i$ for $A = 1$.

Despite its widespread use, Bruggeman's effective medium theory is only an approximation, which involves hidden assumptions about the morphology. Especially if one of the media exhibits infinite ε, the effective dielectric function also becomes infinite as soon as the corresponding volume fraction reaches the threshold value $f = A$. This tells us that any of the phases becomes percolating when its volume fraction exceeds A [105]. This is not suitable, for example, for a highly porous material (where the solid phase remains connected though its volume fraction may be very small). More elaborate theories have tried to overcome this limitation. A rigorous formulation has been derived by Bergman [106]. For a two-component medium, it amounts more or less to a generalization of Eq. (7), with the 1/3 factor replaced by a variable A, itself distributed according to a function g (with, among other conditions, $\int_0^1 g(A)dA = 1$). The practical problem of determining the function g for a given morphology is as yet unsolved, but reasonable guesses for g can be made. Typically, parametrized functions made of a broadened distribution around a mean value of A (typically 1/3) plus a contribution near $A=0$ (associated with a "percolating strength") give a fair representation of many morphologies [105]. Unfortunately, the link between g and the morphology is not explicit, and up to now this practical version of the general Bergman theory has been worked out only for a two-phase material.

IR absorption in the presence of composite films   In Section 6.4, we noticed that the presence of a thin film over the semiconductor surface may lead to a decreased IR absorption from the electrolyte in ATR geometry. A quantitative analysis of this effect requires going beyond the treatment leading to Eqs. (1)–(2), which did not take absorption of medium 2 into account. The result actually depends on the refractive index of the film. If it is lower than that of the semiconducting substrate n_1, a loss in electrolyte absorption is indeed to be expected (most frequent case for an oxide film). If it is equal or close to n_1, total reflection takes place at the outer edge of the film, and no change in electrolyte absorption is to be expected.

More interestingly, if the film is porous and made from a high-index material, the opposite behavior may occur, that is, electrolyte absorption may be increased (rather than decreased) by the presence of the film, as observed in Fig. 6.16 in the potential region where a porous-silicon film is present at the interface [47, 57, 107]. This increase in electrolyte absorption can be rationalized by using effective medium theory. Here, the medium can be described as a silicon-electrolyte mixture, and the electrolyte absorption bands appear enhanced because of the high index of silicon. Intuitively, for a given Maxwell (spatially averaged) field, the high dielectric function of silicon tends to concentrate the field in the electrolyte regions where the local field comes out enhanced, leading to increased absorption although the volume fraction is lower than in the bulk electrolyte. A similar effect has been observed when a porous arsenic hydride film is formed on GaAs during H_2O_2 reduction [68]. An especially strong effect is to be expected for thick layers (thickness $\gg \delta$) with an effective index such that reflection takes place at the outer surface of the layer, which is then crossed all the way through by the IR beam.

The effective-medium description may be applied to the case of rough surfaces (the limit of a "thin" porous film made of a substrate-electrolyte mixture) or to discontinuous metal deposits on a semiconductor surface. The latter case is of special interest, as the local-field enhancement may then become quite strong, leading to the effect called Surface-Enhanced InfraRed Absorption (SEIRA) [108–114]. The origin of the effect is similar to the enhancement of electrolyte absorption in porous semiconductor films. However, for metals, the dielectric function in the IR region is essentially large and negative, which makes the enhancement still larger. Optimum enhancement is obtained for de-

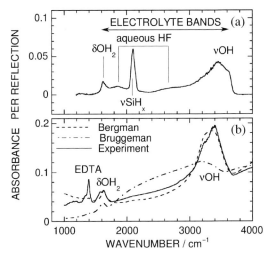

Fig. 6.18 Examples of IR spectra in the presence of a composite layer on a silicon electrode. (a) Porous-silicon layer in 15% HF aqueous solution [107] and (b) discontinuous metal deposit (Ag deposited in 0.1 M EDTA (pH 4) + 0.01 M AgNO₃) [109]. In both cases, note the increased absorption from the electrolyte and the surface species (SiH in (a), adsorbed EDTA in (b)). The electrolyte absorption in (b) can be rationalized only through Bergman effective medium theory.

posits made of closely packed islets, just below the percolation threshold. Figure 6.18 shows that the absorption of such a medium can be analyzed using effective-medium theory, but the Bergman form is necessary to account for the shape of the observed spectra. The enhancement is present for the absorption of the electrolyte between the islets and also for that of species adsorbed on the islet surface, which makes the effect of high interest for the detection of species at the metal-electrolyte interface [108–114]. The application of effective-medium theory to such systems may however be criticized, as the layer thickness is of the same order of magnitude as the scale of the inhomogeneities, and a true quantitative theoretical analysis of the enhancement effect in SEIRA remains as yet to be worked out.

6.8
Conclusion

IR spectroscopy has been turned into a widespread technique for the investigation of semiconductor/liquid interfaces. Multiple-internal-reflection geometry can generally be used, offering an increased sensitivity, together with other advantages: possibility of using a conventional electrochemical cell, allowing for a potential modulation up to higher frequencies, interest in changing the polarization of the IR beam to gain information on the orientation of interface species. Also, semiconductors appear as unavoidable substrates for IR studies on metallic-particle electrodes or other (for example organic) layers in contact with an electrolyte. Though vibrational spectroscopy appears to represent the most common use of the technique, one should keep in mind that electronic absorption is able to bring valuable information on the electronic structure (interface states) or on the energetics of the interface (free-carrier absorption, flat-band potential). Also, careful examination of a spectrum may bring *in-situ* information hardly available from other techniques, for example on interface roughening (from diffusion losses), on the formation of either compact or porous interface films (from changes in electrolyte absorption), or on the dynamic response and lifetime of intermediate species (from an analysis of the response of IR absorption to a potential modulation). These manifold possibilities have been little used in the past, and it is plausible that great benefits will result from their wider use.

References

1 A.J. Bard, L.R. Faulkner, *Electrochemical Methods, Fundamentals and Applications*, Wiley, New York, USA, **1980**.

2 C.A. Melendres, A. Tadjeddine (Eds.), *Synchrotron Techniques in Interfacial Electrochemistry*, Kluwer, Dordrecht, The Netherlands, **1993**.

3 K. Itaya, *Prog. Surf. Sci.* **1998**, *58*, 121.

4 I. Zudans, C.J. Seliskar, W.R. Heineman, *Thin Solid Films* **2003**, *426*, 238.

5 B.H. Erné, M. Stchakovsky, F. Ozanam, J.-N. Chazalviel, *J. Electrochem. Soc.* **1998**, *145*, 447.

6 Z.Q. Tian, J.S. Gao, X.Q. Li, B. Ren, Q.J. Huang, W.B. Cai, F.M. Liu, B.W. Mao, *J. Raman Spectrosc.* **1998**, *29*, 703.

7 A. Bewick, S. Pons, in: R.J.H. Clark, R.E. Hester (Eds.), *Advances in Infrared and Raman Spectroscopy*, vol. 2, Wiley-Heyden, Chichester, UK, **1985**, pp. 1–63.

8 K. Ashley, S. Pons, *Chem. Rev.* **1988**, *88*, 673.

9 T. Iwasita (Ed.), *Infrared Spectroscopy in Electrochemistry*, Electrochim. Acta **1996**, *41*, 621–781.

10 R. Holze, *J. Solid-State Electrochem.* **2004**, *8*, 982.

11 J.-N. Chazalviel, B.H. Erné, F. Maroun, F. Ozanam, *J. Electroanal. Chem.* **2001**, *509*, 108.

12 I. Yagi, S. Nakabayashi, K. Uosaki, *J. Phys. Chem. B* **1997**, *101*, 7414 and references therein.

13 A. Tadjeddine, A. Peremans, P. Guyot-Sionnest, *Surf. Sci.* **1995**, *335*, 210.

14 M. Buck, M. Himmelhaus, *J. Vac. Sci. Technol. A* **2001**, *19*, 2717.

15 M. Born, E. Wolf, *Principles of Optics*, 4th ed., Pergamon, Bath, UK, **1970**, pp. 23–51.

16 D.W. Lynch, W.R. Hunter, in: E.D. Palik (Ed.), *Handbook of Optical Constants of Solids*, Academic, Orlando, USA, **1985**, pp. 333–341.

17 D.F. Edwards, in: E.D. Palik (Ed.), *Handbook of Optical Constants of Solids*, Academic, Orlando, USA, **1985**, pp. 547–569.

18 J.E. Bertie, Z. Lan, *Appl. Spectrosc.* **1996**, *50*, 1047.

19 B. Aurian-Blajeni, M.A. Habib, I. Taniguchi, J. O'M. Bockris, *J. Electroanal. Chem.* **1983**, *157*, 399.

20 J.I. Pankove, *Optical Processes in Semiconductors*, Prentice Hall, Englewood Cliffs, USA, **1971**, pp. 67–76.

21 (a) N.J. Harrick, *Internal Reflection Spectroscopy*, 3rd ed., Harrick Scientific Corporation, Ossining, USA, **1987**; (b) F.M. Mirabella, Jr., N.J. Harrick, *Internal Reflection Spectroscopy: Review and Supplement*, Harrick Scientific Corporation, Ossining, USA, **1985**.

22 R. Heineman, J.N. Burnett, R.W. Murray, *Anal. Chem.* **1968**, *40*, 1974.

23 C. da Fonseca, F. Ozanam, J.-N. Chazalviel, *Surf. Sci.* **1996**, *365*, 1.

24 Y.J. Chabal, *Surf. Sci. Rep.* **1988**, *8*, 211.

25 E. P. Boonekamp, J. J. Kelly, J. van der Ven, A. H. M. Sondag, *J. Electroanal. Chem.* **1993**, *344*, 187.

26 H. Neugebauer, T. Yohannes, N. S. Sariciftci, *Synth. Met.* **2001**, *119*, 379.

27 R. Nakamura, A. Imanishi, K. Murakoshi, Y. Nakato, *J. Am. Chem. Soc.* **2003**, *125*, 7443.

28 H. B. Martin, P. W. Morrison, Jr., *Electrochem. Solid-State Lett.* **2001**, *4*, E17.

29 D. K. Lambert, *Electrochim. Acta* **1996**, *41*, 623.

30 H. Neugebauer, G. Nauer, N. Brinda-Konopik, G. Gidaly, *J. Electroanal. Chem.* **1981**, *122*, 381.

31 A. Venkateswara Rao, J.-N. Chazalviel, F. Ozanam, *J. Appl. Phys.* **1986**, *60*, 696.

32 F. Ozanam, J.-N. Chazalviel, *Rev. Sci. Instr.* **1988**, *59*, 242.

33 C. J. Manning, P. R. Griffiths, *Appl. Spectrosc.* **1993**, *47*, 1345.

34 K. Ataka, Y. Hara, M. Osawa, *J. Electroanal. Chem.* **1999**, *473*, 34.

35 J. Daschbach, D. Heisler, S. Pons, *Appl. Spectrosc.* **1986**, *40*, 489.

36 P. R. Griffiths, B. L. Hirsche, C. J. Manning, *Vibr. Spectrosc.* **1999**, *19*, 165.

37 F. Maroun, F. Ozanam, J.-N. Chazalviel, *J. Phys. Chem. B* **1999**, *103*, 5280.

38 S. Fellah, A. Teyssot, F. Ozanam, J.-N. Chazalviel, J. Vigneron, A. Etcheberry, *Langmuir* **2002**, *18*, 5851.

39 P. Allongue, C. Henry de Villeneuve, S. Morin, R. Boukherroub, D. D. M. Wayner, *Electrochim. Acta* **2000**, *45*, 4591.

40 A. Venkateswara Rao, F. Ozanam, J.-N. Chazalviel, *J. Electrochem. Soc.* **1991**, *138*, 153.

41 L. M. Peter, D. J. Blackwood, S. Pons, *Phys. Rev. Lett.* **1989**, *62*, 308.

42 L. M. Peter, D. J. Blackwood, S. Pons, *J. Electroanal. Chem.* **1990**, *294*, 111.

43 J. Rappich, H.-J. Lewerenz, H. Gerischer, *J. Electrochem. Soc.* **1993**, *140*, L187.

44 J. Rappich, H.-J. Lewerenz, *J. Electrochem. Soc.* **1995**, *142*, 1233.

45 J. Rappich, H.-J. Lewerenz, *Electrochim. Acta* **1996**, *41*, 675.

46 J. C. Flake, M. M. Rieger, G. M. Schmid, P. A. Kohl, *J. Electrochem. Soc.* **1999**, *146*, 1960.

47 R. Outemzabet, M. Cherkaoui, F. Ozanam, N. Gabouze, N. Kesri, J.-N. Chazalviel, *J. Electroanal. Chem.* **2004**, *563*, 3.

48 H. Fukidome, O. Pluchery, K. T. Queeney, Y. Caudano, K. Raghavachari, M. K. Weldon, E. E. Chaban, S. B. Christman, H. Kobayashi, Y. J. Chabal, *Surf. Sci.* **2002**, *502/503*, 498.

49 B. H. Erné, F. Ozanam, J.-N. Chazalviel, *J. Phys. Chem. B* **1999**, *103*, 2948.

50 R. Nakamura, Y. Nakato, *J. Am. Chem. Soc.* **2004**, *126*, 1290.

51 Q. Fan, L. M. Ng, *J. Electroanal. Chem.* **1995**, *398*, 151.

52 Q. Fan, L. M. Ng, *J. Electrochem. Soc.* **1994**, *141*, 3369.

53 M. Nakamura, M.-B. Song, M. Ito, *Electrochim. Acta* **1996**, *41*, 681.

54 M. Niwano, Y. Kondo, Y. Kimura, *J. Electrochem. Soc.* **2000**, *147*, 1555.

55 M. Niwano, T. Miura, N. Miyamoto, *J. Electrochem. Soc.* **1998**, *145*, 659.

56 K. T. Queeney, H. Fukidome, E. E. Chaban, Y. J. Chabal, *J. Phys. Chem. B* **2001**, *105*, 3903.

57 A. Belaïdi, M. Safi, F. Ozanam, J.-N. Chazalviel, O. Gorochov, *J. Electrochem. Soc.* **1999**, *146*, 2659.

58 J.-N. Chazalviel, C. da Fonseca, F. Ozanam, *J. Electrochem. Soc.* **1998**, *145*, 964.

59 B. H. Erné, F. Ozanam, M. Stchakovsky, D. Vanmaekelbergh, J.-N. Chazalviel, *J. Phys. Chem. B* **2000**, *104*, 5974.

60 K. Uosaki, Y. Shigematsu, H. Kita, K. Kunimatsu, *J. Phys. Chem.* **1990**, *94*, 4623.

61 P. A. Christensen, J. Eameaim, A. Hamnett, *Phys. Chem. Chem. Phys.* **1999**, *1*, 5315.

62 L. Kavan, P. Krtil, M. Grätzel, *J. Electroanal. Chem.* **1994**, *373*, 123.

63 K. E. Shaw, P. A. Christensen, A. Hamnett, *Electrochim. Acta* **1996**, *41*, 719.

64 P. Krtil, L. Kavan, I. Hoskovcova, K. Kratochvilova, *J. Appl. Electrochem.* **1996**, *26*, 523.

65 K. C. Mandal, F. Ozanam, J.-N. Chazalviel, *J. Electroanal. Chem.* **1992**, *336*, 153.

66 F. Maroun, F. Ozanam, J.-N. Chazalviel, *J. Electroanal. Chem.* **1997**, *435*, 225.

67 B. H. Erné, F. Maroun, F. Ozanam, J.-N. Chazalviel, *Electrochem. Solid-State Lett.* **1999**, *2*, 231.

68 B. H. Erné, F. Ozanam, M. Stchakovsky, D. Vanmaekelbergh, J.-N. Chazalviel, *J. Phys. Chem. B* **2000**, *104*, 5961.

69 L. Kavan, K. Kratochvilova, M. Grätzel, *J. Electroanal. Chem.* **1995**, *394*, 93.

70 K. Ogura, M. Nakayama, N. Endo, *J. Electroanal. Chem.* **1998**, *451*, 219.

71 M.-C. Pham, P.-C. Lacaze, *J. Electrochem. Soc.* **1994**, *141*, 156.

72 T. Yohannes, H. Neugebauer, S.A. Jenekhe, N. S. Sariciftci, *Synth. Met.* **2001**, *116*, 241.

73 P. Damlin, C. Kvarnström, H. Neugebauer, A. Ivaska, *Synth. Met.* **2001**, *123*, 141.

74 R.-M. Latonen, C. Kvarnström, A. Ivaska, *J. Electroanal. Chem.* **2001**, *512*, 36.

75 C. Kvarnström, H. Neugebauer, H. Kuzmany, H. Sitter, N.S. Sariciftci, *J. Electroanal. Chem.* **2001**, *511*, 13.

76 K.C. Mandal, F. Ozanam, J.-N. Chazalviel, *Appl. Phys. Lett.* **1990**, *57*, 2788.

77 M.-C. Pham, J. Moslih, P.-C. Lacaze, *J. Electrochem. Soc.* **1991**, *138*, 449.

78 Q. Fan, L.M. Ng, *J. Vac. Sci. Technol. A* **1996**, *14*, 1326.

79 Z.-V. Vardeny, X. Wei, in: T.A. Skotheim, R.L. Elsenbaumer, J.R. Reynolds (Eds.), *Handbook of Conducting Polymers*, 2nd ed., Dekker, New York, USA, **1998**, pp. 639–666.

80 V.M. Dubin, F. Ozanam, J.-N. Chazalviel, *Phys. Rev. B* **1994**, *50*, 14867.

81 P. O'Connor, J. Tauc, *Phys. Rev. B* **1982**, *25*, 2748.

82 K. Rerbal, J.-N. Chazalviel, F. Ozanam, I. Solomon, *Phys. Rev. B* **2002**, *66*, 184209.

83 G. Schlichthörl, H. Tributsch, *Electrochim. Acta* **1992**, *37*, 919.

84 H.O. Finklea (Ed.), *Semiconductor Electrodes. Studies in Physical and Theoretical Chemistry*, vol. 55, Elsevier, Amsterdam, The Netherlands, **1988**.

85 A. Tardella, J.-N. Chazalviel, *Phys. Rev. B* **1985**, *32*, 2439.

86 F. Ozanam, C. da Fonseca, A. Venkateswara Rao, J.-N. Chazalviel, *Appl. Spectrosc.* **1997**, *51*, 519.

87 J.-N. Chazalviel, S. Fellah, F. Ozanam, *J. Electroanal. Chem.* **2002**, *524–525*, 137.

88 B.H. Erné, F. Ozanam, J.-N. Chazalviel, *J. Phys. Chem. B* **2000**, *104*, 11591.

89 A. Venkateswara Rao, J.-N. Chazalviel, *J. Electrochem. Soc.* **1987**, *134*, 1138.

90 A. Venkateswara Rao, J.-N. Chazalviel, *J. Electrochem. Soc.* **1987**, *134*, 2777.

91 J.-N. Chazalviel, F. Ozanam, *J. Electron Spec. Relat. Phenom.* **1990**, *54/55*, 1229.

92 U. Fano, *Phys. Rev.* **1961**, *124*, 1866.

93 C. J. Hirschmugl, G.P. Williams, F.M. Hoffmann, Y.J. Chabal, *Phys. Rev. Lett.* **1990**, *65*, 480.

94 D.W. Berreman, *Phys. Rev.* **1963**, *130*, 2193.

95 F. Ozanam, J.-N. Chazalviel, *J. Electroanal. Chem.* **1989**, *269*, 251.

96 A.M. Bradshaw, E. Schweizer, in: R.J.H. Clark, R.E. Hester (Eds.), *Spectroscopy of Surfaces*, Wiley, Chichester, UK, **1988**, Chap. 8, pp. 413–483.

97 C.T. Kirk, *Phys. Rev. B* **1988**, *38*, 1255.

98 K.T. Queeney, N. Herbots, J.M. Shaw, V. Atluri, Y.J. Chabal, *Appl. Phys. Lett.* **2004**, *84*, 493.

99 J.W. Russell, J. Overend, K. Scanlon, M. Severson, A. Bewick, *J. Phys. Chem.* **1982**, *86*, 3066.

100 R.J. MacDonald, *Impedance Spectrosc.*, Wiley, New York, USA, **1987**.

101 J.-N. Chazalviel, K.C. Mandal, F. Ozanam, *SPIE Conf. Proc.* **1992**, *1575*, 40.

102 M. Lharch, M. Aggour, J.-N. Chazalviel, F. Ozanam, *J. Electrochem. Soc.* **2002**, *149*, C250.

103 D.E. Aspnes, in: E.D. Palik (Ed.), *Handbook of Optical Constants of Solids*, Academic, Orlando, USA, **1985**, Chap. 5, pp. 104–108.

104 D.A. Bruggeman, *Ann. Phys. (Leipzig)* **1935**, *24*, 636.

105 W. Theiß, *Surf. Sci. Rep.* **1997**, *29*, 91.

106 D. Bergman, *Phys. Rep. C* **1978**, *43*, 377.

107 V.M. Dubin, F. Ozanam, J.-N. Chazalviel, *Vibr. Spectrosc.* **1995**, *8*, 159.

108 A. Hartstein, J.R. Kirtley, J.C. Tsang, *Phys. Rev. Lett.* **1980**, *45*, 201.

109 F. Maroun, F. Ozanam, J.-N. Chazalviel, W. Theiß, *Vibr. Spectrosc.* **1999**, *19*, 193.

110 M. Osawa, K. Ataka, K. Yoshii, T. Yotsuyanagi, *J. Electroanal. Chem.* **1993**, *64/65*, 371.

111 M. Osawa, *J. Electron Spectrosc. Relat. Phenom.* **1997**, *70*, 2861.

112 T. Wandlowski, K. Ataka, D. Mayer, *Langmuir* **2002**, *18*, 4331.

113 M. Loster, K.A. Friedrich, *Surf. Sci.* **2003**, *523*, 287.

114 S. Pronkin, T. Wandlowski, *Surf. Sci.* **2004**, *573*, 109.

7

Recent Advances in *in-situ* Infrared Spectroscopy and Applications in Single-crystal Electrochemistry and Electrocatalysis

Carol Korzeniewski

7.1
Introduction

Infrared spectroscopy is frequently used together with electrochemical methods to gain molecular-level information regarding species present at the electrode/ solution interface. Early efforts focused on establishing techniques for *in-situ* sampling and acquisition of spectra with selectivity for molecules at or near the electrode surface (see Refs. [1–9]). The foundations of infrared spectroelectrochemistry were developed mainly through the study of electrocatalytic reactions involving small molecules, such as CO and CH_3OH, on polycrystalline electrodes [1, 4]. Growth in the area of single-crystal electrochemistry led to the application of infrared spectroscopy as a probe of bonding and reactivity at well-defined electrode surfaces [10–12]. Through comparisons to related measurements at single crystals in ultra high vacuum (UHV), experimental approaches and understanding of physical and chemical phenomena that affect the appearance of spectral features has progressed considerably [7, 13, 14]. Applications of infrared spectroelectrochemistry have continued to broaden. The technique is providing support for the study of new materials, such as nanoparticle catalysts [15–19] and organic films [20], and is advancing in capability through sampling innovations [21, 22] and next-generation instrumentation [23, 24].

The principles that govern infrared spectroelectrochemistry have been reported in a number of papers. This chapter highlights recent progress and discusses spectroscopic instrumentation and measurement concepts that have received little coverage in previous reports. Applications in the study of molecules on well-defined surfaces, bimetallic alloys, and nanoparticle catalysts are described.

Advances in Electrochemical Science and Engineering Vol. 9.
Edited by Richard C. Alkire, Dieter M. Kolb, Jacek Lipkowski and Philip N. Ross
Copyright © 2006 WILEY-VCH Verlag GmbH & Co. KGaA, Weinheim
ISBN: 3-527-31317-6

7.2
Experimental

Infrared spectroelectrochemical measurements are typically performed with Fourier transform infrared (FTIR) spectrometers. The traditional approach involves external reflection sampling, but methods based on internal attenuated total reflection (ATR) (Refs. [21, 22, 25–27], and see Chapters 6 and 9 in this volume) are becoming more prevalent. In the external reflection mode, after passing through the interferometer the infrared radiation beam is focused onto the surface of a polished-disk (typically ≥ 0.5 cm^2 area) working electrode positioned in a thin-layer electrochemical cell adjacent to an infrared-transparent window. Radiation reflected from the working electrode surface is collected and directed to the infrared detector. Designs for working electrodes, spectroelectrochemical cells, and optics employed for external reflection sampling have been described in several reports [1–9]. Therefore, no attempt will be made to review either this area or standard methods for spectral acquisition. The emphasis of Section 7.2 is on figures of merit pertaining to modern spectrometer systems. Approaches based on ATR sampling [21, 22, 26, 28] are briefly described, but are discussed more fully in Section 7.3 and Chapters 6 and 9 of this volume.

7.2.1
Spectrometer Systems

Infrared spectroelectrochemistry has been carried out almost exclusively with FTIR spectrometers, although dispersive instruments were employed during the early development stages [1]. Dispersive spectrometers have potential to achieve significantly better detection limits than FTIR systems, because the phase-sensitive detection strategies on which they are based provide excellent noise rejection, and the dynamic range of the modulated signal produced falls within the resolution capabilities of conventional analog-to-digital converter (ADC) electronics. However, infrared spectroelectrochemical measurements performed with dispersive spectrometers are limited to chemically reversible systems, and cost and availability have become issues as FTIR spectrometers have replaced dispersive instruments in the marketplace. Nevertheless, FTIR spectrometers are relatively simple to use and offer high throughput designs that provide for sensitivity in *in-situ* infrared spectroscopy measurements. Current systems can also be configured with capabilities for step-scan interferometry [23, 24, 29–31] and spectrochemical imaging [32–34].

7.2.2
Spectrometer Throughput Considerations

The signal-to-noise ratio (SNR) of FTIR spectrometers is directly proportional to the optical throughput (θ). For measurements at high resolution (approximately ≤ 2 cm^{-1}), throughput is determined by the allowable divergence of the infrared beam as it traverses the interferometer [35]. Under these conditions, the throughput (θ_I) is fixed by the resolution $(\Delta\bar{v})$, the high energy spectral limit $(\bar{v}_{max})$, and the area of the beam on the interferometer mirrors (A_M) according to

$$\theta_I = \frac{2\pi A_M \Delta\bar{v}}{\bar{v}_{max}} \;. \tag{1}$$

For instruments operating at lower resolution, throughput is generally determined by the detector area (A_D) and the solid angle of the beam being focused onto the detector. The latter can be computed from the half angle of convergence of the detector mirror (a_M). The detector area-limited throughput (θ_D) is given by [35]

$$\theta_D = 2\pi A_D [1 - \cos a_M] \;. \tag{2}$$

Scattering, reflection from window materials, and absorption by solvent contribute to strong radiation losses during infrared spectroelectrochemical measurements. Therefore, high-throughput instrument designs are generally recommended for these types of measurements [35].

The radiant power that reaches the detector depends in part upon the properties of the source. The spectral radiance of the source $(B_{\bar{v}})$ can be approximated by Planck's radiation law:

$$B_{\bar{v}} = \frac{c_1 \bar{v}^3}{\exp(c_2 \bar{v}/T) - 1} \;. \tag{3}$$

In Eq. (3), $c_1 = 2\,h\,c^2 = 1.19\times10^{-12}$ W cm^{-2} sr^{-1} (cm^{-1})$^{-4}$, $c_2 = h\,c/k = 1.439$ K cm, T is temperature in degrees Kelvin (K), and $B_{\bar{v}}$ has units of W sr^{-1} cm^{-2} (cm^{-1})$^{-1}$. The symbols h and k are Planck's constant and the Boltzmann constant, respectively. For an ideal black body, $B_{\bar{v}}$ is a function of temperature. FTIR spectrometers operating at mid-infrared spectral frequencies typically hold the source within the range 1200–1500 K. Little advantage is gained by operating at higher temperatures because of the corresponding shift in spectral emission to higher energies [36]. Gains may be achieved by increasing the size of the source (area of the emitting surface), but only up to the point where the collection optics become limiting.

7.2.3
Detectors

The noise equivalent power (NEP) and specific detectivity (D*) are figures of merit that express the sensitivity of infrared detectors. The NEP is the root-mean-square (rms) power in a sinusoidally modulated radiation signal incident on the detector that gives a response equal to the rms dark noise in a 1-Hz bandwidth. NEP has units of $W \cdot Hz^{-1/2}$. The D* is proportional to the NEP and the detector area, as follows: $D* = A_D^{1/2} \times NEP^{-1}$. For infrared spectroelectrochemistry measurements, high-sensitivity mercury-cadmium-telluride (MCT) detectors are typically employed, with the narrow band MCT being the most common. The D* for a narrow band MCT detector is just above 10^{10} cm $Hz^{1/2}$ W^{-1} across the mid-infrared spectral region. MCT detectors for FTIR use are available with areas of 1 mm square and 2 mm square, with the larger-area element enabling higher throughput and thus advantages for spectroelectrochemistry measurements. MCT detectors require cooling to liquid nitrogen temperatures. The detection element is housed in a dewar and sealed behind an infrared transparent window, and the window can determine the lower wavenumber spectral cutoff. MCT detector response is also sensitive to the interferometer mirror speed. Detectors typically operate with mirror velocities above 1 cm s^{-1}, or speeds that correspond to modulation of the 632.8 nm HeNe laser reference signal at frequencies near 20 kHz.

7.2.4
Signal-to-Noise Ratio Considerations

The SNR is a figure of merit that is frequently cited in product literature. Values are often reported as peak-to-peak absorbance measured for a given spectrometer configuration over a specified time period and spectral range and at a specified spectral resolution. The SNR varies in direct proportion to the instrument throughput, source spectral radiance, detector D*, and spectral resolution [35]. SNR also increases with the square root of measurement time ($t^{1/2}$), which depends upon the number of scans averaged, the mirror velocity, and the spectral resolution setting. Peak-to-peak noise levels near 10^{-5} absorbance units measured in a region of the spectrum free of absorption by atmospheric species (~ 2000 cm^{-1}) are typical for FTIR spectrometers operating with MCT detectors, spectral resolution of 4 cm^{-1}, and data collection periods of a few seconds.

7.2.5
Signal Digitization

Spectral SNR is also affected by the resolution of the ADC employed for interferogram digitization. To benefit from ensemble averaging of repeated interferogram scans, ADC resolution must be sufficient to capture signals at the detector noise level. However, the signal at the point of zero mirror retardation (the centerburst) is several orders of magnitude greater than the detector noise level, placing challenging constraints on ADC electronics. Gain ranging techniques are commonly employed to address the problem. A gain-ranging amplifier selectively boosts the signal in regions of the interferogram at a distance from the centerburst where the signal attains low levels and thus enables the amplified segments of the interferogram to be digitized with greater resolution. However, gain-ranging amplifiers introduce additional noise into the interferogram, and re-scaling of the amplified regions is required following digitization. With advances in integrated circuit technology in recent years, ADCs with sufficient dynamic range for FTIR measurements are becoming affordable. A few high-throughput FTIR systems that incorporate 24-bit or larger ADC electronics are available.

7.2.6
Signal Modulation and Related Data Acquisition Methods

An approach that has been employed to overcome interferogram dynamic range limitations involves introducing modulation into the analytical signal at a frequency that is fast compared to the highest Fourier frequency [37, 38]. Upon demodulation, the centerburst in the resulting differential interferogram is removed or greatly attenuated [37, 38]. For detection of molecules on surfaces, the approach has been applied mainly through the use of polarization modulation (PM), which takes advantage of the different sensitivities of the p- and s-polarization states of the infrared beam to species at the surface [37, 39–42]. PM-FTIR was adapted to infrared spectroelectrochemistry early in its development (cf. Refs. [1, 2, 43–46]). A primary motivation behind the use of PM-FTIR in electrochemistry has been the ability to discriminate against background interferences such as atmospheric water vapor and CO_2. Advances have been made in recent years by adapting fast, real-time analog sampling electronics for the direct measurement of differential and average interferograms [40–42, 47–49].

With the emergence of step-scan interferometers, it has become possible to employ electrode potential modulation as a means to enhance spectral SNR in measurements with FTIR systems (cf. Refs. [23, 24]). Potential modulation was originally employed with dispersive infrared spectrometers in a technique known as electrochemically modulated infrared spectroscopy (EMIRS) [1]. In EMIRS experiments, the electrode potential is switched between two states at a frequency of about 10 Hz as the monochromator is slowly scanned. A differential spectral signal is recovered with phase-sensitive detection. In extending the potential modulation strategy to measurements with Fourier transform instruments, conventional rapid-scan FTIR spectrometers have not been suitable. The electrode potential should be modulated at a frequency at least ten times [38] faster than the highest Fourier frequency (8 kHz for 4000 cm^{-1} radiation at a mirror velocity of 1 cm s^{-1}) [23, 29], and measurements under these conditions become challenged by the typically slow time constants of spectroelectrochemical cells (cf. Ref. [24]). Early demonstrations employed a slow-scanning interferometer [50]. More recently, however, success has been achieved with step-scan spectrometers, as these systems operate in a manner that reduces the Fourier frequencies effectively to zero [23, 29]. Step-scan FTIR spectrometers have been applied to perform time-resolved [51–54] and kinetic measurements [24, 31]. Modulation times from a few Hz up to 100 kHz have been attained through the use of spectroelectrochemical cells that enable ATR sampling [24, 31]. Step-scan FTIR measurements are discussed in greater detail in Section 7.3.5 in the context of applications.

7.3
Applications

7.3.1
Adsorption and Reactivity at Well-defined Electrode Surfaces

Experiments that applied infrared spectroscopy to investigate the properties of CO and other small molecules adsorbed on surfaces of single-crystal electrodes led to significant advances in understanding molecular structure and chemical bonding at metal/solution interfaces. The use of well-defined surfaces provided two notable advantages. First, the arrangement of molecules in an adlayer could be suggested by drawing upon results for related metal-adsorbate systems in UHV, where infra-red spectral measurements were frequently accompanied by more direct probes of structure, such as low-energy electron diffraction (LEED). Second, by controlling adsorbate coverages on electrodes, structural information could be derived from comparisons to corresponding infrared and high-resolution electron energy loss spectroscopy (HREELS) measurements on analogous UHV-prepared systems. For CO on Pt(*hkl*) electrodes, in particular, the methods afforded greater insights into several phenomena. A few of these include effects of electrode potential on adlayer structure [13, 55, 56] and adsorbate chemical bonding [13, 57, 58], the pro-motion of CO island formation through hydrophobic interactions with water [57, 59, 60], the adsorption of CO preferentially at different structural sites on stepped electrode surfaces [61–65], and optical effects that influence the appearance of ad-sorbate infrared spectral bands [55, 56, 63].

Infrared spectroscopy has continued to support the study of adsorption and re-activity at well-defined electrode surfaces. Single crystals are employed to probe ac-tive site models for catalytic reactions and as templates for the deposition and growth of other phases. Infrared spectroscopy has played an important role in en-abling *in-situ* detection and molecular-level characterization of species present at these surfaces. The sections below highlight some recent areas of application.

7.3.1.1 Adsorption on Pure Metals

Infrared spectroscopy is frequently applied to investigate CO adsorption on elec-trodes, because CO is important as an intermediate and surface poison in many electrocatalytic reactions and the C-O stretching vibrational modes of the adlayer are sensitive to the chemical environment at the metal/solution interface. Infra-red spectra of CO adsorbed on low-index surface planes of Pt single-crystal elec-trodes have become a benchmark for use in understanding the behavior of CO on other surfaces. Related approaches have been extended to bulk single-crystal metal electrodes that include Pd [66, 67], Ir [68–71], Rh [13, 70], Ru [72–74], Ni [75, 76] and Au [77].

In the study of Pt(111)/CO, key information that advanced understanding of relationships between infrared spectral features and specific adlayer structures was provided by *in-situ* scanning tunneling microscopy (STM) measurements

[55, 56]. STM images showed that CO forms a hexagonal closed-packed (hcp) (2×2)–3 CO adlattice with a coverage, θ_{CO}, of 0.75 on Pt(111) in 0.1 M $HClO_4$ at potentials just positive of 0.0 V_{RHE} (where the subscript RHE indicates that the voltage is reported with respect to the reversible hydrogen electrode reference). CO molecules in the adlayer occupy atop and three-fold bridging sites in a 1:2 (atop:bridging) ratio. The STM data enabled the vibrational spectra to be modeled theoretically [56]. Until that point, simulations had been based on adlattices known to form in UHV, and close agreement between experiment and theory was not achieved [55, 56]. In addition to predicting the experimental spectra, the modeling shed light on the complex relationship between band intensities and the population of adsorbates in different coordination environments (i.e., atop, two-fold bridging, etc.). The studies also provided insights into the physical interactions, such as dipole-dipole coupling, which have a strong influence on the appearance of spectra and are thus of wide-ranging importance for applications in infrared spectroelectrochemistry [7, 14, 55–57, 60, 68]. As potential is scanned positive from 0.0 V_{RHE}, the adlayer was observed by STM to transform from the (2×2)–3 CO state to a lower coverage ($\theta_{CO}=13/19$) hcp ($\sqrt{19} \times \sqrt{19}$) R23.4°–13 CO structure dominated by CO in asymmetric (near-atop and near-bridging) sites. The transformation was recently confirmed by *in-situ* surface X-ray scattering (SXS) measurements [78]. (See Chapter 1 of this volume for detailed discussions of *in-situ* SXS and its applications.) The SXS study also showed that CO molecules assigned to asymmetric binding in the ($\sqrt{19} \times \sqrt{19}$) R23.4° structure adopt a tilted configuration, but are otherwise likely present in normal atop and two-fold bridging sites [78]. This finding is important, since the adlayer shift leads to marked spectral changes that simulations were not able to fully reproduce [56]. The added information from SXS will enable refined models to be tested in the future. In general, *in-situ* SXS is a valuable complement to infrared spectral measurements. Its use is mentioned further below in the context of bimetallic electrodes.

More recently, the experimental strategy that combines *in-situ* infrared spectroscopy and STM has been used to characterize CO on Ni(111) [75] and Ru(0001) [72] electrodes in aqueous solution. Both surfaces are capable of supporting a stable CO coverage (θ_{CO}) of 0.33, producing a ($\sqrt{3} \times \sqrt{3}$) R30° structure in which CO molecules occupy exclusively atop sites on Ru and bridging sites on Ni. Adsorbed CO was detected on each surface as a single infrared spectral band at energies appropriate for the respective coordination environments. A higher coverage state ($\theta_{CO}=0.5$) is also stable on the surfaces, which leads to the appearance of two C-O stretching spectral bands, reflecting the presence of CO in both atop and multi-fold coordination environments. For the Ru(0001) surface, infrared measurements also indicate that the high coverage adlayer is oxidation resistant [73]. In the case of Ni(111), subsequent experiments moved beyond adsorbed CO to examine stages of surface oxide formation. Changes in the coverage of adsorbed water, electrolyte anions, and surface oxides could be detected in infrared spectra and correlated to the formation of oxide patches as potential was scanned [76].

Pt-group metals are most often the targets of investigations concerning CO adsorption. However, in recent work that examined CO electrochemistry on Au single crystal electrodes, it was revealed that CO forms adsorbed states on the low-index surface planes which can be detected by infrared spectroscopy [77, 79]. Because of weak chemical bonding interactions between CO and Au, the C-O stretching vibrational frequencies are not strongly red-shifted relative to gas phase CO, particularly in comparison to bands for CO on Pt-group metals. On reconstructed surfaces of Au(111), Au(100), and Au(110), C-O stretching vibrational bands were observed in the range 1940–1990 cm^{-1} for 3-fold bridging, 2005–2070 cm^{-1} for 2-fold bridging, and 2115–2140 cm^{-1} for atop-coordinated CO. Representative spectra are shown in Fig. 7.1. The modes assigned to bridge-bonded species are 150–200 cm^{-1} higher in energy on Au than on Pt single crystals. The differences for atop-coordinated CO are smaller (~ 50 cm^{-1}) but still substantial. Furthermore, the spectra, and hence the adlayer structures formed, were found to be strongly sensitive to the presence of dissolved CO in solution. The results have been explained in terms of limitations in the extent of back-donation from the metal d-orbital to the CO $2\pi^*$-orbital in Au-CO complexes [77]. For CO bonding to Au surfaces, the effect leads to a small metal-CO bond energy and only weak perturbation of the C-O bond. The study also ob-

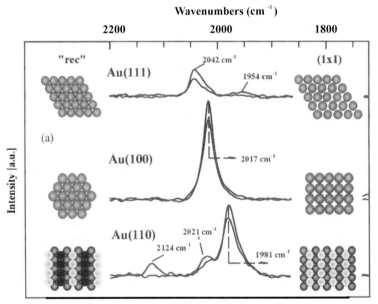

Fig. 7.1 Representative infrared spectra of the C-O stretching mode for CO adsorbed on reconstructed surfaces of Au single-crystal electrodes in 0.1 M HClO$_4$ saturated with CO (black curves) and in Ar-purged solution (grey curves). Models showing the reconstructed Au(111)-(1×23), Au(100)-"hex", and Au(110)-(1×2) surfaces and unreconstructed surfaces are included. See Chapter 1 in this volume for additional details. Adapted with permission from Ref. [77].

served the rate of CO oxidation to be structure sensitive and fastest on the highly stepped [Au(110)–(1×2)] or strained [Au(100)–hex] surfaces, which also support the greatest CO coverages [77].

Anions and neutral organic compounds are among other species whose adsorption on single-crystal electrodes has been probed with sensitivity by infrared spectroscopy. Because of their importance as common electrolytes, there has been long-standing interest in the behavior of simple oxoanions such as perchlorate, sulfates, and phosphates, at the electrode/solution interface (cf. Refs. [6, 80–89]). These ions undergo only weak perturbations upon adsorption. Thus, it can be difficult to discriminate species in solution, or in the diffuse double-layer, from ions at the electrode surface [54]. Better selectivity for adsorbed electrolyte anions has been achieved through use of the surface-enhanced infrared absorption spectroscopy (SEIRAS) technique [22, 54, 89, 90]. Methods for the preparation of quasi-single crystalline thin films are enabling the study of electrolyte adsorption on structurally well-defined surface sites by SEIRAS [22, 89, 90]. Applications of the SEIRAS technique are discussed more fully in Section 7.3.2. Carbonate and bicarbonate are other interesting oxoanions that have received attention recently as adsorbates on single-crystal electrodes [91, 92]. The ions can be generated at the electrode/solution interface from CO_2 through potential-induced acid-base reactions. Adsorbed carboxylate ions are also of interest, and, in contrast to oxoanions, they are more easily detected. Strong bands associated with symmetric and asymmetric stretching modes of the surface-coordinated carboxylate group appear in the region between 1500 and 1300 cm^{-1} [93–98]. The bands provide a signature for the transformation between carboxylic acids and adsorbed carboxylate species [93, 94]. Recent studies of oxalic acid [96, 97] and citric acid [98] chemistry on single-crystal electrodes demonstrate the development of these features as probes of adsorption. In the study of neutral organic compounds, applications of infrared spectroscopy are increasing (cf. Ref. [20]). Growth has been somewhat slow, because the absorption cross-sections for organic adsorbates are often weak, and the detectable vibrational modes can be limited because of screening of transition dipoles that lie along the electrode surface plane. During the last few years, infrared spectroscopy has been employed fruitfully to probe organic systems that included small molecules [99–107], self-assembling species [42, 108, 109], and oriented lipids [110, 111] on single-crystal electrodes.

7.3.1.2 Electrochemistry at Well-defined Bimetallic Electrodes

Bimetallic electrodes have been of continuing interest for their importance in catalysis and relevance to understanding metal electro-deposition processes. Infrared spectroscopy has played a supporting role by providing insights into the involvement of solvent, coordinating anions, and adsorbed neutral molecules in the case of metal deposition (cf. [112–114]), and surface-mediated reaction pathways in the case of catalysis [115–119]. An interesting application involving metal deposition on single-crystal electrodes is motivated by the desire to produce well-defined surfaces for materials that are not easily cleaned and ordered by bench-top methods like flame annealing. An approach has been to deposit a thin film of the material of interest by epitaxial growth on the surface of a Pt single-crystal electrode [113, 120]. The physical and electronic structure of the film can then be characterized *in situ* by infrared spectroscopy through detection of adsorbed CO probe molecules [113, 114, 120]. The system Pt(*hkl*)/Pd has received a great deal of attention recently (see Refs. [113, 117–119, 121] and Chapter 1 in this volume). The preference of CO for bridging sites on Pd results in the appearance of strong C-O stretching features for Pd/CO that can be easily distinguished from the typically intense bands for atop CO on Pt sites. By monitoring the positions of C-O stretching bands as a function of Pd coverage, Feliu and coworkers demonstrated that insights into the structure and growth mechanism of Pd islands can be gained [113, 121]. Furthermore, it was shown that deposition of Pd multilayers on Pt(111), or a complete monolayer on stepped-Pt single-crystal electrodes results in additional bands characteristic of CO bonding to low-coordination atomic sites along step edges of Pd. Examining the position and intensity of the bands as a function of terrace width for Pd films on stepped Pt surfaces provided a means to probe multilayer growth and estimate the sizes of Pd terraces [121]. Surfaces formed by modification of Pt(*hkl*) by Pd also have demonstrated enhanced catalytic activity toward the oxidation of the C_1 species formic acid [117, 118] and CO [119]. *In-situ* infrared spectra indicate that Pd sites are considerably more active than Pt in the low-potential region below about 0.4 V. Adsorbed CO was observed to form only on Pt sites, and the presence of surface Pd atoms lowered the onset potential for the oxidation reactions and increased the rate for CO_2 formation.

The capabilities of infrared spectroscopy as a probe of reaction pathways are demonstrated more fully in the example below for CO oxidation over Pt_3Sn single-crystal electrodes. Pt_3Sn is one of a number of binary metal intermetallic electrodes that have been employed lately for the study of electrocatalytic reactions [115, 116, 122–124]. The spectra in Fig. 7.2 indicate that dynamic changes occur in the CO adlayer on the surface of $Pt_3Sn(111)$ in CO-saturated electrolyte solution as the potential is scanned through a region where water activation is catalyzed by the surface [115, 124]. At the lowest potential (0.04 V_{RHE}), a pair of peaks is present at about 2080 cm^{-1}. The energy is typical for C-O stretching modes of atop CO on Pt. Although the band splitting is not well understood, it has been suggested that it may reflect the presence of two different structural regions on the surface which support different CO packing densities [115, 116]. As potential is stepped positive, the formation of CO_2 is detected by the characteristic band indicated at 2345 cm^{-1}. In parallel, the atop CO band splitting becomes less apparent, and by 0.25 V_{RHE} a feature typical of CO in two-fold bridging sites on Pt is present. The spectral changes suggest that CO is adsorbed on $Pt_3Sn(111)$ at high packing densities near 0.04 V_{RHE} and that a lower coverage state is attained with the onset of CO_2 formation and as potential is scanned increasingly positive. The integrated intensity of the CO_2 band provides an estimate of the quantity of CO_2 produced over a measurement period. The

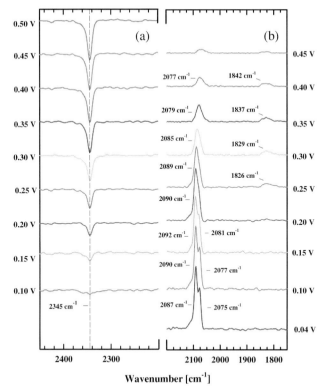

Fig. 7.2 Series of infrared spectra showing the formation of CO$_2$ (a) and the progressive oxidation of adsorbed CO (b) on a Pt$_3$Sn(111) electrode in 0.5 M H$_2$SO$_4$ saturated with CO. The electrode was initially held at 0.04 V$_{RHE}$ and stepped positive. Each spectrum is accumulated from 50 interferometer scans at the potential indicated. Adapted with permission from Ref. [115].

growth of the 2345 cm^{-1} band in Fig. 7.2 shows that the rate of CO$_2$ formation becomes faster with increasing potential and lowering of adsorbed CO coverage.

General models for the CO adlayer on Pt$_3$Sn(111) were proposed with the aid of *in-situ* SXS measurements (see Chapter 1 in this volume for additional details). As discussed earlier, *in-situ* SXS provides valuable structural information to guide understanding of infrared spectral data. In this case, *in-situ* SXS established the structure of metal atoms at the surface of Pt$_3$Sn(111) as a function of potential [115]. While it was not possible to obtain adlattice structures for CO, characterization of the metal lattice established the potential range for its stability and enabled models for CO to be proposed based on infrared spectra. From knowledge of the position of Sn in the lattice, it was suggested that OH species on Sn influence the formation of CO clusters with high local coverage through repulsive interactions.

7.3.2
SEIRAS

SEIRAS is emerging as a powerful technique for interfacial studies in electro-chemistry and electrochemical catalysis. Chapter 9 of this volume is devoted to SEIRAS and describes experimental aspects and many applications. Highlights are presented below, and the reader is referred to Chapter 9 for details regarding cell design, working electrode characteristics, and preparation techniques and applications.

SEIRAS measurements are typically performed in an ATR sampling config-uration with the working electrode prepared as a thin metal film on the surface of the internal reflection element [22, 26, 90, 125, 126]. The metal films employed are composed of nanometer-scale islands (cf. Refs. [90, 127]). The islands can be polarized by electromagnetic radiation in the mid-infrared region [28, 128]. An oscillating electric dipole is produced which gives rise to strong electro-magnetic fields at the metal surface [28, 126, 128–130]. The electromagnetic fields set up by the particle stimulate molecules in the vicinity of the metal sur-face to undergo enhanced infrared absorption. In addition to an electromagnetic mechanism, metal-adsorbate chemical bonding interactions are also thought to play a role through their effect on the vibrational polarizability [90, 128].

In infrared spectroelectrochemistry, the ATR-based SEIRAS technique has sev-eral advantages over traditional external reflection sampling. First, infrared absor-bance by interfacial species is enhanced by an order of magnitude or more [22, 28, 89, 90, 126, 128–130]. The improvements in sensitivity and detection limit that re-sult have led to some unexpected findings. For example, in a recent study of metha-nol electrochemical oxidation on platinum, formate was detected as an adsorbed intermediate [21]. The results furnished evidence for the existence of reaction path-ways that parallel the frequently studied direct dehydrogenation route to CO and have stimulated new thinking on the mechanism for methanol electrochemical oxi-dation [21]. Second, the electric fields produced at the metal surface are short range, so infrared absorption is preferentially enhanced in molecules at the metal/solution interface. The selectivity of SEIRAS for interfacial species has enabled adsorbed water molecules and electrolyte ions to be detected without interference from the bulk solution [22, 89, 90]. The external reflection method especially suffers in this regard, since the infrared beam must traverse a thin layer of solution (~ 1–5 µm) before reaching the metal electrode. Finally, the ATR sampling geom-etry overcomes limitations from restricted mass transport and uncompensated re-sistance in the thin-layer cells required for external reflection measurements. Prob-lems that may arise from reactant depletion [131–134] and pH changes [133, 135, 136] in thin-layer cells are avoided. Furthermore, the rapid time response of cells employed for SEIRAS has enabled changes following a potential step to be recorded spectrally with a time resolution near 100 µs [51–54] and potential modulation FTIR to be performed in a manner similar to ac voltammetry measurements [24, 53].

Figure 7.3 shows an example of the sensitivity and selectivity of SEIRAS to-ward interfacial water. The spectral bands in the region 1600–1700 cm^{-1} and

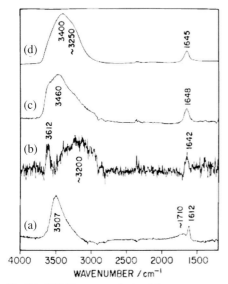

Fig. 7.3 SEIRAS spectra obtained from an Au(111) film electrode in 0.5 M HClO₄. Spectra were recorded at potentials of 0.12 V_{RHE} (a), 0.77 V_{RHE} (b), and 1.22 V_{RHE} (c). A spectrum recorded near the potential of zero charge (0.62 V_{RHE}) was used as the reference. Spectrum d is the transmission spectrum of 0.5 M HClO₄. Adapted with permission from Ref. [22].

2800–3700 cm^{-1} arise from HOH bending and OH stretching modes of water, respectively [22, 89, 90]. The spectra demonstrate that the structure of water at the Au/solution interface is highly sensitive to the electrode potential. The spectrum recorded at the lowest potential (Fig. 7.3a) displays features that have been interpreted as indicating that water is present at the surface with the hydrogen atoms toward the metal and the oxygen extending into solution [22]. Under the conditions of the experiment, the excess surface charge is negative at 0.12 V_{RHE}, and therefore the water dipole is stabilized in this manner. The downshift of the HOH bending mode and upshift and narrowing of the OH stretching mode in comparison to water in the bulk electrolyte solution (Fig. 7.3d) is reported to be characteristic of a decrease in hydrogen bonding between water molecules [22]. Furthermore, the band at about 1710 cm^{-1} is associated with a protonated state of water and is thought to form through proton donation from second-layer water molecules to oxygen atoms that are directed away from the surface in the first layer (see Fig. 7.4a).

Upon increasing the potential (Fig. 7.3b and c), the water bending mode shifts toward higher energy, and a high wavenumber feature appears on the broad envelope of the OH stretching bands. The feature near 3600 cm^{-1} is characteristic of free (or non-hydrogen-bonded) OH. It begins to grow coincident with the start of ClO₄$^-$ adsorption near 0.77 V_{RHE}. Other features in the spectra in Fig. 7.3b and c are characteristic of an ice-like structure, and it has been sug-

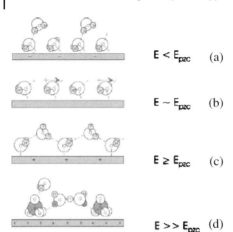

Fig. 7.4 Proposed models for the structure of water at the Au/solution interface at various potential regions relative to the potential of zero charge, E_{pzc} (A–D). In D, adsorbed sulfate ions are included in the drawing. Adapted with permission from Ref. [90].

$E < E_{pzc}$ (a)

$E \sim E_{pzc}$ (b)

$E \geq E_{pzc}$ (c)

$E \gg E_{pzc}$ (d)

gested that anion adsorption induces some disruption in the ice-like network enabling the formation of non-hydrogen bonded OH groups [22]. An interesting observation is that the free OH band is suppressed in the presence of anions that are proton acceptors, such as sulfate [22, 89, 90]. Thus, it appears that adsorbed ClO_4^- ions weaken hydrogen bonds between water molecules and, being only a weak acceptor of protons, produce –OH groups with free motion.

Structural models for water at the Au-aqueous solution interface at various potentials relative to the potential of zero charge (pzc) are summarized in Fig. 7.4. Negative excess surface charge induces water to orient with the oxygen atom toward the bulk, with the possibility for hydrogen bonding with second layer water molecules. Near the pzc, the absence of detail in spectra has led to the proposed model in which the water molecules lie with the dipole in the plane of the surface. Since spectra recorded at potentials near the pzc (0.62 V_{RHE} for spectra shown in Fig. 7.3) are essentially featureless, they are sometimes employed as the background for measurements at other potentials. Just positive of the pzc (Fig. 7.4 C), water dipoles appear to orient with the oxygen atom toward the surface and develop a strongly hydrogen-bonded network with molecules in the second layer. These water-water interactions become disrupted by anions, which can develop high coverages at positive potentials far from the pzc (Fig. 7.4 D). The models in Fig. 7.4 are based on structures proposed in the study of Au in contact with 0.5 M $HClO_4$ [22], as reported in Fig. 7.3, but were further developed in experiments performed in sulfate-containing aqueous electrolytes [89, 90]. Thus, adsorbed sulfate is the anion included in Fig. 7.4 D.

Another area in which SEIRAS is providing important new information is electrocatalysis [21, 137–142]. In recent studies of methanol oxidation on Pt, adsorbed formate [21] and formate-like species [142] were identified as by-products. The findings showed that, in addition to the frequently studied reaction leading to adsorbed CO, methanol oxidation can progress through a parallel pathway to formate, which avoids surface poisoning by CO and provides a po-

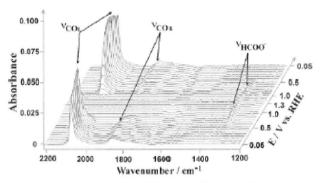

Fig. 7.5 SEIRAS spectra obtained from a Pt film electrode in 0.1 M HClO$_4$ containing 0.5 M CH$_3$OH. Spectra were recorded in sequence starting from 0.05 V$_{RHE}$ and progressing positive. The scan rate was 5 mV s^{-1}. Adapted with permission from Ref. [21].

tential route to CO$_2$ that does not involve CO as an intermediate [143, 144]. Fig. 7.5 shows representative SEIRAS spectra recorded during a slow scan of Pt in a solution containing 0.1 M HClO$_4$ and 0.5 M CH$_3$OH. The scan was initiated from 0.05 V$_{RHE}$ and advanced positive, while spectra were collected in sequence, as shown. Expected bands for CO bonded to the Pt surface in linear and bridging configurations are present within the range 2080–1700 cm^{-1}. The adsorbed CO is produced by dissociative chemisorption at the initial potential of 0.05 V$_{RHE}$, and the dominant spectral features for CO persist up to about 0.7 V$_{RHE}$. The band for formate (1320 cm^{-1}) begins to become apparent as the scan approaches 0.6 V$_{RHE}$. The spectra suggest that the formation of reactive oxides near 0.6 V$_{RHE}$ not only depletes the adsorbed CO population, but also promotes the conversion of other carbon-containing fragments to formate. The intensity of the formate band follows the current in the cyclic voltammetry. The current reaches a peak at about 1.0 V$_{RHE}$ and falls at more positive potentials because of the formation of passivating surface oxides. On the reverse scan, the formate band appears again between 1.0 V$_{RHE}$ and 0.6 V$_{RHE}$, before being replaced by adsorbed CO at the lower potentials. Since the initial reports, adsorbed formate has been shown to form during the electrochemical oxidation of methanol and related C$_1$ species under other experimental conditions [138, 139]. With its high sensitivity and selectivity for interfacial species, SEIRAS will continue to be valuable for the study of electrocatalytic reactions. Most recently, SEIRAS has been extended to probe molecular-scale processes involved in hydrogen evolution [145].

In addition to electrocatalysis and double-layer phenomena, SEIRAS has proved particularly useful in the study of organic adsorption on electrodes [99, 146–149]. This is an area where SEIRAS can have an especially important impact, as traditional *in-situ* infrared methods are often limited by the low absorption cross-sections of organic adsorbates. Along these same lines, SEIRAS is being fruitfully applied to probe properties of biological molecules at electrodes [150, 151]. In recent work, interfacial hydrogen bond pairing in complementary nucleic acid bases was probed by SEIRAS [150]. A thiol-derivatized adenine moiety immobilized on an Au electrode was observed to orient toward the surface normal in a geometry favorable to base pairing with thymidine under electrode potential control. SEIRAS is also contributing to the understanding of protein electrochemistry by serving as a real-time, *in-situ* probe of the macromolecule immobilization process and the subsequent redox behavior [151]. In this regard, SEIRAS has promise for developing into a centrally important probe of interactions between biological macromolecules and charged surfaces in a manner similar to that achieved by surface-enhanced Raman spectroscopy (SERS) [152], but with the added advantage of enabling measurements at quasi-single-crystalline surfaces [22, 89, 90]. As an example, an application involving the immobilization of the redox enzyme cytochrome *c* oxidase onto an Au electrode followed by its reconstitution into a lipid bilayer structure on the electrode is highlighted.

Figure 7.6 A shows SEIRAS spectra that follow the adsorption of the protein cytochrome *c* oxidase onto the surface of a chemically modified Au electrode. Bands of the amide I and amide II modes of the protein backbone appear at 1658 cm^{-1} and 1550 cm^{-1}, respectively. The peak intensity of the bands increases with time. The amide II band was shown to grow according to an exponential rate law with a time constant of 213 s [151]. The amide I band position is characteristic of an α-helical protein, consistent with the structure of cytochrome *c* oxidase. The protein was adsorbed onto the Au surface via affinity for a chemical layer that was constructed in a step-wise fashion upon exposure of the surface to a series of reagents. Initially, the metal surface was modified by self-assembly of DTSP [dithiobis-(suc-

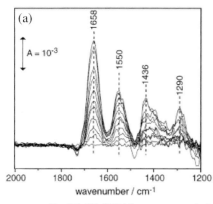
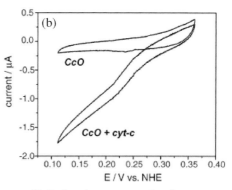

Fig. 7.6 (A) SEIRAS spectra recorded during the adsorption of cytochrome *c* oxidase onto an NTA-Ni^{2+}-modified Au surface via the affinity for its histidine tag introduced genetically into the protein. The bands shown are increasing in intensity with time.

(B) Cyclic voltammogram of surface-immobilized cytochrome *c* oxidase in the absence and presence of the natural cytochrome *c* (cyt. *c*) electron donor, as indicated. The scan rate was 2 mV s^{-1}. Adapted with permission from Ref. [151].

cinimidylpropionate)] on the Au. The next step involved exposure of the layer to amino-nitrilotriacetic acid (ANTA), which underwent reaction with DTSP through the amino group to form a carboxamide bond and leave the layer terminated by the NTA moiety. NTA contains three carboxylate groups and one tertiary amine site in close proximity. Incubation of the layer with a solution containing Ni^{2+} produced a surface with high affinity for short peptide segments containing consecutive histidine residues. Such a peptide segment was engineered into the cytochrome c oxidase used in the study. A peptide with six histidine units was linked to the carboxy-terminus of subunit I on the protein. The strategy enabled the protein to be immobilized to the Au surface with a specific orientation relative to the plane of the Au. The spectra in Fig. 7.6 A track the oriented binding of the histidine-tagged cytochrome c oxidase to the NTA-Ni^{2+} affinity sites. SEIRAS was employed as an *in-situ* monitor of progress in the reaction steps that accompanied each stage of surface modification. Thus, SEIRAS was applied in the same way as reflectance infrared spectroscopy had been applied earlier to probe steps in the modification of surfaces for use in areas such as chemical sensing [153, 154]. In a final step, the oriented, surface-attached cytochrome c oxidase was reconstituted into a lipid environment formed at the surface by spontaneous assembly from exposure to the vesicles, in order to enhance the stability and functionality of the membrane protein.

In Fig. 7.6 A, the bands at 1436 cm^{-1} and 1290 cm^{-1} are not associated with protein, but are thought to arise from changes in the structure and conformation of the NTA-Ni^{2+} layer. The band at 1436 cm^{-1} can be assigned to the symmetric stretching vibrational mode of the carboxylate groups on NTA. It is likely that the band reflects a perturbation to the NTA-Ni^{2+} interaction during coordination of histidine residues onto the protein. Cyclic voltammograms of the cytochrome c oxidase bound to the Au electrode are included in Fig. 7.6 B. The orientation of the enzyme does not permit direct electron transfer between the active site and the Au. Thus, the voltammogram displays only background charging current in the absence of a mediator. Since the histidine linker orients the enzyme such that the binding site for the natural electron donor cytochrome c is freely accessible to the solution phase, the addition of cytochrome c enables current flow. The electrode appears to facilitate cycling of cytochrome c into the reduced state, which is capable of donating electrons to the enzyme during the conversion of O_2 to water [151].

7.3.3
Infrared Spectroscopy as a Probe of Surface Electrochemistry at Metal Catalyst Particles

An area of increasing importance is the study of surface electrochemistry at nanometer-scale metal catalyst particles. An effective approach for applying *in-situ* infrared spectroscopy has been to adsorb catalyst onto the surface of a metal substrate that is highly reflective and also inert over the potential range of interest [15, 17, 155–161]. Polycrystalline Au electrodes have been employed frequently. Catalyst is supported as a thin film and held in place by physical adsorption [15, 17, 156–161], or sometimes through adhesion to Nafion [157]. Infrared radiation is reflected from the Au, and molecules adsorbed on the metal particles can be detected through their interaction with the standing wave electromagnetic field that develops at the Au surface. By keeping the catalyst loading below a few monolayers, spectral distortions arising from diminished reflectivity of the bulk substrate are avoided [18, 158].

The approach has been demonstrated with free [15, 156] as well as carbon-supported metal particles [17, 158–160]. Another method that has a number of advantages, but has not been used extensively thus far, involves anchoring metal nanoparticles to Au via an ultra-thin silane film [18]. A scheme for film formation by self-assembly chemistry, cross-linking, and capping by amine ligands is summarized in Fig. 7.7. Adsorption of metal particles to the amine-terminated layer was shown to reduce particle agglomeration and improve film optical properties [18]. At the same time, the films possessed sufficient electronic conductivity to enable electrochemical measurements. Figure 7.8 displays voltammograms of an ultra-thin silane layer on Au-supporting PtRu nanoparticles. The responses prior to and just after exposure of catalyst to CO are similar to those of bulk metals. Infrared spectra recorded following adsorption of CO are also included in Fig. 7.8. The bands have high SNR and are free of distortions caused by anomalous dispersion effects [18, 158]. The spectra reflect the core-island structure of the PtRu particles. The two main features present are associated with CO bound in atop positions on

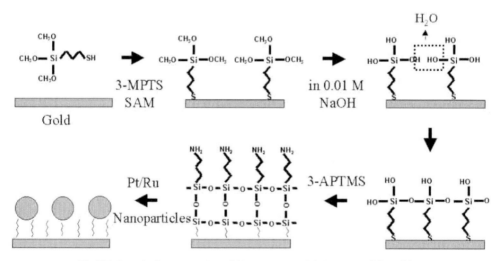

Fig. 7.7 Steps in the preparation of thin, amine-terminated silane films on Au. The Au substrate was immersed in a solution of 3-mercaptopropyltrimethoxysilane (3-MPTS) to produce a self-assembled monolayer (SAM). The film was cross-linked in 0.01 M NaOH followed by immersion in a solution containing 3-aminopropyltrimethoxysilane (3-APTMS) to introduce terminal amine groups capable of binding metal nanoparticles. Adapted with permission from Ref. [18].

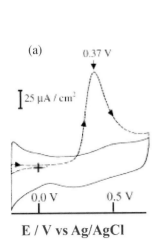

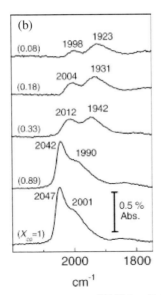

Fig. 7.8 Voltammograms (A) and *in-situ* infrared spectra (B) recorded from PtRu catalyst particles immobilized on a poly-crystalline Au electrode according to the procedure outlined in Fig. 7.7. Measurements were performed in 0.05 M H_2SO_4. PtRu particles were prepared by electroless deposition of Ru on Pt-black to an Ru atom coverage of 66 % (see Ref. [18] for details). The voltammograms were recorded at a scan rate of 50 mV s^{-1} in the absence (solid line) and presence (dashed line) of a CO monolayer. Infrared spectra were recorded for catalyst with different coverages of CO, as indicated. Adapted with permission from Ref. [18].

areas of the particles that are rich in the different metals. The particles were pre-pared by electroless deposition of Ru onto Pt-black to form a catalyst consisting of a Pt core decorated with islands of Ru on the surface [18, 162, 163]. The lower energy band of the pair is assignable to CO in Ru-rich areas of the particles, while the higher energy band is consistent with CO in regions containing mainly Pt. The spectra were recorded at different CO coverages ranging from 8% to 100% of the available metal sites. The high spectral sensitivity attained enabled the surface electrochemistry of CO on the catalyst particles to be probed in a manner similar to that employed for bulk PtRu electrodes [16, 164].

In another recent study, *in-situ* infrared spectroscopy was applied to investigate the properties of a carbon-supported Pt electrocatalyst sample containing two different metal nanoparticle populations [19]. The nanoparticles differed in size, with one population averaging approximately 3.6 nm and the other 1.7 nm. The surface chemistry of the particles was investigated by adsorption of the catalyst onto the surface of a bulk Au electrode. In CO-stripping voltammo-grams, two waves, a dominant peak near 0.7 V_{RHE} and a smaller peak at about 1.0 V_{RHE}, were present. Infrared spectra were consistent with a model in which CO on the smaller particles was somewhat resistant to oxidation and hence comprised the dominant reactant in the high-potential stripping wave. Figure

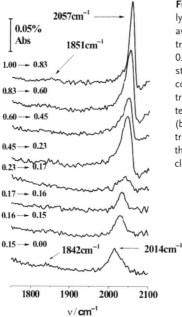

Fig. 7.9 Infrared spectra of CO adsorbed on Pt catalyst containing two populations of particles, one of average size 3.6 nm and the other 1.7 nm. The spectra were collected with the electrode at 0.1 V_{RHE} in 0.1 M $HClO_4$. The absorbance bands represent the state of the surface at the higher of the two fractional coverages indicated toward the left side of each spectrum. The coverage was lowered by stepping the potential to 0.7 V_{RHE} (top four spectra) or 0.95 V_{RHE} (bottom 4 spectra) for a fixed time period. The spectra shown were recorded in sequence starting with the CO coverage at saturation on the catalyst particles. Adapted with permission from Ref. [19].

7.9 shows example spectra from the study. A difference technique was used to uncover bands of CO molecules that underwent oxidation during a brief positive potential excursion to a point coincident with the first oxidation wave. The change in CO coverage that occurred is indicated beside each spectrum in the plot. Starting from full CO coverage (Fig. 7.9, top spectrum), the reactive species has the spectral signature of atop CO molecules in large islands at high coverage – conditions of strong dipole-dipole coupling. Thus, the oxidation current is believed to arise from CO adsorbed on the ~3.6-nm Pt particles. To reduce the CO coverage below 0.17, a larger positive potential excursion reaching into the second oxidation wave was required. The bottom four spectra in Fig. 7.9 reflect the chemical environment of CO molecules reacting at potentials in the high-potential CO stripping peak. The energy of the band for atop CO species is typical of molecules bound to low coordination surface atoms on Pt. The reacting molecules are thought to reside on the smaller, ~1.7-nm crystallites, which are expected to have a higher density of edge sites than the ~3.6-nm particles.

Results of the Maillard study [19] additionally suggested that barriers to CO diffusion existed *between* particles on the carbon support. Hence, CO molecules could become trapped on small particles, being unable to reach active sites on the ~3.6-nm Pt. In their pioneering work, Weaver and coworkers also observed spectral features and reactivity behavior dependent upon the size of Pt deposits on carbon-supported catalyst materials [17, 160]. Working with homogeneous particle size distributions, they showed that responses typical of low-coordination surface atoms become dominant for Pt particles smaller than about 4 nm

[17, 160], and spectra had several features in common with CO on surfaces of high step density single-crystal Pt electrodes [17, 61, 63, 165].

Another example of where insights derived from early *in-situ* infrared spectroscopy studies of CO on extended single-crystal surfaces have been fruitfully applied to understand the reactivity and bonding of CO on metal nanoparticle catalyst is in a recent study by Arenz and coworkers [166]. Plots of atop CO band position versus applied potential provided information about the density of defects on the particles and rearrangements of the CO adlayer during slow potential scans. For particles in the range of 1–30 nm, the average slopes of the plots were similar when CO was adsorbed to reach coverages near saturation. However, deviations from the average were observed upon oxidation of CO, and the direction and onset potential of the change in (Stark tuning [13, 58, 166, 167]) slope was particle size dependent. Drawing upon studies of CO adsorption on bulk single-crystal electrodes [7, 13, 14, 57, 68], the deviations signaled structure changes in the CO adlayer to either a more dense (blue-shift) or a less dense (red-shift) arrangement as potential was scanned [166]. In general, by combining *in-situ* infrared spectroscopy with electrochemistry and high-resolution transmission electron microscopy (HRTEM) measurements, the group was able to advance understanding of atomic-scale factors, including CO mobility, surface defect density, and OH adsorption strength, that limit the rate of CO electrochemical oxidation on nanometer dimension catalyst particles [166, 168, 169].

7.3.4
Nanostructured Electrodes and Optical Considerations

In early *in-situ* infrared spectroscopy studies of catalyst supported on reflective Au substrates, it was recognized that spectral bands develop distortions due to anomalous optical effects as the catalyst coverage becomes greater than a few monolayers [17, 18, 158]. Different procedures were adopted to ensure that catalyst coverages on Au remained low [15, 17, 18, 157, 158]. Related to these efforts are experiments that examine the electrochemical and optical properties of nanostructured metal electrodes prepared by chemical, vapor, or electrochemical deposition methods (cf. Refs. [79, 137, 141, 170–179] and references therein). Much of the work in this area has targeted SEIRAS measurements.

The understanding of factors that lead to enhanced band intensities and dispersive band shapes is of central interest in studies with nanostructured electrodes. Effective medium theory has often been employed to identify mechanisms for enhanced infrared absorption [28, 128, 172, 174, 175]. Osawa and coworkers applied Maxwell-Garnett and Bruggeman effective medium models in early SEIRAS work [28, 128]. Recently, Ross and Aroca overviewed effective medium theory and discussed the advantages and disadvantages of different models for predicting characteristics of SEIRAS spectra [174]. When infrared measurements on nanostructured electrodes are performed by ATR sampling, as is typically the case in SEIRAS experiments, band intensity enhancements occur, but the band shapes are usually not obviously distorted. In contrast, external reflection sampling often produces derivative-like and inverted band shapes, in addition to intensity enhancements [79, 141, 170, 171, 173, 175, 176, 178, 179]. The band distortions have been explained in terms of the relationship between reflectivity at the metal/solution interface and the dielectric properties of the nanostructured layer [172, 175]. Mechanisms involving electronic damping and interparticle interactions have also been proposed [170, 171]. The nanostructured electrodes behave similarly to dielectric films in that optical dispersion effects are more prominent in reflectance spectra than ATR spectra [180].

7.3.5
Emerging Instrumental Methods and Quantitative Approaches

7.3.5.1 Step-scan Interferometry

Experimentation with step-scan interferometry in electrochemistry began in the early 1990s (cf. Ref. [23]), and interest has grown steadily [24, 29–31, 51–54]. Step-scan FTIR spectroscopy provides a means to investigate time- and frequency-dependent processes. Measurements are limited to reversible systems. However, a great deal of insight can be gained into the molecular transformations that accompany the external perturbation [181–183].

Step-scan measurements can be performed through different modes of spectrometer operation. Probably the simplest implementation involves changing the mirror retardation in fixed steps and recording the detector transient response to an external perturbation that is synchronized to the mirror stepping motion. In electrochemistry, this approach has been used to advantage to monitor reversible processes on timescales of a few hundred microseconds [51–54]. Another method involves applying a modulated perturbation to the sample and recovering in-phase and quadrature components of the signal at each mirror position [23, 29, 184]. For a complete mirror scan, separate interferograms for the in-phase and quadrature channels can be stored and further processed to spectra. In electrochemistry, this type of sample modulation has been performed by continuously varying the electrochemical cell potential via a sinusoidal waveform [23, 31]. More often, however, for electrochemical studies, the latter scheme has been employed with the spectrometer operating under conditions of "phase modulation", whereby one of the interferometer mirrors is dithered sinusoidally (ca. 400 Hz) between steps to produce a high-frequency periodic variation in the detector signal [23, 24, 29, 30]. In its early applications, the detector signal was fed to a lock-in amplifier referenced to the 400 Hz mirror dither frequency. The time constant was kept short (~ 40 ms) such that the output carried the low-frequency signal induced by the electrochemical cell potential modulation (typically 2–25 Hz). The signal was then sent to a second lock-in amplifier, where changes induced in the detector by the electrode potential modulation were recovered. After a settling period, the in-phase and quadrature signals at the second lock-in amplifier and the output from the first lock-in amplifier were digitized and stored. The data acquisition process was repeated at every step of the moving mirror, and an interferogram from each of the three signals was reconstructed from the stored data at the end of a complete mirror scan.

The single-beam spectrum obtained by Fourier transformation of the interferogram produced from the output of the first lock-in amplifier was used as a background for spectra generated by signals from the second lock-in amplifier. In more recent applications, the steps described above are carried out with the use of digital signal processing (DSP) electronics on board the spectrometer, so that external lock-in amplifiers are not required [24]. However, systems that employ DSP are relatively new, and limitations in the software-based processing of signals produced by electrochemical modulation have been brought to light recently [24].

The example in Fig. 7.10 demonstrates an application of step-scan interferometry in electrochemistry. Electrode potential modulation was employed, with the

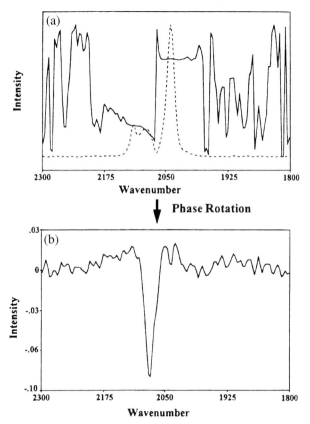

Fig. 7.10 Step-scan FTIR spectra of 10 mM ferrocyanide in 1.0 M KCl. (A) Power spectrum (dashed line) and phase spectrum (solid line) recorded by modulating the potential between the limits of 0.02 V_{SCE} and 0.42 V_{SCE} at 1 Hz. The electrode had been pretreated by cycling between -0.3 V_{SCE} and $+0.8$ V_{SCE} for 1 h to form an adsorbed hexacyanoferrate complex. The phase delay is $+1.44$ rad across the 2040 cm^{-1} band and -1.65 rad across the 2115 cm^{-1} band. (B) In-phase spectrum after phase rotation of solution ferricyanide and ferrocyanide bands into the quadrature channel. Reprinted from Ref. [30] with permission.

spectrometer operating in phase modulation mode. The method was able to uncover bands of surface species that are buried by strong features from reactive ferro- and ferricyanide molecules in solution [30]. Figure 7.10A shows the power spectrum determined from the in-phase and quadrature spectra. Three main bands are present. The strong feature at 2040 cm^{-1} is from a C≡N stretching mode of ferrocyanide in solution. The phase delay is positive and constant across the wavenumber region spanned by this band, which indicates that ferrocyanide is consumed during the potential step. The highest energy band is at 2115 cm^{-1} and arises from a C≡N stretching mode of the ferricyanide produced from the oxidation of ferrocyanide at +0.42 V$_{SCE}$ (volts versus the saturated calomel electrode (SCE) reference). As a reaction product that diffuses into solution, the phase delay is negative and constant across the band. The center feature in the power spectrum in Fig. 7.10A appears at about 2090 cm^{-1} and can be ascribed to a surface species [30]. The phase signal varies across the spectral region near 2090 cm^{-1}, a sign that the position of this band is sensitive to the potential modulation. Figure 7.10B shows the in-phase spectrum that results after the spectral signals of the solution ferro- and ferricyanide species are rotated into the quadrature channel. Since the phase across the bands of the solution species remains constant during the potential modulation, the signals from the solution species can be rotated into the quadrature spectrum, leaving only bands with a different phase delay in the in-phase spectrum [30]. Hence, the method allows strong bands of molecules in solution to be removed from *in-situ* spectra and the adsorbate bands of interest to be isolated. The strategy is likely to be employed more frequently as step-scan FTIR spectrometers gain greater use in electrochemistry.

7.3.5.2 Two-dimensional Infrared Correlation Analysis

Two-dimensional infrared (2D-IR) correlation spectroscopy is a technique for uncovering time-dependent relationships between different peaks in a dynamic spectrum (cf. Refs. [52, 183, 185–187]). The correlation analysis is often performed using the in-phase and quadrature components of the differential spectra that result from signal demodulation in response to a sinusoidal perturbation [185, 186]. Two-dimensional correlation spectra are derived from differential spectra by computing a correlation intensity function [52, 185, 186]. The function correlates the spectral intensity between two wavenumbers, $\bar{v}_1$ and $\bar{v}_2$, and is comprised of a real ($\Phi(\bar{v}_1, \bar{v}_2)$) and an imaginary ($\Psi(\bar{v}_1, \bar{v}_2)$) component, which are referred to as the synchronous and asynchronous correlations, respectively. Plotting $\Phi(\bar{v}_1, \bar{v}_2)$ on a plane defined by the two independent wavenumber axes generates a synchronous 2D correlation spectrum. A similar plot of $\Psi(\bar{v}_1, \bar{v}_2)$ gives the asynchronous 2D correlation spectrum (see Fig. 7.11B). The 2D-IR technique has been used in the manner described to investigate dynamic properties of materials on the molecular scale [182, 183, 185, 187]. For example, in the study of thin polymer films under the influence of a small-amplitude oscillatory strain, 2D-IR correlation analysis of infrared spectra measured with polarized light revealed different functional groups in the materials undergoing separate dynamic motions [182, 183, 185, 187]. Peaks in 2D spectra bring to light synchronous and asynchronous motions of functional groups in response to an external perturbation. In addition, by spreading peaks out over two dimensions, complex spectral regions containing overlapping bands can sometimes be resolved [183, 185, 187]. 2D correlation analysis is not limited to sinusoidal perturbations [52, 186]. Theory in relation to other types of waveforms has been presented [186].

In electrochemistry, Osawa and coworkers have demonstrated the usefulness of 2D-IR correlation analysis. Figure 7.11 gives an overview of an instructive application. Film formation upon reduction of heptylviologen at an Ag electrode was investigated with the SEIRAS technique. The spectra shown (Fig. 7.11 A) were recorded with a step-scan FTIR. The sequence tracks the increase in the intensity of bands arising from the heptylviologen radical cation following the onset of the reaction upon increasing the potential to –0.55 V (versus an Ag-AgCl reference electrode) from –0.2 V, where the dication is stable. The synchronous 2D correlation plot indicated that all six of the major bands between 1700 cm^{-1} and 1100 cm^{-1} increase with time simultaneously. However, the asynchronous 2D plot (Fig. 7.11 B) reveals more detailed insights into the dynamics of the reduction reaction. The correlation analysis was performed on the spectrum computed by time-averaging over the series of individual spectra that record the development of the system in response to the potential step [52]. The time-averaged spectrum is included along the top and left side of the 2D plot in Fig. 7.11 B for reference. The bands at 1636 cm^{-1} and 1184 cm^{-1} do not have an asynchronous peak connecting them. Therefore, the time-dependent intensity increase in these two bands is completely correlated. In contrast, both of the bands possess correlation peaks that connect them to the four other major bands in the spectrum at 1593 cm^{-1}, 1504 cm^{-1}, 1331 cm^{-1}, and 1161 cm^{-1}. This finding indicates that the former two and latter four bands are associated with processes that have a somewhat different time dependence. A vibrational analysis indicated that the spectral features at 1636 cm^{-1} and 1184 cm^{-1} can be assigned to dipyridine ring modes of the heptylviologen cation monomer, while the remaining bands are associated with a radical dimer [52]. The asynchronous 2D-IR spectra indicate that the monomer and dimer are formed at different rates. Also notable is that the 2D spectral peaks have positive or negative signs. The shaded regions in the plot indicate negative-intensity regions. In an asynchronous correlation plot, negative intensity in a peak means that the intensity fluctuation at $\bar{v}_1$ occurs after that at $\bar{v}_2$ [52, 185]. Thus, the correlation analysis indicates that the monomer, with bands at 1636 cm^{-1} and 1184 cm^{-1}, forms ahead of the dimer species. This finding is consistent with a nucleation and growth model in which the monomer forms rapidly during the nucleation period, after which the dimer becomes the dominant species as film growth proceeds [52].

The asynchronous correlation analysis also uncovered several minor spectral peaks. Shoulders on the main bands in the region between 1600 cm^{-1} and

(a)

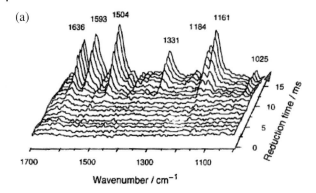

(b)

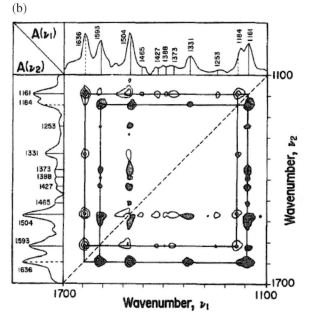

Fig. 7.11 (a) Series of time-resolved spectra showing the generation of heptylviologen radical cation at an Ag film electrode in 0.3 M KBr containing 1 mM heptylviologen dibromide. Spectra were recorded with a step-scan FTIR spectrometer after stepping the potential from –0.2 V to –0.55 V (versus an Ag-AgCl reference electrode). (b) Asynchronous 2D correlation spectra of electrochemically generated heptylviologen. The shaded areas represent negative intensity regions. Adapted from Ref. [52] with permission.

1450 cm^{-1} were resolved and found to be associated with two different time-dependent processes. These weaker bands could be assigned to changes in the crystal structure of the film [52].

Osawa and coworkers also applied 2D-IR correlation analysis in the study of interfacial water and electrolyte at Au electrodes [89]. It was possible to enhance resolution in the O-H stretching region and show a relationship between the movement of interfacial water and sulfate ions with changes in potential. While its use has been somewhat limited, 2D-IR correlation analysis has demonstrated the capability to become a valuable tool for the study of electrode reactions.

7.3.5.3 Quantitation of Molecular Orientation

The ability to detect monolayer and sub-monolayer quantities of organic molecules on electrodes with high SNR is enabling quantitative methods to be applied to determine the average orientation of species at the electrode/solution interface (Refs. [42, 104], and see Chapter 8 of this volume). The approaches being used were established through studies of adsorbates at gas/solid interfaces by infrared reflectance measurements (cf. Refs. [3, 93]). A general strategy involves simulating the reflectance across the wavelength range of interest by employing the Fresnel equations and a multi-phase model of the interface [93, 104]. The method has proved to be powerful for gaining insights into spectral band distortions and predicting conditions for high detectability [3, 93, 104, 188]. Comparison of measured and simulated spectra provides information about molecular orientation. For molecules on reflective metal surfaces, the angle of a transition dipole moment relative to the surface normal can be calculated from the integrated intensities of the experimental and simulated bands for the mode [42, 93, 103].

In the spectroelectrochemical experiment, the simulation of reflectivity responses can become quite complicated owing to the multiple interfaces and phases through which the optical beam must pass [104]. However, the models employed have gained in sophistication over the years as they have been needed to assist in the design of cells and optical systems and in the interpretation of spectra [23, 80, 104, 189–193]. An expanded discussion of the computational approach and its application to a four-phase model of the infrared spectroelectrochemical experiment was presented by Pettinger and coworkers recently [104]. An example of its use and the type of detailed structural information that can be gained is described for the case of pyridine adsorption on an Au electrode [103].

Figure 7.12 shows an *in-situ* infrared spectrum of pyridine on Au(110) together with a transmission infrared spectrum of neat pyridine for reference. The vibrational mode symmetry assignments for some of the peaks are indicated. The *in-situ* spectrum is displayed as the reflectivity difference between the background and sample potentials. The background spectrum was collected at a potential (–0.75 V_{SCE}) where pyridine is desorbed from the surface and is present only in solution. At the sample potential of 0.4 V_{SCE}, pyridine is adsorbed at near-saturation coverages on Au(110). Sharp downward bands in the *in-situ* spectrum labeled with a_1 symmetry arise from ring mode vibrations of adsorbed pyridine, while bands pointing upward arise from excess pyridine in solution following desorption [103].

The vibrational modes detected for adsorbed pyridine are of a_1 symmetry. The inset to Fig. 7.12 shows that the transition dipole moments are directed along

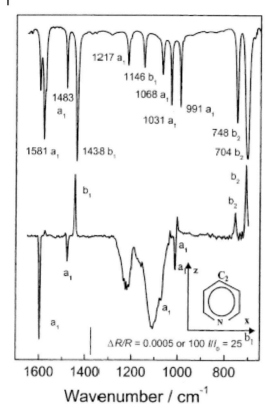

Fig. 7.12 Transmission infrared spectrum of neat pyridine (top) and an *in-situ* infrared spectrum for an Au(110) electrode in D_2O containing 0.1 M $KClO_4$ + 1 mM pyridine (bottom) are shown. The y-axis scale on the *in-situ* spectrum reports the differential reflectivity of the system; upward bands are associated with species present at the reference potential of –0.75 V_{SCE} and downward bands reflect the system at the sample potential of 0.40 V_{SCE}. Symmetry assignments for vibrational modes of certain bands are indicated. The inset shows directions of pyridine transition dipole moments with a_1 and b_1 symmetry. Reprinted from Ref. [103] with permission.

the C_2 axis of pyridine for a_1 modes and in the plane of the ring, but perpendicular to the C_2 axis for b_1 modes. For b_2 modes, the transition dipole moments are orthogonal to those for both the a_1 and b_1 modes. With consideration of the dipole selection rule for molecules adsorbed at the surface of a reflective metal [1, 2, 93, 104], the spectrum suggests that pyridine adopts an orientation whereby the C_2 axis has a significant component along the surface normal. Furthermore, small shifts in band position relative to the solution-free molecule indicate that pyridine coordinates to the Au surface through its nitrogen atom [103].

The relative intensities of the upward and downward bands in the *in-situ* infrared spectrum in Fig. 7.12 can be interpreted by considering general relationships that govern band intensity and infrared absorption strength. Equation

(7.4) shows that the absorbance (A) of a vibrational band integrated over its wavenumber range is proportional to the square of the dot product of the transition dipole moment (μ) and the electric field of the radiation (E):

$$\int A d\bar{v} \propto |\mu \cdot E|^2 \propto \cos^2 \theta |\mu|^2 \langle E^2 \rangle \tag{4}$$

where $\langle E^2 \rangle$ is the mean squared electric field (MSEF) of the photon and θ is the angle between the vectors for the radiation electric field and the molecular dipole moment. At a metal surface, the resulting electric field vector for reflected radiation lies along the normal, which is defined here as the z direction. The MSEF at the surface (z=0) can be written as $\langle E^2(z=0) \rangle$, and at an arbitrary position a distance into solution the MSEF in the z-direction can be given by $\langle E^2(z) \rangle$. Bands in the reflectance difference spectrum recorded with p-polarized radiation will have upward or downward directions depending upon the orientation of transition dipoles with respect to the z-component of the MSEF at the reference and sample potentials. Li and coworkers showed that the following relationship can be used to understand and predict the appearance of these spectra [103]:

$$\frac{\Delta R}{R} \propto \left[\frac{1}{3} \frac{\partial |\mu(E_{\mathrm{Ref}})|^2}{\partial \bar{v}} \langle E^2_{\mathrm{E\,Ref}}(z) \rangle - \cos^2 \theta \frac{\partial |\mu(E_{\mathrm{samp}})|^2}{\partial \bar{v}} \langle E^2_{\mathrm{E\,samp}}(z = 0) \rangle \right] \tag{5}$$

where E_{Ref} and E_{samp} refer to the background and sample potentials, respectively. The relationship assumes that molecules are adsorbed on the surface at the sample potential and desorbed into homogeneous solution at the reference potential. The factor of 1/3 accounts for the fact that in homogeneous solution the molecules adopt a random orientation. Furthermore, the MSEF is expected to be a few times greater at z=0 than in solution because of the well-known enhancement effect at the metal surface [1, 2, 93]. Since it is reasonable to assume for pyridine that the magnitude of the transition dipole moment derivative term is the same for the molecule whether it is in free solution or adsorbed on Au, this term in Eq. (5) can be ignored. Thus, positive bands are expected when $\cos^2\theta \ll 1/3$, which will be the case when the transition dipole moment of the adsorbate is far from the surface normal. On the other hand, negative bands are expected when $\cos^2\theta > 1/3$, indicating that the angle between the transition dipole and the surface normal is small.

In Fig. 7.12, the appearance of positive bands in the *in-situ* spectrum implies that these modes are not strongly sensed by the infrared radiation field when the molecules are in the adsorbed state. These modes, which are of b_1 and b_2 symmetry, are expected to lie along the surface plane, for the most part, when the angle between the C_2 axis of pyridine and the surface normal is small. Thus, the appearance of positive bands for b_1 and b_2 symmetry species and negative bands for modes having a_1 symmetry is consistent with the picture of pyridine bonded to the surface through the ring nitrogen atom with the C_2 axis perpendicular to the surface plane.

Li and coworkers developed a more quantitative description of the adsorption geometry by comparing the measured spectrum to spectra simulated using their Fresnel model for the spectroelectrochemical experiment [103]. Results of the analysis indicate that pyridine molecules on Au(110) adopt a tilt angle relative to the surface normal of approximately 15° at 0.4 V_{SCE}. Further study showed that the tilt angle for pyridine on Au(110) increases essentially linearly with potential between 0.4 V_{SCE} and –0.3 V_{SCE} and reaches about 35° at –0.3 V_{SCE}.

The quantitative infrared studies of pyridine orientation on Au electrodes outline details of procedures that can be applied generally to gain insights into molecular structure at electrode/solution interfaces [103, 104]. These and related ideas have been extended to other systems more recently [42, 109–111]. The methods provide a powerful means to probe molecular orientation at electrodes.

7.4
Summary

Infrared spectroscopy has long been regarded as valuable for the study of interfacial electrochemistry. The earliest work focused on molecular adsorption and electrocatalysis of CO and small organic molecules [1, 2, 4, 14, 194]. Methods for data acquisition were established, and in addition to spectral band assignments, connections were made between the positions and shapes of bands and general properties of the environment at the electrode/solution interface. Features were identified as characteristic of CO adsorbed on electrodes in high coverage, densely packed structures versus small, dispersed islands, for example [14, 45, 46, 57, 194]. With developments in bench-top methods for the preparation and handling of single-crystal electrodes [195, 196], the ability to perform *in-situ* infrared spectroscopy measurements on well-defined surfaces leads to greater understanding of relationships between spectral properties and interfacial structure. Advances in this direction were accelerated by the demonstration of *in-situ* STM, which provided an ability to determine real-space structures of adlattices [55]. The application of *in-situ* STM to CO adsorption on Pt(111) electrodes revealed adlattices that had not been predicted, and the findings enabled theoretical analyses to explain quantitative aspects of spectra, such as non-ideal Beer's law effects [55, 56, 60, 68]. The progress stimulated additional theoretical work, and significant strides were made in understanding the origins of spectral perturbations that result from changes in electrode potential, including those derived from surface-adsorbate chemical bonding and external electric fields at the electrode/solution interface [58].

The present chapter aims to highlight recent applications of these approaches in the study of adsorption and catalysis at well-defined electrode surfaces and to describe related areas where important developments are occurring. Infrared spectroscopy continues to be widely used as a probe of CO surface electrochemistry, both when CO is the starting reactant and when it is formed as an intermediate. Investigations involving CO electrochemistry are targeting bimetallic as well as pure metal single-crystal surfaces, and experiments are combining additional *in-situ* structural techniques, such as STM and SXS. Also coming into greater use are methods for obtaining infrared spectra of species on metal nano-particle electrocatalysts.

With the emergence of SEIRAS, there have been improvements in sensitivity and selectivity toward surface species in infrared spectroelectrochemistry. Factors that limit traditional external reflection sampling methods, such as interferences from bulk solution species and low sensitivity toward organic adsorbates, are often overcome through SEIRAS measurements. Advances have been made in detecting the structure of solvent and electrolyte ions within the first few layers of the electrode surface, and improved sensitivity toward organic adsorbates and intermediates of electrocatalytic reactions has been demonstrated. Time-resolved studies on the 100-µs scale have also been performed with SEIRAS. Many more applications of SEIRAS are expected in the future. Another area where significant gains are being made in infrared spectroelectrochemistry is in quantitation of molecular orientation on single-crystal electrodes. Reports that demonstrate Fresnel calculations on multi-phase models of the thin-layer optical region in the spectroelectrochemical cell are providing guidance for determining tilt angles and general information about molecular orientation. In terms of instrumentation, the availability of step-scan FTIR systems is renewing interest in electromodulation and stimulating investigations into the use of 2D correlation spectroscopy for data analysis. Further development of these approaches will be valuable for approaching more complex problems in areas such as electrocatalysis.

Acknowledgments

Support from the U.S. National Science Foundation for the author's work in the area of infrared spectroelectrochemistry is gratefully acknowledged.

References

1 A. Bewick, B.S. Pons, In *Advances in Infrared and Raman Spectroscopy*, R.J.H. Clark and R.E. Hester (Eds.), Wiley, London, Vol. 12, p. 1 (1985).

2 B. Beden, C. Lamy, In *Spectroelectrochemistry: Theory and Practice*, R.J. Gale, (Ed.) Plenum Press, New York, p. 189 (1988).

3 S.M. Stole, D.D. Popenoe, M.D. Porter, In *Electrochemical Interfaces. Modern Techniques for In-Situ Interface Characterization*, H.D. Abruna (Ed.), VCH Publishers, Inc., New York, p. 339 (1991).

4 B. Beden, J.-M. Leger, C. Lamy, In *Modern Aspects of Electrochemistry*, J.O.M. Bockris, B.E. Conway and R.E. White (Eds.), Plenum, New York, Vol. 22, p. 97 (1992).

5 R.J. Nichols, In *Adsorption of Molecules at Metal Electrodes*, J. Lipkowski and P.N. Ross (Eds.), VCH Publishers, New York, Chapter 7, p. 347 (1992).

6 T. Iwasita, F.C. Nart, In *Advances in Electrochemical Science and Engineering*, H. Gerischer and C. Tobias, (Eds.), VCH

Publishers, New York, Vol. 4, p. 123 (1995).

7 C. Korzeniewski, *Critical Reviews in Analytical Chemistry,* **1997**, *27*, 81.

8 T. Iwasita and E. Pastor, In *Interfacial Electrochemistry: Theory, Experiment and Applications,* A. W. Wieckowski (Ed.), Marcel Dekker, New York, p. 353 (1999).

9 C. Korzeniewski, In *Handbook of Vibrational Spectroscopy,* J. M. Chalmers and P. R. Griffiths (Eds.), Wiley, New York, Vol. 4, p. 2699 (2002).

10 F. Kitamura, M. Takeda, M. Takahashi, M. Ito, *Chem. Phys. Lett.,* **1987**, *142*, 318.

11 F. Kitamura, M. Takahashi, M. Ito, *J. Phys. Chem.,* **1988**, *92*, 3320.

12 S.-C. Chang, L.-W. H. Leung, M. J. Weaver, *J. Phys. Chem.,* **1989**, *93*, 5341.

13 S. C. Chang, M. J. Weaver, *J. Phys. Chem.,* **1991**, *95*, 5391.

14 C. Korzeniewski, In *Interfacial Electrochemistry: Theory, Experiment and Applications,* A. Wieckowski (Ed.), Dekker, New York, p. 345 (1999).

15 K. A. Friedrich, F. Henglein, U. Stimming, W. Unkauf, *Colloids Surf. A.: Physiochem. Eng. Asp.,* **1998**, *134*, 193.

16 K. A. Friedrich, K. P. Geyzers, A. J. Dickinson, U. Stimming, *J. Electroanal. Chem.,* **2002**, *524/525*, 261.

17 S. Park, S. A. Wasileski, M. J. Weaver, *J. Phys. Chem. B,* **2001**, *105*, 9719.

18 S. Park, A. Wieckowski, M. J. Weaver, *J. Am. Chem. Soc.,* **2003**, *125*, 2282.

19 F. Maillard, E. R. Savinova, P. A. Simonov, V. I. Zaikovskii, U. Stimming, *J. Phys. Chem. B,* **2004**, *108*, 17893.

20 (Special Volume on Thin Organic Films) *J. Electroanal. Chem.,* J. S. E. Lipkowski (Ed.), **2003**; Vol. 550/551.

21 Y. X. Chen, A. Miki, S. Ye, H. Sakai, M. Osawa, *J. Am. Chem. Soc.,* **2003**, *125*, 3680.

22 K. Ataka, T. Yotsuyanagi, M. Osawa, *J. Phys. Chem.,* **1996**, *100*, 10664.

23 B. O. Budevska, P. R. Griffiths, *Anal. Chem.,* **1993**, *65*, 2963.

24 D. A. Brevnov, E. Hutter, J. H. Fendler, *Appl. Spectrosc.,* **2004**, *58*, 184.

25 I. T. Bae, M. Sandifer, Y. W. Lee, D. A. Tryk, C. N. Sukenik, D. A. Scherson, *Anal. Chem.,* **1995**, *67*, 4508.

26 M. Osawa, K. Yoshii, *Appl. Spectrosc.,* **1997**, *51*, 512.

27 S. M. Moon, C. Bock, B. MacDougall, *J. Electroanal. Chem.,* **2004**, *568*, 225.

28 M. Osawa, K. Ataka, K. Yoshi, Y. Nishikawa, *Appl. Spectrosc.,* **1993**, *9*, 1497.

29 C. M. Pharr, P. R. Griffiths, *Anal. Chem.,* **1997**, *69*, 4665.

30 C. M. Pharr, P. R. Griffiths, *Anal. Chem.,* **1997**, *69*, 4673.

31 K. Ataka, Y. Hara, M. Osawa, *J. Electroanal. Chem.,* **1999**, *473*, 34.

32 E. N. Lewis, P. J. Treado, R. C. Reeder, G. M. Story, A. E. Dowrey, C. Marcott, I. W. Levin, *Anal. Chem.,* **1995**, *67*, 3377.

33 R. Bhargava, I. W. Levin, *Anal. Chem.,* **2001**, *73*, 5157.

34 C. Pellerin, C. M. Snively, D. B. Chase, J. F. Rabolt, *Appl. Spectrosc.,* **2004**, *58*, 639.

35 P. R. Griffiths, J. A. de Haseth, *Fourier Transform Infrared Spectrometry,* Wiley, New York, 1986, Vol. 83.

36 J. D. Ingle, Jr., S. R. Crouch, *Spectrochemical Analysis,* Prentice Hall, New Jersey, 1988.

37 A. E. Dowrey, C. Marcott, *Appl. Spectrosc.,* **1982**, *6*, 414.

38 L. A. Nafie, D. W. Vidrine, In *Fourier Transform Infrared Spectroscopy,* J. R. Ferraro and L. J. Basile (Eds.), Academic Press, New York, Vol. 3, p. 83 (1982).

39 W. G. Golden, In *Fourier Transform Infrared Spectroscopy,* J. R. Ferraro and L. J. Basile (Eds.), Academic Press, New York, Vol. 4, p. 315 (1985).

40 B. J. Barner, M. J. Green, E. I. Saez, R. M. Corn, *Anal. Chem.,* **1991**, *63*, 55.

41 M. J. Green, B. J. Barner, R. M. Corn, *Rev. Sci. Instrum.,* **1991**, *62*, 1426.

42 V. Zamlynny, I. Zawisza, J. Lipkowski, *Langmuir,* **2003**, *19*, 132.

43 J. W. Russell, J. Overend, K. Scanlon, M. Severson, A. Bewick, *J. Phys. Chem.,* **1982**, *86*, 3066.

44 J. W. Russell, M. Severson, K. Scanlon, J. Overend, A. Bewick, *J. Phys. Chem.,* **1983**, *87*, 293.

45 K. Kunimatsu, W. G. Golden, H. Seki, M. R. Philpott, *Langmuir,* **1985**, *1*, 245.

46 K. Kunimatsu, H. Seki, W. G. Golden, J. G. Gordon, II, M. R. Philpott, *Langmuir,* **1986**, *2*, 464.

47 E. I. Saez, R. M. Corn, *Electrochim. Acta*, **1993**, *38*, 1619.

48 W. N. Richmond, P. W. Faguy, R. S. Jackson, S. C. Weibel, *Anal. Chem.*, **1996**, *68*, 621.

49 P. W. Faguy, W. N. Richmond, *J. Electroanal. Chem.*, **1996**, *410*, 109.

50 F. Ozanam, J.-N. Chazalveil, *Rev. Sci. Instrum.*, **1988**, *59*, 242.

51 M. Osawa, K. Yoshii, K. Ataka, T. Yotsuyanagi, *Langmuir*, **1994**, *10*, 640.

52 M. Osawa, K. Yoshii, Y. Hibino, T. Nakano, I. Noda, *J. Electroanal. Chem.*, **1997**, *426*, 11.

53 H. Noda, K. Ataka, L.-J. Wan, M. Osawa, *Surf. Sci.*, **1999**, *427/428*, 190.

54 A. Rodes, J. M. Orts, J. M. Perez, J. M. Feliu, A. Aldaz, *Electrochem. Commun.*, **2003**, *5*, 56.

55 I. Villegas, M. J. Weaver, *J. Chem. Phys.*, **1994**, *101*, 1648.

56 M. W. Severson, C. Stuhlmann, I. Villegas, M. J. Weaver, *J. Chem. Phys.*, **1995**, *103*, 9832.

57 S. C. Chang, M. J. Weaver, *J. Chem. Phys.*, **1990**, *92*, 4582.

58 S. A. Wasileski, M. T. M. Koper, M. J. Weaver, *J. Chem. Phys.*, **2001**, *115*, 8193.

59 S. C. Chang, J. D. Roth, M. J. Weaver, *Surf. Sci.*, **1991**, *244*, 113.

60 M. W. Severson, M. J. Weaver, *Langmuir*, **1998**, *14*, 5603.

61 C. S. Kim, C. Korzeniewski, W. J. Tornquist, *J. Chem. Phys.*, **1994**, *100*, 628.

62 C. S. Kim, W. J. Tornquist, C. Korzeniewski, *J. Chem. Phys.*, **1994**, *101*, 9113.

63 C. S. Kim, C. Korzeniewski, *Anal. Chem.*, **1997**, *69*, 2349.

64 J. Shin, C. Korzeniewski, *J. Phys. Chem.*, **1995**, *99*, 3419.

65 N. P. Lebedeva, A. Rodes, J. M. Feliu, M. T. M. Koper, R. A. van Santen, *J. Phys. Chem. B*, **2002**, *106*, 9863.

66 S. Z. Zou, R. Gomez, M. J. Weaver, *Surf. Sci.*, **1998**, *399*, 270.

67 S. Z. Zou, R. Gomez, M. J. Weaver, *Langmuir*, **1999**, *15*, 2931.

68 C. Tang, S. Zou, M. W. Severson, M. J. Weaver, *J. Phys. Chem.*, **1998**, *102*, 8796.

69 R. Gomez, M. J. Weaver, *Langmuir*, **1998**, *14*, 2525.

70 R. Gomez, J. M. Feliu, A. Aldaz, M. J. Weaver, *Surf. Sci.*, **1998**, *410*, 48.

71 R. Gomez, M. J. Weaver, *J Phys. Chem. B*, **1998**, *102*, 3754.

72 N. Ikemiya, T. Senna, M. Ito, *Surf. Sci.*, **2000**, *464*, L681.

73 N. S. Marinkovic, J. X. Wang, H. Zajonz, R. R. Adzic, *Electrochem. Solid State Lett.*, **2000**, *3*, 508.

74 W. F. Lin, P. A. Christensen, A. Hamnett, M. S. Zei, G. Ertl, *J. Phys. Chem. B*, **2000**, *104*, 6642.

75 N. Ikemiya, T. Suzuki, M. Ito, *Surf. Sci.*, **2000**, *466*, 119.

76 M. Nakamura, N. Ikemiya, A. Iwasaki, Y. Suzuki, M. Ito, *J. Electroanal. Chem.*, **2004**, *566*, 385.

77 B. B. Blizanac, M. Arenz, P. N. Ross, N. M. Markovic, *J. Am. Chem. Soc.*, **2004**, *126*, 10130.

78 Y. V. Tolmachev, A. Menzel, A. V. Tkachuk, Y. S. Chu, H. D. You, *Electrochem. Solid State Lett.*, **2004**, *7*, E23.

79 S.-G. Sun, W.-B. Cai, L.-J. Wan, M. Osawa, *J. Phys. Chem. B*, **1999**, *103*, 2460.

80 P. W. Faguy, N. S. Marinkovic, *Anal. Chem.*, **1995**, *67*, 2791.

81 P. W. Faguy, N. S. Marinkovic, R. R. Adzic, *Langmuir*, **1996**, *12*, 243.

82 A. Lachenwitzer, N. Li, J. Lipkowski, *J. Electroanal. Chem.*, **2002**, *532*, 85.

83 N. Hoshi, M. Kuroda, T. Ogawa, O. Koga, Y. Hori, *Langmuir*, **2004**, *20*, 5066.

84 N. Hoshi, A. Sakurada, S. Nakamura, S. Teruya, O. Koga, Y. Hori, *J. Phys. Chem. B*, **2002**, *106*, 1985.

85 I. R. Moraes, F. C. Nart, *J. Electroanal. Chem.*, **2004**, *563*, 41.

86 T. Senna, N. Ikemiya, M. Ito, *J. Electroanal. Chem.*, **2001**, *511*, 115.

87 N. S. Marinkovic, J. X. Wang, H. Zajonz, R. R. Adzic, *J. Electroanal. Chem.*, **2001**, *500*, 388.

88 Y. Shingaya, M. Ito, *J. Electroanal. Chem.*, **1999**, *467*, 299.

89 K. Ataka, M. Osawa, *Langmuir*, **1998**, *14*, 951.

90 T. Wandlowski, K. Ataka, S. Pronkin, D. Diesing, *Electrochim. Acta*, **2004**, *49*, 1233.

91 K. Arihara, F. Kitamura, T. Ohsaka, K. Tokuda, *J. Electroanal. Chem.*, **2001**, *510*, 128.

92 A. Berna, A. Rodes, J. M. Feliu, F. Illas, A. Gil, A. Clotet, J. M. Ricart, *J. Phys. Chem. B*, **2004**, *108*, 17928.

93 M. D. Porter, *Anal. Chem.*, **1988**, *60*, 1143A.

94 D. S. Corrigan, E. K. Krauskopf, L. M. Rice, A. Wieckowski, M. J. Weaver, *J. Phys. Chem.*, **1988**, *92*, 1596.

95 J. Shin, W. J. Tornquist, C. Korzeniewski, C. S. Hoaglund, *Surf. Sci.*, **1996**, *364*, 122.

96 A. Berna, A. Rodes, J. M. Feliu, *Electrochim. Acta*, **2004**, *49*, 1257.

97 A. Berna, A. Rodes, J. M. Feliu, *J. Electroanal. Chem.*, **2004**, *563*, 49.

98 R. J. Nichols, I. Burgess, K. L. Young, V. Zamlynny, J. Lipkowski, *J. Electroanal. Chem.*, **2004**, *563*, 33.

99 S. Pronkin, T. Wandlowski, *J. Electroanal. Chem.*, **2003**, *550*, 131.

100 R. Albalat, J. Claret, A. Rodes, J. M. Feliu, *J. Electroanal. Chem.*, **2003**, *550*, 53.

101 V. Climent, A. Rodes, J. M. Perez, J. M. Feliu, A. Aldaz, *Langmuir*, **2002**, *16*, 10376.

102 V. Climent, A. Rodes, R. Albalat, J. Claret, J. M. Feliu, A. Aldaz, *Langmuir*, **2001**, *17*, 8260.

103 N. Li, V. Zamlynny, J. Lipkowski, F. Henglein, B. Pettinger, *J. Electroanal. Chem.*, **2002**, *524/525*, 43.

104 B. Pettinger, J. Lipkowski, M. Hoon-Khosla, *J. Electroanal. Chem.*, **2001**, *500*, 471.

105 M. Hoon-Khosla, W. R. Fawcett, J. D. Goddard, W.-Q. Tian, J. Lipkowski, *Langmuir*, **2000**, *16*, 2356.

106 A. Chen, D. Yang, J. Lipkowski, *J. Electroanal. Chem.*, **1999**, *475*, 130.

107 M. Hoon-Khosla, W. R. Fawcett, A. Chen, J. Lipkowski, B. Pettinger, *Electrochim. Acta*, **1999**, *45*, 611.

108 I. Zawisza, J. Lipkowski, *Langmuir*, **2004**, *20*, 4579.

109 T. Doneux, C. Buess-Herman, J. Lipkowski, *J. Electroanal. Chem.*, **2004**, *564*, 65.

110 X. M. Bin, I. Zawisza, J. D. Goddard, J. Lipkowski, *Langmuir*, **2005**, *21*, 330.

111 I. Zawisza, X. M. Bin, J. Lipkowski, *Bioelectrochemistry*, **2004**, *63*, 137.

112 M. Nakamura, K. Matsunaga, K. Kitahara, M. Ito, O. Sakata, *J. Electroanal. Chem.*, **2003**, *554*, 175.

113 A. Gil, A. Clotet, J. M. Ricart, F. Illas, B. Alvarez, A. Rodes, J. M. Feliu, *J. Phys. Chem. B*, **2001**, *105*, 7263.

114 J. Inukai, M. Ito, *J. Electroanal. Chem.*, **1993**, *358*, 307.

115 V. R. Stamenkovic, M. Arenz, C. A. Lucas, M. E. Gallagher, P. N. Ross, N. M. Markovic, *J. Am. Chem. Soc.*, **2003**, *125*, 2736.

116 V. Stamenkovic, M. Arenz, B. B. Blizanac, K. J. J. Mayrhofer, P. N. Ross, N. M. Markovic, *Surf. Sci.*, **2005**, *576*, 145.

117 M. Arenz, V. Stamenkovic, P. N. Ross, N. M. Markovic, *Surf. Sci.*, **2004**, *573*, 57.

118 M. Arenz, V. Stamenkovic, T. J. Schmidt, K. Wandelt, P. N. Ross, N. M. Markovic, *Phys. Chem. Chem. Phys.*, **2003**, *5*, 4242.

119 M. Arenz, V. Stamenkovic, P. N. Ross, N. M. Markovic, *Electrochem. Commun.*, **2004**, *5*, 809.

120 R. Gomez, J. M. Feliu, *Electrochim. Acta*, **1998**, *44*, 1191.

121 B. Alvarez, A. Rodes, J. M. Perez, J. M. Feliu, *J. Phys. Chem. B*, **2003**, *107*, 2018.

122 E. Casado-Rivera, D. J. Volpe, L. Alden, C. Lind, C. Downie, T. Vazquez-Alvarez, A. C. D. Angelo, F. J. DiSalvo, H. D. Abruna, *J. Am. Chem. Soc.*, **2004**, *126*, 4043.

123 V. Stamenkovic, T. J. Schmidt, P. N. Ross, N. M. Markovic, *J. Phys. Chem. B*, **2002**, *106*, 11970.

124 B. E. Hayden, M. E. Rendall, O. South, *J. Am. Chem. Soc.*, **2003**, *125*, 7738.

125 A. Hatta, T. Ohshima, W. Suetaka, *Appl. Phys. A*, **1982**, *29*, 71.

126 A. Hartstein, J. R. Kirtley, J. C. Tsang, *Phys. Rev. Lett.*, **1980**, *45*, 201.

127 H. Miyake, S. Ye, M. Osawa, *Electrochem. Commun.*, **2002**, *4*, 973.

128 M. Osawa, K. Ataka, *Surf. Sci.*, **1992**, *262*, L118.

129 A. Hatta, Y. Suzuki, W. Suetaka, *Appl. Phys. A*, **1984**, *35*, 135.

130 M. Osawa, In *Handbook of Vibrational Spectroscopy*, J. M. Chalmers and P. R. Griffiths (Eds.), Wiley, Chichester, Vol. 1, p. 785 (2002).

131 D. S. Corrigan, M. J. Weaver, *J. Phys. Chem.*, **1986**, *90*, 5300.

132 D. S. Corrigan, M. J. Weaver, *Langmuir,* **1988**, *4*, 599.

133 D. S. Corrigan, M. J. Weaver, *J. Electroanal. Chem.,* **1988**, *239*, 55.

134 J. D. Roth, M. J. Weaver, *Anal. Chem.,* **1991**, *63*, 1603.

135 I. T. Bae, D. A. Scherson, E. B. Yeager, *Anal. Chem.,* **1990**, *62*, 45.

136 F. Maroun, F. Ozanam, J.-N. Chazalviel, *J. Electroanal. Chem.,* **1997**, *435*, 225.

137 A. Miki, S. Ye, M. Osawa, *Chem. Commun.,* **2002**, 1500.

138 A. Miki, S. Ye, T. Senzaki, M. Osawa, *J. Electroanal. Chem.,* **2004**, *563*, 23.

139 T. Yajima, H. Uchida, M. Watanabe, *J. Phys. Chem. B,* **2004**, *108*, 2654.

140 M. Watanabe, Y. Zhu, H. Uchida, *J. Phys. Chem. B,* **2000**, *104*, 1762.

141 Y. Zhu, H. Uchida, M. Watanabe, *Langmuir,* **1999**, *15*, 8757.

142 Y. M. Zhu, H. Uchida, T. Yajima, M. Watanabe, *Langmuir,* **2001**, *17*, 146.

143 R. Parsons, T. VanderNoot, *J. Electroanal. Chem.,* **1988**, *257*, 9.

144 T. D. Jarvi, E. M. Stuve, In *Electrocatalysis,* J. Lipkowski and P. N. Ross (Eds.), Wiley-VCH Publishers, New York, Vol. 3, p. 75 (1998).

145 K. Kunimatsu, T. Senzaki, M. Tsushima, M. Osawa, *Chem. Phys. Lett.,* **2005**, *401*, 451.

146 Y. Sato, F. Mizutani, *Phys. Chem. Chem. Phys.,* **2004**, *6*, 1328.

147 D. Mayer, T. Dretschkow, K. Ataka, T. Wandlowski, *J. Electroanal. Chem.,* **2002**, *524*, 20.

148 T. Wandlowski, K. Ataka, D. Mayer, *Langmuir,* **2002**, *18*, 4331.

149 N. Goutev, M. Futamata, *Appl. Spectrosc.,* **2003**, *57*, 506.

150 Y. Sato, H. Noda, F. Mizutani, A. Yamakata, M. Osawa, *Anal. Chem.,* **2004**, *76*, 5564.

151 K. Ataka, F. Giess, W. Knoll, R. Naumann, S. Haber-Pohlmeier, B. Richter, J. Heberle, *J. Am. Chem. Soc.,* **2004**, *126*, 16199.

152 T. M. Cotton, In *Spectroscopy of Surfaces,* R. J. H. Clark and R. E. Hester (Eds.), Wiley, New York, Vol. 16, p. 91 (1988).

153 C. E. Jordan, B. L. Frey, S. Kornguth, R. M. Corn, *Langmuir,* **1994**, *10*, 3642.

154 C. E. Jordan, A. G. Frutos, A. J. Thiel, R. M. Corn, *Anal. Chem.,* **1997**, *69*, 4939.

155 P. A. Christensen, A. Hamnett, J. Munk, G. L. Troughton, *J. Electroanal. Chem.,* **1994**, *370*, 251.

156 K. A. Friedrich, F. Henglein, U. Stimming, W. Unkauf, *Electrochim. Acta,* **2000**, *45*, 3283.

157 K. A. Friedrich, F. Henglein, U. Stimming, W. Unkauf, *Electrochim. Acta,* **2001**, *47*, 689.

158 S. Park, Y. Y. Tong, A. Wieckowski, M. J. Weaver, *Electrochem. Commun.,* **2001**, *3*, 509.

159 S. Park, Y. Y. Tong, A. Wieckowski, M. J. Weaver, *Langmuir,* **2002**, *18*, 3233.

160 S. Park, Y. Xie, M. J. Weaver, *Langmuir,* **2002**, *18*, 5792.

161 V. Stamenkovic, M. Arenz, P. N. Ross, N. M. Markovic, *J. Phys. Chem. B,* **2004**, *108*, 17915.

162 P. Waszczuk, J. Solla-Gullon, H. S. Kim, Y. Y. Tong, V. Montiel, A. Aldaz, A. Wieckowski, *J. Catal.,* **2001**, *203*, 1.

163 Y. Y. Tong, H. S. Kim, P. K. Babu, P. Waszczuk, A. Wieckowski, E. Oldfield, *J. Am. Chem. Soc.,* **2002**, *124*, 468.

164 G. Q. Lu, J. O. White, A. Wieckowski, *Surf. Sci.,* **2004**, *564*, 131.

165 C. S. Kim, W. J. Tornquist, C. Korzeniewski, *J. Phys. Chem.,* **1993**, *97*, 6484.

166 M. Arenz, K. J. J. Mayrhofer, V. Stamenkovic, B. B. Blizanac, T. Tomoyuki, P. N. Ross, N. M. Markovic, *J. Am. Chem. Soc.,* **2005**, *127*, 6819.

167 M. J. Weaver, *Appl. Surf. Sci.,* **1993**, *67*, 147.

168 K. J. J. Mayrhofer, B. B. Blizanac, M. Arenz, V. R. Stamenkovic, P. N. Ross, N. M. Markovic, *J. Phys. Chem. B,* **2005**, *109*, 14433.

169 V. Stamenkovic, K. C. Chou, G. A. Somorjai, P. N. Ross, N. M. Markovic, *J. Phys. Chem. B,* **2005**, *109*, 678.

170 Y.-J. Chen, S.-G. Sun, S.-P. Chen, J.-T. Li, H. Gong, *Langmuir,* **2004**, *20*, 9920.

171 C.-X. Wu, H. Lin, Y.-J. Chen, W.-X. Li, S.-G. Sun, *J. Chem. Phys.,* **2004**, *121*, 1553.

172 C. Pecharroman, A. Cuesta, C. Gutierrez, *J. Electroanal. Chem.,* **2004**, *563*, 91.

173 W. Chen, S.-G. Sun, Z.-Y. Zhou, S.-P. Chen, *J. Phys. Chem. B*, **2003**, *107*, 9808.

174 D. Ross, R. Aroca, *J. Chem. Phys.*, **2002**, *117*, 8095.

175 A. E. Bjerke, P. R. Griffiths, *Anal. Chem.*, **1999**, *71*, 1967.

176 G.-Q. Lu, S.-G. Sun, L.-R. Cai, S.-P. Chem, Z.-W. Tian, *Langmuir*, **2000**, *16*, 778.

177 T. R. Jensen, R. P. Van Duyne, S. A. Johnson, V. A. Maroni, *Appl. Spectrosc.*, **2000**, *54*, 371.

178 A. Bo, S. Sanicharane, B. Sompalli, Q. Fan, B. Gurau, R. Liu, E. S. Smotkin, *J. Phys. Chem. B*, **2000**, *104*, 7377.

179 G.-Q. Lu, S.-G. Sun, S.-P. Chen, L.-R. Cai, *J. Electroanal. Chem.*, **1997**, *421*, 19.

180 N. Harrick, *Internal Reflection Spectroscopy*, Interscience Publishers, New York, 1967.

181 C. J. Manning, In *Handbook of Vibrational Spectroscopy*, J. M. Chalmers and P. R. Griffiths (Eds.), Wiley, West Sussex, UK, Vol. 1, p. 283 (2002).

182 J. N. Wilking, C. J. Manning, G. A. Arbuckle-Keili, *Appl. Spectrosc.*, **2004**, *58*, 304.

183 C. Marcott, A. E. Dowrey, I. Noda, *Anal. Chem.*, **1994**, *66*, 1065A.

184 B. O. Budevska, C. J. Manning, P. R. Griffiths, *Appl. Spectrosc.*, **1994**, *48*, 1556.

185 I. Noda, *Appl. Spectrosc.*, **1990**, *44*, 550.

186 I. Noda, *Appl. Spectrosc.*, **1993**, *47*, 1329.

187 I. Noda, A. E. Dowrey, C. Marcott, *Appl. Spectrosc.*, **1993**, *47*, 1317.

188 D. L. Allara, A. Baca, C. A. Pryde, *Macromolecules*, **1978**, *11*, 1215.

189 H. Seki, K. Kunimatsu, W. G. Golden, *Appl. Spectrosc.*, **1985**, *39*, 437.

190 D. S. Bethune, A. C. Luntz, J. K. Sass, D. K. Roe, *Surf. Sci.*, **1988**, *197*, 44.

191 D. D. Popenoe, S. M. Stole, M. D. Porter, *Appl. Spectrosc.*, **1992**, *46*, 79.

192 P. W. Faguy, W. R. Fawcett, *Appl. Spectrosc.*, **1990**, *44*, 1309.

193 P. W. Faguy, N. S. Marinkovic, *Appl. Spectrosc.*, **1996**, *50*, 394.

194 C. Korzeniewski, M. W. Severson, *Spectrochim. Acta*, **1995**, *51A*, 499.

195 D. Zurawski, L. Rice, M. Hourani, A. Wieckowski, *J. Electroanal. Chem.*, **1987**, *230*, 221.

196 M. Wasberg, L. Palaikis, S. Wallen, M. Kamrath, A. Wieckowski, *J. Electroanal. Chem.*, **1988**, *256*, 51.

8
In-situ Surface-enhanced Infrared Spectroscopy of the Electrode/Solution Interface

Masatoshi Osawa

8.1
Introduction

Conventional electrochemical methods provide a vast amount of kinetic and mechanistic information about heterogeneous redox processes. However, it is desirable to supplement this with the molecular structural information that can now be provided by several *in-situ* surface analytical techniques [1, 2]. Of the techniques available, infrared spectroscopy is well suited for this task since the spectral data can yield valuable information on the identity as well as the reactivity of the interfacial species. This is especially true when examining multistep reactions involving adsorbed intermediate.

So far, infrared reflection-absorption spectroscopy (IRAS) has been used most frequently in observing both polycrystalline and single-crystal electrode surfaces. IRAS has greatly contributed to electrochemistry, as is borne out by several review articles [3–5], but it should be noted that this technique has several serious drawbacks, when used for the examination of electrochemical surface reactions, which are inherent to the external reflection geometry. Infrared radiation is reflected at the electrode surface through an infrared transparent window such as CaF_2 and the electrolyte solution, as schematically shown in Fig. 8.1a. Since solutions (especially water) strongly absorb infrared radiation, the solution layer between the electrode and the window must be made very thin ($1 \sim 10$ μm) by pushing the electrode against the window to reduce the background absorption from the solution. The thin-layer structure of the cell severely prevents the diffusion of reactants and products between the thin layer and the reservoir. When the reaction products are gases, they are trapped in the thin-layer cavity and disturb both spectral and electrochemical measurements. In addition, the high resistance of the thin-layer cell prevents quick response of the system to externally applied potential changes. Furthermore, the reaction is not uniform over the electrode: initial reactions begin at the edge of the electrode and propagate gradually toward its center. Since infrared radiation probes a relatively wide area of the electrode surface, IRAS is not suitable for studying fast surface reactions.

Advances in Electrochemical Science and Engineering Vol. 9.
Edited by Richard C. Alkire, Dieter M. Kolb, Jacek Lipkowski and Philip N. Ross
Copyright © 2006 WILEY-VCH Verlag GmbH & Co. KGaA, Weinheim
ISBN: 3-527-31317-6

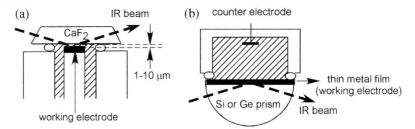

Fig. 8.1 *In-situ* monitoring of the electrochemical interface by IRAS (a) and ATR-SEIRAS (b).

Another serious problem with IRAS is that the infrared absorption by the solution is much greater than that by the adsorbates. If the thickness of the adsorbed layer is assumed to be 1 nm, a simple calculation reveals that the absorption by the solution is three to four orders of magnitude greater than that by the adsorbate (1–$10 \ \mu m/1 \ nm = 10^3$–10^4). Although the solution background can be reduced greatly by employing potential difference tactics such as EMIRS (electrochemical modulated IRS) [6] and SNIFTIRS (subtractively normalized interfacial FT-IR spectroscopy) [7], complete removal of the solution background is not easy in practice.

These problems introduced by IRAS can be solved by the use of the attenuated-total-reflection (ATR) configuration shown in Fig. 8.1b, in which a thin metal film deposited on an infrared-transparent prism with high refractive index, such as Si and Ge, is used as the working electrode. IR radiation from a light source is focused at the interface from the back of the electrode through the prism, and the totally reflected radiation is detected with a detector, which is most commonly a liquid-N_2-cooled MCT (HgCdTe) detector. If the electrode surface has nanometer-scale roughness, the absorption by the molecules adsorbed on the surface is significantly enhanced ($10^1 \sim 10^3$ times compared to normal measurements without the metal) [8–10]. The observed absorption is typically more than ten times larger than that observed by IRAS (the sensitivity of IRAS is one order of magnitude higher than normal measurements without the metal). The so-called surface-enhanced infrared absorption (SEIRA) effect is significant at the surface and fades away within a distance of a few monolayers from the surface. Although the infrared radiation penetrates into the solution in the order of the wavelength as an evanescent wave and gives the signals from the bulk solution, the background absorption by the solution is much less compared with that in the case of IRAS. As a result of surface enhancement and the reduction of the absorption by the solution achieved by the use of the ATR configuration, the signals from the surface and the solution are comparable [11]. Therefore, the solution background can be cancelled out much more easily than in the case of IRAS. Because of the high surface sensitivity and the free mass transport, SEIRA spectroscopy in the ATR mode (ATR-SEIRAS) enables time-resolved monitoring of surface electrochemical reactions with time resolutions down to the microsecond scale (depending on the spectrometer used) [12–

16]. The SEIRA effect can be observed in transmission and external modes as well [8–10, 17], but the ATR configuration is more favorable for electrochemistry as mentioned above.

SEIRA is similar to surface-enhanced Raman scattering (SERS) [18] in nature: both SEIRA and SERS are characterized by a remarkable enhancement of signals from adsorbed molecules on rough metal surfaces. However, a remarkable difference exists between them with respect to the metal available. SERS can be observed mainly on free-electron metals such as Ag, Au, and Cu, whereas SEIRA can occurs on many transition metals as well. This broad range of available metals makes SEIRAS more useful in electrochemistry. In addition, the mechanism of SEIRA is simpler than that of SERS. It is generally accepted that at least two mechanisms, electromagnetic (EM) and chemical, contribute to SERS, although the details are not yet fully understood. The EM mechanism assumes that there is an enhancement of the excitation and Raman scattered light at the surface, while the chemical mechanism assumes that some kind of resonance Raman effect is associated with a photo-driven charge transfer (CT) between the adsorbed molecule and the surface [18]. The contributions from these two mechanisms differ from system to system, which makes the interpretation of the observed SERS spectra difficult. On the other hand, an EM mechanism similar to that for SERS is the main contributor in SEIRA [8–10, 19]. The increase in the absorption coefficients of molecules by adsorption on metal surfaces was suggested to give an additional enhancement [20, 21], but such a chemical effect should be effective also in IRAS on bulk metals and is not specific to SEIRA. Since the chemical mechanism that involves electronic excitations of the metal surface and/or adsorbed molecules (photo-driven CT) as in SERS is ruled out in SEIRAS, the interpretation of the obtained spectra is much easier.

This chapter is organized as follows. First, the EM mechanism of SEIRA is briefly described for a better understanding of the technique, and this is followed by experimental aspects of the studies. Then several electrochemical systems studied by ATR-SEIRAS are described. Since SEIRA studies before 1997 have already been summarized in some review articles [8–10], this chapter deals mainly with the work performed after 1998.

8.2
Electromagnetic Mechanism of SEIRA

Figure 8.2 shows typical AFM images of SEIRA-active Au thin films deposited on Si by vacuum evaporation (a) and electroless (or chemical) deposition [22]. The vacuum-evaporated Au film (20 nm in mass thickness measured with a quartz crystal microbalance) consists of Au particles of average diameter 70 nm. On the other hand, larger particles with an average diameter of 300 nm and a height of 50 nm are found on the chemically deposited Au film. Detailed analysis of the latter film suggests that the particles exist on top of a thin (<50 nm), continuous Au underlayer which is in direct contact with the Si substrate. It is well established

(a) (b)

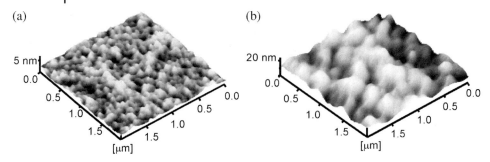

Fig. 8.2 AMF images of vacuum-evaporated (a) and chemically deposited (b) Au thin films on Si ($2 \times 2\ \mu m^2$) [22].

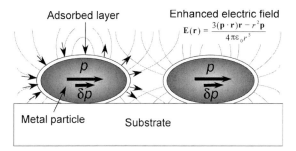

Fig. 8.3 Schematic representation of the EM mechanism of SEIRA on an island metal film. Incident photon field polarizes the metal particles and the dipole p generates an enhanced local electric field around the islands and excites adsorbed molecules. The molecular vibration induces an additional dipole δp in the metal islands and perturbs the *effective* optical properties of the metal island film. The thin curves and arrows represent electrostatic potential and field at the surface, respectively.

that such nanometer-scale surface roughness plays a crucial role in SEIRA [19, 23, 24]. By modeling the metal film by a set of rotating metal ellipsoids as shown in Fig. 8.3, the EM mechanism explains the SEIRA effect as follows.

Metal particles, much smaller than the wavelength of light, are polarized by the incident photon field through which an EM field stronger than that of the incident filed is generated around the particles [25] and excites the vibrations of adsorbed molecules. The EM field at the surface is estimated to be 10 times larger than that of the incident field [20]. Another EM effect gives an additional enhancement. A molecular vibration induces a dipole δp in the metal and perturbs the polarization of the metal particles p. The perturbation induced by the adsorbed molecules is significant at the frequencies of the molecular vibrations and is negligible at other frequencies. Consequently, the molecular vibrations can be observed through the change in absorption or reflectance of the metal film. Since the volume fraction of the metal is larger than that of the adsorbed molecules in the composite layer consisting of both metal and molecules and, further, the absorption coefficients of the most metals are larger than those of molecules in the infrared region, the change in absorbance or reflectance of the metal is larger than the IR absorption by the molecule itself, resulting in an *apparently* large absorption. In other words, the metal particles work as an amplifier of the infrared absorption in SEIRA. A large surface area of the rough surface (that is, a larger amount of adsorbed molecules) can also contribute to the larger absorption. The roughness factors of SEIRA-active surfaces are typically within the range of 2–7.

Two other important characters of SEIRA should be addressed. First, the enhanced surface EM field around the metal particles is essentially polarized along the surface normal at any point on the particles, as shown by the small arrows in Fig. 8.3 [25]. Consequently, only the vibrations that give transition dipole components perpendicular to the local surface can be excited. The surface selection rule can be explained also by the interaction of the adsorbate dipole with its image induced in the metal: the adsorbate dipole that is perpendicular to the surface constructively interacts with its image dipole to enhance the absorption, while the adsorbate dipole that is parallel to the surface destructively interacts with its image dipole to reduce the absorption [26]. The surface selection rule is identical to that in IRAS [27] and the orientations of adsorbed molecules can be elucidated by using this rule, as will be described in more detail later.

The second issue is that the distance dependence of the enhanced electric field is approximately proportional to $1/r^3$ (where r stands for the distance from the point dipole induced in a metal particle p) [25]. The interaction between adsorbed molecules and metal particles is also a function of $1/r^3$. Therefore, the enhancement is expected to be a function of $1/r^6$. That is, the enhancement is relatively short ranged and is mainly confined to the adsorbed molecules.

The SEIRA spectra can be simulated by using effective medium theories and the Fresnel equations [8–11, 17, 19, 24]. Such simulations predict that (i) the SEIRA effect is not limited to coinage metals and can also occur on most transition metals, (ii) the enhancement factor is primarily a function of the size, shape, and proximity between the metal particles, and (iii) the band shapes of the enhanced absorption depend on the morphology of the metal film. These predictions have been well confirmed by experiments.

8.3
Experimental Procedures

8.3.1
Electrochemical Cell and Optics

Typical electrochemical cells used for ATR-SEIRAS are shown in Fig. 8.4 [28, 29]. The cell can be made of either glass or plastics such as Kel-F. The prism works as the cell window as well as the substrate on which a thin-film electrode is deposited. Infrared-transparent materials with high reflective indices such as

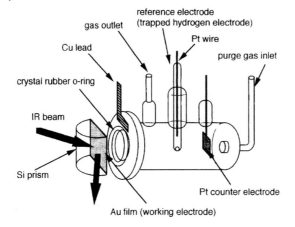

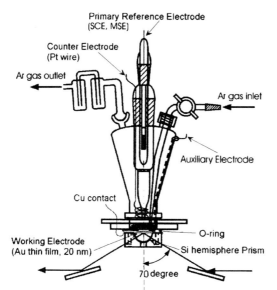

Fig. 8.4 Horizontal- [28] and vertical-type [29] spectroelectrochemical cells for ATR-SEIRAS.

Si, Ge, and ZnSe can be used as the window material. Among these materials, Si is used most frequently because of its high chemical stability, although the spectral range available is limited to above $\sim 1000\,\mathrm{cm^{-1}}$ for non-doped Si (the spectral range of doped Si is further limited). Ge has a wider spectral window ($>700\,\mathrm{cm^{-1}}$), but is easily dissolved at high potentials (in the double-layer potential regions of Pt and Au in acid), and pits are formed on the surface even if its surface is coated with a thin metal film. However, it can be used safely for less noble metals with lower double-layer potential regions such as Ag [11] and Cu [30]. Hemicylindrical, hemispherical, and triangular prisms are often used, of which the first two are convenient if one wants to change the incident angle

of infrared radiation for special purposes (for example, in the study of the incident-angle dependence of the spectral intensity).

The optimum optical conditions (the angle of incidence and polarization of
the incident infrared radiation) for ATR-SEIRAS depend on the thickness and
morphology of the thin metal film [8–10, 31]. Figure 8.5 shows the simulated
incident-angle dependence of the band intensity of a model molecule adsorbed
on an Ag film for p-polarization, where the metal film was assumed to be a bilayer consisting of a continuous underlayer and a nanoparticle overlayer [11].
The sharp peak around 20° corresponds to the surface plasmon excitation, but
this arises from the modeling and does not actually occur. The intensity for s-
polarization is negligibly small and thus omitted in the figure. The polarization
dependence arises from the constructive and destructive interactions of multiple
reflected p- and s-polarized radiations, respectively, within the metal layer [23].
This plot shows that the maximum sensitivity is obtained by using p-polarized
radiation at a high incident angle around 80°. The use of the high incident angle is also favorable for reducing the absorption by the solution due to the decrease in the penetration depth of the evanescent wave. Nevertheless, two issues
should be noted about the optical conditions in the practical measurements.
First, the calculation assumes that the electrode is large enough. If its size is
limited to less than 2 cm, the sensitivity does not differ so greatly at 60–80°.
Therefore, a reflection accessory for IRAS (typically with an incident angle of
60–65° for CaF_2 windows) can be used also in ATR-SEIRAS. The second issue
is with respect to the polarization used in the measurements. Although only
the p-polarized component of the incident infrared radiation can give the absorption of the adsorbed molecule as shown in Fig. 8.5, the polarizer reduces
the infrared radiation reaching the detector to one half or less. As a result, better S/N can be obtained without a polarizer (e.g., unpolarized radiation),
although the peak intensity itself is reduced to one half or less [32, 33].

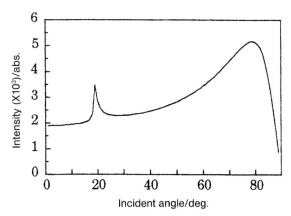

Fig. 8.5 The incident angle dependence of the band intensity
in ATR-SEIRAS for p-polarization (a model calculation). Band
intensity is negligibly small for s-polarization [11].

8.3.2
Preparation of Thin-film Electrodes

Thin-film electrodes that exhibit the SEIRA effect can be deposited on the total-reflecting plane of the prism by vacuum evaporation or electroless plating of the desired metal. Electrochemical deposition cannot be used for non-doped (and thus non-conducting) Si windows, but is possible for Ge.

Vacuum evaporation has been used in earlier experiments [8–10]. Metals with low melting points such as Au, Ag, and Cu can be deposited by resistive thermal heating of a tungsten basket in which the source metal is placed [10]. Electron-beam heating is also available [29]. In the case of metals with high melting points such as Pt, electron-beam heating or Ar-sputtering is more favorable [34, 35]. Thin-film electrodes of alloys can be prepared by co-deposition of the component metals [36–38]. Very slow deposition (0.1 nm s^{-1} or less) is absolutely essential to obtain SEIRA-active film electrodes [39]. The thickness of the film used for electrochemistry is typically 20 nm. Both the thickness of the film and the deposition rate can be monitored during the deposition by using a quartz microbalance incorporated into the vacuum chamber. Since the adhesion of vacuum-evaporated Au and Ag films is rather poor and they are easily peeled off from the substrate by wiping, they must be handled with care. Potential excursions to strong oxygen and hydrogen evolution regions should also be avoided.

Recently, electroless deposition techniques have been developed for preparing stable SEIRA-active thin-film electrodes on Si and Ge. So far, Pt [32, 33], Au [22], Cu [30], and Ag [16] electrodes have been reported in the literature. Electroless deposition is a more convenient and cost-effective method than vacuum deposition. One can prepare a thin-film electrode just by dropping a plating solution, in which a complex of the desired metal, reducing agent, and some additives are dissolved, onto the total reflecting plane of the prism. Good films can be obtained at relatively high temperatures (60–70 °C) [22, 30, 32, 33]. Careful cleaning of the substrate surface is required before the deposition. For Si, the surface is contacted with 40% NH$_4$F for a few minutes to remove oxide layer on the surface and to terminate it with hydrogen, which gives good adhesion of the deposited metal layer to the substrate.

Recently, a two-step wet process that can be applied to the preparation of SEIRA-active films of many metals including Pt, Pd, Rh, and Ru has been reported [40]. This technique includes the initial electroless deposition of an Au film on the prism followed by an electrochemical deposition of several monolayers of the desired metal on the Au film. This technique is similar to that used for observing SERS on transition metals [41, 42], but the strategy is somewhat different. In SERS, the underlying Au film works as an amplifier of Raman scattering. Since transition metals strongly damp the electromagnetic field enhanced by the underlying Au film, the overlayer of the desired metal must be a few monolayers thick. On the other hand, in SEIRAS the underlying Au film is used as a template for controlling the morphology of the desired metal overlayer. Therefore, relatively thick overlayers can be deposited in the latter, which is favorable when preparing

pinhole-free overlayers. Quasi-Au(111) electrodes prepared by electrochemical annealing of evaporated Au films (see below) are also used as templates for preparing a Pd(111) surface by electrochemical deposition of Pd [43].

Vacuum-evaporated very thin (~ 20 nm) Au and Ag electrodes suitable for
SEIRAS show pale blue to purple colors, while chemically deposited film electrodes show the colors of the corresponding massive metals and are as shiny as a well-
polished surface. Both thin-film electrodes have enough conductivity for electrochemistry. These film electrodes are usually polycrystalline. In the case of Au films,
(111) single-crystal-like surfaces can be obtained by flame annealing [44] or potential cycling in the double-layer region (electrochemical annealing) [29]. The trace in
Fig. 8.6 shows a typical cyclic voltammogram (CV) for an Au thin-film electrode
recorded in 0.1 M H_2SO_4 at 50 mV s^{-1} after flame annealing. The voltammogram
is close to that of an Au(111) single crystal with a miscut of $4°$ [29]. The peaks characteristic of Au(111) are found at 0.6 and 1.1 V (vs RHE), and are ascribed to the
phase transition of the surface and the disorder-order transition of adsorbed sulfate. The STM images of the electrode shown in Fig. 8.6 reveal that the round surfaces of each particle are composed of (111)-terraces and steps [44].

SEIRAS usually gives spectral features of the adsorbed molecules the same as
those observed by IRAS on massive bulk electrodes, but the band intensity and

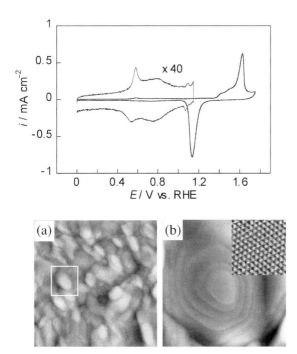

Fig. 8.6 Cyclic voltammograms of a vacuum-
evaporated Au film in 0.1 M H_2SO_4 at
50 mV s^{-1} after flame annealing (top) and
STM images of the electrode (bottom). B is
the expanded image of the area shown by
the square in A (100×100 nm^2). The inset in
B is a high-resolution image of the terrace at
the center (6×6 nm^2) [44].

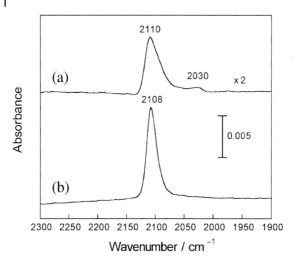

Fig. 8.7 SEIRA spectra of CO adsorbed on vacuum-evaporated (a) and chemically deposited (b) Au film electrodes measured in CO-saturated H_2SO_4 at 0 V (vs SCE). The reference spectrum of each spectrum was acquired at the same potential in the pure supporting electrolyte without CO [22].

spectral features greatly depend on the morphology of the thin-film electrode. An example is shown in Fig. 8.7, in which SEIRA spectra of CO adsorbed on a vacuum evaporated and chemically deposited Au film electrodes recorded in CO-saturated H_2SO_4 at 0 V (vs SCE) are compared [22]. A single peak ascribed to CO adsorbed at an atop site is observed at 2108 cm^{-1} on the chemically deposited electrode (b), which is consistent with an SRES study [45]. On the other hand, an additional weak band is observed at 2030 cm^{-1} on the vacuum-evaporated Au film electrode (a). The latter peak was ascribed to CO adsorbed at Au silicide that could be formed between the substrate and the deposited Au particles [44]. In the case of the chemically deposited Au film, the Au particles that enhance infrared absorption are separated from the substrate surface by a thin continuous Au underlayer. Therefore, the interference from the silicide is absent for this film. It is also noteworthy that the main CO band is broader and weaker on the vacuum-evaporated film than on the chemically deposited film. Since the integrated band intensities on the different electrodes are almost the same, the broader band shape on the vacuum-evaporated film may reflect greater inhomogeneity of the surface.

Transition metal electrodes prepared by sputtering and electrochemical deposition often show derivative-like (i.e., bipolar) [17, 34–38] or negative absorption (anti-absorption) bands [46–50]. Spectral features change from normal absorption to anti-absorption through bipolar shapes with increasing amount of the metal deposited [17]. The bipolar band shape was ascribed to a Fano-type resonance of electronic interactions between molecular vibrations and the metal [34, 35]. The anti-absorption was claimed to be a novel effect characteristic of nanoparticles and is called the abnormal infrared effect (AIRE) [46–48]. Although the origins of AIRE and the bipolar band shape are still a matter of discussion [17, 51–53], it should be noted that the distorted spectral patterns can be simulated by physical optics without taking into account any chemical interactions between the surface and the molecules [17, 51, 53] and can be removed by tuning the morphology of the metal film [30]. Since the distortion of the band shapes makes the interpretation of the observed spectra difficult and prevents quantitative analysis of the data, precise control of the morphology is essential.

8.4
General Features of SEIRAS

8.4.1
Comparison of SEIRAS with IRAS

To show the general features of the ATR-SEIRAS and to highlight some of the advantages of the technique over IRAS, the IR spectra of pyridine adsorbed on Au(111) electrode surfaces measured by IRAS [54] and ATR-SEIRAS [55] under nearly identical conditions are compared in Fig. 8.8. Figure 8.8a shows p-polarized IRAS spectra of pyridine adsorbed on the well-defined Au(111) single-crystal surface in a 1 mM pyridine + 0.05 M $KClO_4$ solution in D_2O. The spectra were collected by the SNIFTIRS technique with the reference potential of –0.75 V (SCE) at which pyridine is totally desorbed from the electrode surface. The use of D_2O as the solvent was aimed at avoiding the interference from the bending mode of H_2O (1645 cm^{-1}). Up- and down-going bands are observed in the spectra. The down-going bands are ascribed to pyridine adsorbed on the electrode surface in the normalized reflectance change units defined as $\Delta R/R = (R-R_0)/R_0$, where R and R_0 represent sample and reference spectra, respectively. On the other hand, the up-going bands are ascribed to pyridine in the bulk solution. Since mass transport between the thin layer and the reservoir is restricted in the IRAS set-up, the adsorption of the molecule on the electrode surface reduces its concentration in the thin layer and gives up-going bands of the dissolved species in the potential difference spectrum. In fact, the intensity of the up-going bands at 1445 cm^{-1} is proportional to the Gibbs surface excess as determined by chronocoulometry [54].

The corresponding set of spectra obtained by ATR-SEIRAS with a vacuum-evaporated Au(111) thin-film electrode and 1 mM pyridine + 0.1 M $NaClO_4$ solution in H_2O is shown in Fig. 8.8b [55]. The vacuum-evaporated film electrode was subjected to flame annealing before use to give a (111) quasi-single crystal surface. The spectra were recorded with unpolarized radiation and by a simple potential difference tactic with the reference potential of –0.8 V. These spectra are represented in the absorbance units [$A = -\log(R/R_0)$], and thus the up-going bands correspond to pyridine adsorbed on the electrode. Largely different from

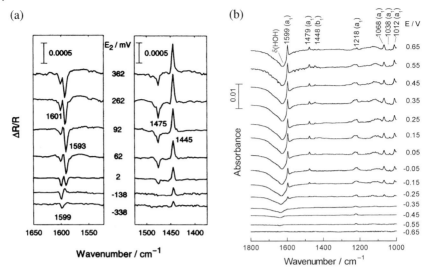

Fig. 8.8 (a) IRA spectra of pyridine adsorbed on Au(111) single-crystal electrode [54]. (b) ATR-SEIRA spectra of pyridine adsorbed on a vacuum-evaporated and flame-annealed Au film electrode (20 nm in mass thickness) [55]. Reference potential is –0.75 V (vs SCE) in both measurements. Solution: (a) 1 mM pyridine + 0.05 M KClO$_4$ D$_2$O; (b) 1 mM pyridine + 0.1 M NaClO$_4$ H$_2$O.

the IRA spectra, only the bands of adsorbed pyridine are observed in the ATR-SEIRA spectra except for the broad band peaked at 1620 cm^{-1} assigned to the bending mode of water. The band is red-shifted by $\sim$25 cm^{-1} from the frequency of bulk water and is ascribed to water molecules at the interface [28, 56]. Since water is repelled from the interface by pyridine adsorption, this band is observed as the negative peak.

The IRA and SEIRA spectra of adsorbed pyridine are nearly identical, but a small difference is found around 1600 cm^{-1}. Two peaks are observed at 1601 and 1593 cm^{-1} in the IRA spectra, while a single peak is observed at 1599 cm^{-1} in the ATR-SEIRA spectra. Since a single peak was observed at 1593 cm^{-1} in the IRAS measurement on the Au(210) electrode, and this surface has steps and kinks, the two bands at 1601 and 1593 cm^{-1} observed on Au(111) were ascribed to pyridine adsorbed at (111) terraces and defect sites, respectively [54]. Nevertheless, the assignment is still ambiguous because pyridine in the bulk solution has a medium strong band at 1594 cm^{-1} and the superposition of the solution spectrum can modify the surface spectrum of the adsorbate. A numerical subtraction of the solution spectrum (s-polarized spectrum, for example) is necessary for obtaining the true spectra [57]. It is especially true when the spectrum is changed only slightly by adsorption. On the other hand, ATR-SEIRAS selectively probes the electrochemical interface and is free from such ambiguity. The absence of the symmetric Cl-O vibration of the supporting electrolyte ClO$_4^-$ in the spectral range around 1100 cm^{-1} ensures the complete subtraction of the solution signal.

By the success of the complete subtraction of the perchlorate band, the weak pyridine bands at 1012–1168 cm^{-1} are clearly identified in the SEIRA spectra. The 1012-cm^{-1} band is assigned to the symmetric ring-breathing mode of pyridine. The corresponding band is located at 992 cm^{-1} for neat pyridine and 1001 cm^{-1} for aqueous pyridine solution [55]. The breathing mode is known to be sensitive to coordination at the N atom on the ring and the blue shift of this mode suggests the adsorption of pyridine via the N atom.

The other important issue to be noted is the great difference in band intensities observed by the two different techniques. The peak intensity of the 1479 cm^{-1} band in the ATR-SEIRA spectrum is approximately 1.5×10^{-3} absorbance, while that in the IRA spectrum is 2×10^{-4} in the $\Delta R/R$ units. The latter value corresponds to $\sim 1 \times 10^{-4}$ absorbance and is about 15 times smaller than that observed by SEIRAS. If p-polarization is used in the ATR-SEIRAS measurements as in the IRAS measurements, the band intensities should be twice as strong (or more) as that for unpolarized radiation [32, 33]; that is, the sensitivity of SEIRAS is about 30 times higher than that of IRAS. Even if the larger surface area of the electrode used in the SEIRAS measurement (roughness factor of ~ 2.5) is taken into account, the surface signal in SEIRAS is still higher than that in IRAS. Since the signal-to-noise ratio (S/N) of a spectrum is proportional to the square root of the co-added number of spectra, high-quality spectra can be obtained by ATR-SEIRAS with much less spectral acquisition time than IRAS, which facilitates time-resolved IR monitoring of electrode dynamics, as described later.

8.4.2
Surface Selection Rule and Molecular Orientation

The SEIRA spectra are often much simpler than the corresponding normal spectra of the precursor molecules, as in IRA spectra. This can be ascribed to specific orientations of the adsorbed molecule in most cases on the basis of the surface selection rule. Since the electrode surfaces used for SEIRAS are rough, the average orientation is random on the macroscopic scale even if adsorbed molecules are oriented with respect to local surfaces. Therefore, one may wonder why the surface selection rule in SEIRAS on a rough metal surface is the same as that in IRAS on flat surfaces. This puzzle can be solved by recalling that the surface EM field that excites adsorbed molecules is normal to the surface at any point on the rough surface and/or by taking image dipoles into account, as discussed in Section 8.2.

As in IRAS, the surface selection rule is represented by the mathematical formula as

$$I \propto \Gamma \left| \frac{d\vec{\mu}}{dQ} \cdot \vec{E} \right| = \Gamma \left| \frac{d\vec{\mu}}{dQ} \right| |\vec{E}|^2 \cos^2 a \propto \Gamma \left| \frac{d\vec{\mu}}{dQ} \right|^2 \cos^2 a \,, \tag{1}$$

where I is the band intensity, Γ the surface concentration of the molecule, $d\vec{\mu}/dQ$ the dipole moment derivative of the mode, E amplitude of the surface

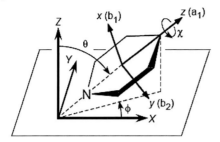

Fig. 8.9 Definition of experimental (XYZ) and molecular (xyz) systems for pyridine adsorbed on a metal surface. The two axis systems can be connected with three Eulerian angles θ, χ and ϕ. Symmetry species a_1, b_1 and b_2 have dipole derivatives along z, x and y directions, respectively.

electric field that excites the adsorbed molecule, and a the angle between E and $d\bar{\mu}/dQ$. Since E is essentially perpendicular to surface, a is the angle between $d\bar{\mu}/dQ$ and the surface normal.

The procedure to determine the orientations quantitatively is described below using pyridine as an example. Pyridine has C_{2v} symmetry and its vibrational modes are classified into three symmetry classes a_1, b_2, and b_1. The a_1 and b_2 modes are in-plane vibrations giving transition dipoles along C_2 rotation axis (z-axis) and perpendicular to it (y-axis), while b_1 modes are out-of-plane vibrations giving transition dipoles perpendicular to the molecular plane (x-axis). The molecular (xyz) and experimental (XYZ) systems are defined as shown in Fig. 8.9, where θ is the tilt angle of z-axis of the molecular system from the surface normal (Z), χ the twist angle of the molecular plane around the z-axis (which is $0°$ when y is parallel to the surface), and ϕ the rotation angle of the molecular system around the Z-axis. The molecular fixed-axis system xyz can be correlated with the experimental axis system XYZ by the three Eulerian angles θ, χ, and ϕ [58]. By averaging over ϕ, Eq. (1) can be rewritten for the a_1, b_1 and b_2 modes as

$$I(a_1) \propto \cos^2 \theta \times I°(a_1) \tag{2}$$

$$I(b_1) \propto \sin^2 \theta \cos^2 \chi \times I°(b_1) \tag{3}$$

$$I(b_2) \propto \sin^2 \theta \sin^2 \chi \times I°(b_2) , \tag{4}$$

where $I°$ represents the intrinsic intensity of the corresponding mode for the free molecule ($\propto |d\bar{\mu}/dQ|^2$). From these three equations, the following two equations can be introduced.

$$\tan^2 \theta = \frac{I(b_1)}{I(a_1)} \times \frac{I°(a_1)}{I°(b_1)} \times \frac{1}{\cos^2 \chi} \tag{5}$$

$$\tan^2 \chi = \frac{I(b_2)}{I(b_1)} \times \frac{I°(b_1)}{I°(b_2)} \tag{6}$$

The molecular orientation of benzenethiol adsorbed on a quasi-Au(111) surface was analyzed from a comparison of a SEIRA spectrum with a KBr spec-

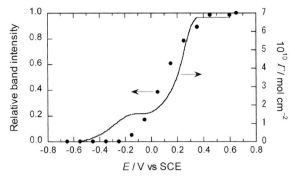

Fig. 8.10 Comparison of the intensity of the 1599 cm^{-1} band of pyridine adsorbed on a highly ordered Au(111) surface in 0.1 M NaClO$_4$+0.1 mM pyridine [55] and the Gibbs surface excess (Γ) of pyridine on a single-crystal Au(111) electrode measured by chronocoulometry [62].

trum of Au(I)-benzenethiolate by this protocol. The SEIRA spectrum was obtained in the transmission mode by using a thin Si wafer as the substrate for observing the low frequency region below 1000 cm^{-1}. The result ($\theta \approx 30°$ and $\chi \approx 0°$) is close to those on Mo(110) ($\theta \sim 23°$) and on Ag ($\theta = 14°$ and $\chi = 0°$) determined by NEXAFS [59, 60] and SERS [61], respectively.

In the SEIIRA and IRAS spectra of pyridine shown in Fig. 8.8, the a_1 modes dominate the spectra and b_2 modes are hardly observed except for the very weak band at 1448 cm^{-1} (the strongest band in the normal IR spectrum of neat pyridine). The result indicates that pyridine is oriented with the y-axis nearly parallel to the surface ($\chi \approx 0°$). Since the spectral range below 1000 cm^{-1} where b_1 modes are located could not be measured because of the low transmittance of the silicon prism, the tilt angle θ cannot be determined from the in-situ spectra in Fig. 8.8 b. An alternative way to determine θ in such a case is to examine the relationship between the intensity of an infrared absorption band and the surface concentration of the adsorbate on the basis of Eq. (1). Figure 8.10 compares the integrated intensity of the 1599-cm^{-1} band of adsorbed pyridine and the Gibbs surface excess determined by chronocoulometry [62]. The Gibbs surface excess data shows that pyridine is adsorbed at potentials more positive than –0.6 V and its coverage increases as the potential is made more positive, while the pyridine bands are observed at potentials more positive than –0.3 V. Assuming that the absorption coefficient of the band is independent of the orientation and coverage, the results suggest that pyridine adopts a flat orientation at low potentials and reorients to a more perpendicular orientation as the potential is made more positive.

It is well known that the adsorption coefficients of adsorbed molecule are a function of coverage (for example, CO on Ru(001) in UHV [63]). This has been explained by dipole-dipole coupling or by coupling through metal electrons [27]. On the other hand, in an electrochemical environment, a linear relationship between band intensity and coverage has been reported for linearly bonded CO on

Pt electrodes [64, 65]. Since the dipole-dipole coupling between adsorbates with weaker absorption than CO is small, this effect may be neglected in the most cases.

Molecular orientations of many organic molecules have been discussed on the basis of the surface selection rule, but the rule must be modified slightly in some special cases. A theoretical consideration suggests that surface spectra of adsorbed molecules are largely modified when charge transfer (CT) occurs between the surface and adsorbed molecules [66]. For centro-symmetric molecules, for example, Raman-active (and thus IR-inactive) modes are activated and significantly enhanced in surface IR spectra through the vibronic coupling of molecular vibrations (particularly the totally symmetric modes) with CT, while the original IR-active modes are quenched by the coupling. This significant modification is commonly observed in infrared spectra of CT complexes [67]. Since the CT is perpendicular to the surface, molecular vibrations that give dipole transitions parallel to the surface can be enhanced selectively in this case. Very recently, such an effect was observed for diprotonated 4,4′-bipyridine adsorbed flat on a Cu electrode [68]. This modified surface selection rule holds also for the π–π stacked monocation radical dimer of heptylviologen (N,N'-heptyl-4,4′-bipyridium) deposited on an Ag electrode [69]. In both of these spectra, Raman-active (originally IR-inactive) modes dominate the SEIRA spectra.

8.4.3
Comparison of SEIRA and SERS

As in usual IR and Raman spectroscopy, SEIRAS and SERS are complementary to each other, and thus a comparison of SEIRA and SER spectra is valuable for a better understanding of the both spectra. Comparisons of SEIRA and SER spectra were reported for *p*-nitroaminothiol and pyrazine adsorbed on polycrystalline Ag and Au surfaces, respectively [70, 71].

Pyrazine is a good sample to examine because it has D_{2h} symmetry, and thus Raman-active (*gerade*) modes are totally IR-inactive and IR-active (*ungerade*) modes are totally Raman-inactive for the free molecule. The SERS of pyrazine adsorbed on metal surfaces has received great attention because normally Raman-forbidden *ungerade* modes are observed in the spectrum [72–79]. This result can be interpreted differently: lowering of the symmetry by adsorption, contribution of quadrature polarizability (field gradient mechanism), and photo-driven CT mechanism. Despite all of the efforts to understand this phenomenon, the mechanism of the activation of the Raman-inactive modes is still controversial.

In Fig. 8.11, the observed SER bands of pyrazine adsorbed on Au electrodes are compared with the normal IR spectrum of an aqueous pyrazine solution (5 M) and the SEIRA spectrum of adsorbed pyrazine [71]. In the SER spectrum (c), the Raman-inactive modes for the free molecule are observed at 1479, 1155, 1123, and 1047 cm^{-1} (shown by dashed lines) together with Raman-active modes (solid lines). The counterparts of these Raman-inactive modes can be found in the solution IR spectrum (a), indicating that these bands are apparently *ungerade* modes. On the other hand, the SEIRA spectrum (b) is essentially identical to the solution

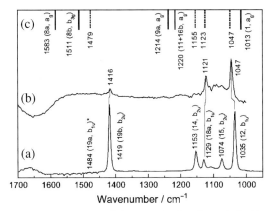

Fig. 8.11 Comparison of the normal infrared spectrum of (a) 5 M aqueous solution of pyrazine and (b) SEIRA spectrum of pyrazine adsorbed on an Au thin-film electrode at 0.0 V (vs SCE) [71]. (c) SER bands of pyrazine adsorbed on a polycrystalline Au electrode taken from Ref. [77]. The solid and dashed bars correspond to normally Raman-active and infrared-active modes, respectively, under D_{2h} symmetry. The 1484 cm^{-1} band denoted by the asterisk is very weak for the solution, but is clearly identified for neat pyrazine.

IR spectrum except for the large difference in relative intensities between the bands. The difference in relative band intensity arises from a specific orientation of pyrazine. The selective observation of b_{1u} modes at 1121 and 1047 cm^{-1} that have transition dipole moments along C_2 rotation axis (N-N direction) in the SEIRA spectrum suggests a perpendicular end-on adsorption via one of the N atoms. The adsorption can reduce the D_{2h} symmetry to C_{2v} or C_s and activate the Raman-active modes, but no Raman-active modes are found in the SEIRA spectrum. The result suggests that pyrazine is adsorbed only weakly and the symmetry reduction does not affect the vibrational properties of pyrazine so greatly. This was supported by the result that adsorbed pyrazine is replaced by the supporting anion, ClO_4^-, at high potentials as revealed by the potential dependence of the SEIRA spectra, chronocoulometric measurements of the Gibbs surface excess, and impedance measurements. From a detailed analysis of allowed vibrations in the three possible mechanisms, the breakdown of the Raman selection rule for adsorbed pyrazine was ascribed to the photo-driven CT mechanism [71]. As can be found in this work and others [80], SEIRAS is useful in interpreting SER spectra.

8.4.4
Baseline Shift by Adsorption of Molecules and Ions

The baseline of a surface spectrum (i.e., the reflectivity of the electrode) is shifted by the adsorption of molecules and ions, and also by the surface oxidation [4, 15, 33, 43, 81–84]. SEIRAS is more sensitive than IRAS for the baseline shift because the enhanced absorption of adsorbed species is observed in SEIRAS through the reflectance change of the metal, as described in Section

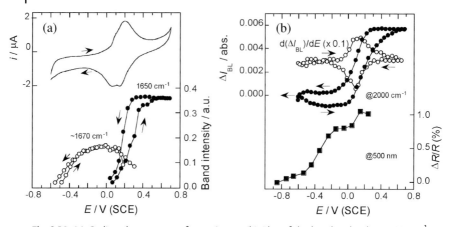

Fig. 8.12 (a) Cyclic voltammogram for an Au film electrode 0.1 M NaClO$_4$ + 1 mM cytosine at a scan rate of 5 mV s^{-1} and the plot of the integrated intensities of the v(C=O) bands for physisorbed (open circles) and chemisorbed (closed curves) cytosine taken from a series of SEIRA spectra collected simultaneously with the voltammogram.

(b) Plot of the baseline level at 2000 cm^{-1} (closed circles) and its derivative (open circles) versus potential [83]. The squares represent the potential dependence of the reflectivity change of a massive Au electrode in 0.1 M NaClO$_4$ + 1.13 mM cytosine measured at 500 nm [88].

8.2. The analysis of the baseline shift can provide some additional information on the interaction between molecules and the electrode as follows.

The adsorption of molecules and anions generally decreases the reflectivity of metal electrodes. For example, the data for cytosine adsorption on an Au electrode [83] is shown in Fig. 8.12. The upper curve of Fig. 8.12a shows a CV for an Au electrode in 0.1 M NaClO$_4$ containing 1 mM cytosine recorded at a sweep rate of 5 mV s^{-1}, in which a pair of anodic and cathodic peaks is found around 0.2 V (vs SCE). A combined STM and impedance study showed that cytosine is adsorbed on Au over a wide potential range above –0.6 V (SCE), and the pair of peaks in the voltammograms arise from a phase transition between disordered physisorbed and ordered chemisorbed cytosine layers [85]. In fact, the frequency of the v(C=O) band at low potentials (1667–1673 cm^{-1}) is close to that for the free molecule and undergoes a red shift to 1650 cm^{-1} at potentials more positive than 0.1 V [83]. The potential dependence of the band intensity at 1650 cm^{-1} (closed circles in Fig. 8.12a) is well correlated with the chemisorption of cytosine. Both physisorbed and chemisorbed cytosine molecules interact with the surface via N(3), O(=C), and NH$_2$ moiety, and are oriented almost perpendicular to the surface [83, 85]. Associated with the chemisorption of cytosine, the baseline of the spectrum at 2000 cm^{-1} at which cytosine has no absorption bands increases (that is, the reflectivity of the electrode decreases) as shown in Fig. 8.12b (closed circles). The first derivative of the baseline-potential curve shown by open circles in the same figure resembles the voltammogram, suggesting that chemisorption of cytosine reduces the reflectivity of the electrode.

In contrast, the adsorption of hydrogen increases the reflectivity of Pt and Pd electrodes [4, 33, 81, 82]. The increase in the reflectivity by hydrogen adsorption is attributed to the increase in free-electron density of the surface [4, 86]. In the case of cytosine adsorption on Au shown above, the charge transfer is from the electrode to the adsorbed molecule as is seen from the CV, and thus the reflectivity of the Au electrode decreases. The baseline is also sensitive to the applied potential and is minimal at the pzc of the electrode [15, 43].

UV-visible spectroscopy is more sensitive to the reflectance change of the electrode surface [87]. Cytosine adsorption on Au [88, 89] and hydrogen adsorption on Pt [86] have been examined in the visible region. It is worth noting that potential dependence of the baseline shift for cytosine adsorption observed in the visible region (at 500 nm, squares in Fig. 8.12 b) [88, 89] is different from that observed in the infrared region. In the visible region, the reflectivity starts to decrease at –0.7 V and saturates at 0.1 V after reaching a plateau around –0.1 V, indicating that the reflectivity change in the visible region is caused by both physisorbed and chemisorbed molecules.

The reflectivity of the interface is a function of the refractive indices (optical constants) of the adsorbed layer and the electrode [87]. Both the replacement of interfacial water molecules by the target molecules and the change in the electronic state of the electrode surface can change the reflectivity. In the case of cytosine adsorption, the molecule is adsorbed over a wide potential range. Therefore, the data in Fig. 8.12 b suggests that the baseline shift in the IR region is sensitive only to the change in electronic state of the electrode surface caused by the chemisorption. The wavelength dependence could be ascribed to the large difference in the optical constant of Au in the two spectral regions (3.748–30.5i at 2000 cm^{-1} and 0.98–1.84i at 500 nm).

8.5
Selected Examples

Owing to the high surface sensitivity, ATR-SEIRAS has been applied to many electrochemical studies including the structure of the double-layer, adsorption/desorption of molecules and ions, characterization of self-assembled monolayers, and real-time monitoring of electrochemical surface reactions. The surface sensitivity (or interface selectivity) advantage has been demonstrated most effectively in the studies of the double-layer structure [28, 29, 56, 90–92]. These studies showed that the vibrational properties of water at the interface are largely different from those in the bulk solution and depend on the applied potential. The interaction of interfacial water molecules with specifically adsorbed anions and molecules was also examined in detail. The recent trend is shifting from simple adsorption/desorption systems to more complicated reaction systems and further to electrode kinetics and dynamics. In this regard, it should be noted that the injection of sample molecules into the cell, gas bubbling, and solution exchange can be done on-line without disturbing spectral acquisition in the ATR mode. This additional advantage of ATR-SEIRAS facilitates the time-resolved studies of reaction mechanisms. Applications directed toward bioelectrochemistry are also growing. In this section, some of the results reported in recent literature are described.

8.5.1
Reactions of a Triruthenium Complex Self-assembled on Au

Self-assembled monolayers (SAMs) of alkanethiols can be used as a platform for fixing functional groups at the electrochemical interface. The oxo-centered triruthenium complex monolayer self-assembled on Au from a disulfide $[Ru_3(\mu_3\text{-}O)(\mu\text{-}CH_3COO)_6(CO)(L_1)(L_2)]$ $(L_1=[(NC_5H_4)CH_2NHC(O)(CH_2)_{10}S\text{-}]_2$ and $L_2=4$-methylpyridine) shows reversible multi-step electron transfer [93, 94]. The Ru complex SAM, in which CO is coordinated to one of the Ru sites (CO-SAM), shows a reversible redox at 0.6 V (vs SCE) arising from the one-electron transfer between $[(Ru^{II}CO)Ru^{III}Ru^{III}]^0$ and $[(Ru^{III}\text{-}CO)Ru^{III}Ru^{III}]^{+1}$ in the CV recorded in 0.1 M NaClO$_4$ (Fig. 8.13a, solid trace). The potential difference IR spectrum recorded at +0.8V with a reference potential of +0.3 V (Fig. 8.13a′, solid trace) shows a downward peak at 1960 cm^{-1} and an upward peak at 2070 cm^{-1}, which correspond to the stretching vibrations of the CO ligand in the $[(Ru^{II}\text{-}CO)Ru^{III}Ru^{III}]^0$ and $[(Ru^{III}\text{-}CO)Ru^{III}Ru^{III}]^{+1}$ states, respectively. The blue shift of the CO peak position reveals a decrease in the electron density on the Ru site caused by the oxidation reaction, which reduces the π-electron back-donation from the Ru to CO and hence strengthens the C–O bond.

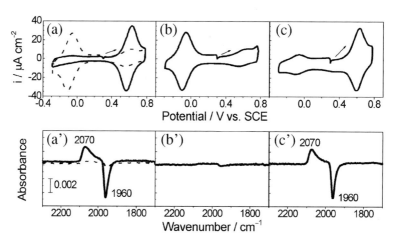

Fig. 8.13 Cyclic voltammograms and IR spectra at +0.8 V for (a, a′): CO-SAM (solid) and solvent-SAM (dotted) in 0.1 M HClO₄. (b, b′): Solvent-SAM after contact with CO at +0.3 V for 1 h. (c, c′): Solvent-SAM after contact with CO at –0.3 V for 30 min. Background IR spectrum for each spectrum was recorded at +0.3 V. Scan rate: 0.5 V s^{-1} [93].

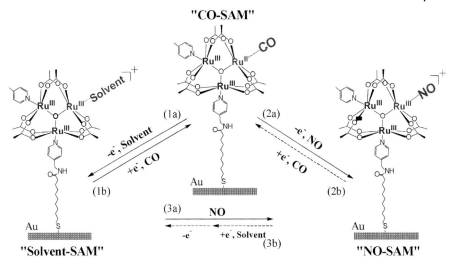

Fig. 8.14 Electrochemically controlled ligand replacement in the Ru complex SAM on Au [93].

After holding the potential at +0.8 V for 20 min, the redox peaks disappear and a new set of redox peaks appear at –0.1 V as shown by the dashed trace in Fig. 8.13 a. The corresponding SEIRAS spectrum (Fig. 8.13 a′, dashed trace) clearly shows the dissociation of the CO ligand. The electron backdonation from the Ru to CO weakens the CO-Ru bonding and results in the CO dissociation. The solvent molecule is believed to coordinate to the Ru center (solvent-SAM), and the redox peaks at –0.1 V are ascribed to the one-electron transfer between $[(Ru^{II}\text{-solvent}) Ru^{III}Ru^{III}]^{+0}$ and $[(Ru^{III}\text{-solvent}) Ru^{III}Ru^{III}]^{+1}$.

As shown in Figs. 13 b and b′, no changes occur in either the CV or IR spectrum of the solvent-SAM after contact with the CO-saturated solution for 1 h at the rest potential (ca. +0.3 V). However, when the potential is changed to –0.3 V where the solvent-SAM is reduced to $[(Ru^{II}\text{-solvent})Ru^{III}Ru^{III}]^{0}$, both the CV (Fig. 8.13 c) and the IR spectrum (Fig. 8.13 c′) recover to the original states (solid lines in Fig. 8.13 a and a′), indicating that CO replaces the solvent ligand and rebinds to the central Ru ion under potential control.

By holding the electrode modified with the CO-SAM at +0.8 V in 1,2-dichloroethane and bubbling NO, the CO-ligand can be replaced by NO (NO-SAM). The NO-SAM can be reconverted to the CO- or solvent-SAM by controlling the potential as well. The reaction scheme found by the combined CV and SEIRAS study is summarized in Fig. 8.14 [93]. As demonstrated in this work, SEIRAS can be used as an on-line monitoring tool in fabrication and characterization of functional surfaces. The rates of the ligand-exchange reactions and their activation energies were also evaluated from time-resolved SEIRAS measurements under temperature control with a double-jacked spectroelectrochemical cell [94].

8.5.2
Cytochrome c Electrochemistry on Self-assembled Monolayers

The redox reactions of metal proteins such as cytochrome c (cyt. c) are known
to be promoted by modifying electrodes with SAMs of nitrogen-containing heterocyclic molecules. SEIRAS studies of the modified electrodes directed toward
the understanding the promotion mechanism have been reported. The systems
reported so far are 2-pyridinethiol, 4-pyridinethiol (4-PySH), bis(2-pyridyl)disul-
fide, bis(4-pyridyl)disulfide, 2-pyrimidinethiol, 6-amino-8-purinethiol, mercaptopropionic acid, mercaptoethanol, 4,4′-dithiodipyridine, and L-cysteine [95–99].

The promotion activities of 4-PySH SAMs on Au greatly depend on the conditions of the self-assembly. The highest activity is obtained when the electrode is
modified in alkaline solution (e.g., 0.1 M KOH) with a low 4-PySH concentration (e.g., 20 μM) by applying a potential more positive than 0.3 V (vs Ag/AgCl)
[96]. The lower activity of the SAMs prepared under different conditions was ascribed to the adsorption of sulfide, a trace impurity in commercially supplied reagents. To clarify the reason why higher activity is achieved under the conditions mentioned above, the kinetics of 4-PySH adsorption on Au was examined
by real-time SEIRAS monitoring [96]. In Fig. 8.15a, the band intensity of a ring
vibration of adsorbed 4-PySH at 1578 cm^{-1}, which shifts to 1621 cm^{-1} in acidic
solution because of the protonation, is plotted as a function of time after injecting the molecule into the cell (0.1 mM) at 0 V. The spectra were recorded at
every 5 s with a reference spectrum recorded before the injection of the molecule. It should be stressed again that the injection of the sample molecule can

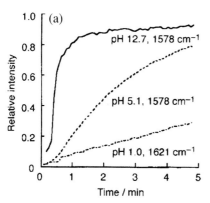

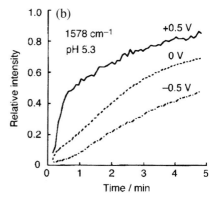

Fig. 8.15 Plots of the integrated intensity of
the 1578-cm^{-1} band (at 1621 cm^{-1} in acidic
solution) from PySH adsorbed on Au
electrodes showing the adsorption kinetics
of this molecule after injecting into the
solutions measured under different conditions (solution pH and applied potential);
(a) pH dependence at the fixed potential of

0 V (vs Ag/AgCl) and (b) potential dependence at the fixed pH of 5.3. The intensity
data were taken from time-resolved SEIRA
spectra collected every 5 s and are given as
relative intensities with respect to that
20 min after the injection of the molecule for
each measurement. The concentration of
4-PySH was fixed at ca. 100 mM [96].

be done quite easily during sequential spectral collections in the ATR-SEIRAS set-up. The plot shows that the adsorption becomes faster as the solution pH is increased. Figure 8.15 b shows the potential dependence of the adsorption in a neutral solution of pH 5.3. As the applied potential is increased, the adsorption becomes faster.

On the basis of the SEIRAS experiments, the difference in the promotion activities of the SAMs prepared under different conditions was explained as follows: the sulfide is strongly adsorbed on Au and prevents the adsorption of 4-PySH. However, since the sulfide concentration is very low, it takes some period of time for its adsorption. Accordingly, by employing the conditions under which 4-PySH can be adsorbed quickly, the electrode surface can be covered with 4-PySH with less interference from sulfide.

The promotion activity also depends on the structure of the precursor molecule. The orientation analyses of the SAMs of 2- and 4-PySH and their disulfides suggest that the N ends of the molecules must be directed toward the solution side for promoting the electrochemistry of cyt. c [95]. It was also found that the ring vibrations of the pyridyl moieties of the SAMs are sensitive to the protonation at the N atom on the ring. From the pH dependence of a ring vibration mode, the pK_a value for a 4-PySH SAM on an Au electrode was determined [100].

Recently, the redox reaction of cytochrome c on Au electrodes modified with mercaptopropionic acid (MPA), mercaptoethanol, 4,4'-dithiodipyridine, and L-cysteine were directly observed with ATR-SEIRAS [98, 99]. Figure 8.16a shows the

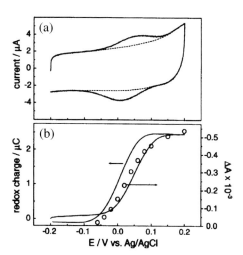

Fig. 8.16 (a) Cyclic voltammogram of horse heart cytochrome c adsorbed onto an MPA-modified Au electrode. Scan rate is 50 mV s^{-1}. The dashed line corresponds to the double-layer capacitance. (b) Total charge pass during oxidation and reduction of cytochrome c (solid curve) calculated by integration of the redox current displayed in (a). Open circles are the intensities of the band at 1692 cm^{-1} in the difference spectra displayed in Fig. 8.17 [98].

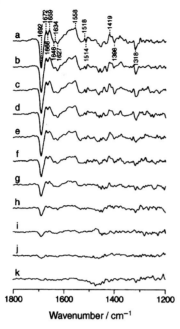

Fig. 8.17 Redox difference spectra of cytochrome *c* on the MPA-modified Au electrode recorded with the ATR-SEIRA technique at (a) +0.20, (b) +0.15, (c) +0.10, (d) + 0.08, (e)+0.06, (f) +0.04, (g) +0.02, (h) 0.00, (i) –0.02, (j) –0.04, (k) –0.06 V (vs Ag/AgCl). The reference spectrum was taken at –0.10 V. The band at 1692 cm^{-1} corresponds to an absorbance change of –5×10^{-4} [98].

CVs of an Au electrode modified with an MPA SAM recorded in a phosphate buffer solution (pH 7.0) with (solid trace) and without (dashed trace) 2 μM horse heart cyt. *c* [98]. Associated with the redox peaks around 0 V (vs Ag/AgCl), very small but clearly noticeable changes appear in the potential difference SEIRA spectra as shown in Fig. 8.17. The spectrum at –0.1 V was used as the reference (i.e., the reduced form of cyt. *c*). The strong negative peak at 1692 cm^{-1} is assigned to the amide I band (mostly ν(C=O) of the peptide bond) of β-turn segment from the reduced form and the decrease of this band by oxidation is indicative of a change in hydrogen bonding of this secondary structure element. Other amide I bands around 1650 cm^{-1} (the negative bands for the reduced form and the positive bands for oxidized form) reflect different types of structural changes of the protein backbone. A comparison of the potential dependence of the 1692-cm^{-1} band intensity with the charge passed in the CVs (Fig. 8.16 b) well demonstrates that the spectral changes are caused by the redox reaction of cyt. *c*.

The spectrum of cyt. *c* greatly depends on the terminal groups (-COOH, -COH, -SH, -NH$_2$) of the SAMs, reflecting different interactions with cyt. *c* [99]. The potential difference tactic used in this measurement is favorable in examining redox behavior of such complex biological compounds, because the changes in the central functional groups induced by charge transfer can be observed selectively.

8.5.3
Molecular Recognition at the Electrochemical Interface

The cyt. *c* redox can be promoted also by the SAMs of 2-amino-6-purinethiol and 6-amino-8-purinethiol. The cyt. *c* redox on the 6-amino-8-purinethiol SAM is largely suppressed by the addition of thymidine, a derivative of thiamine, into the solution, while it is not suppressed on the 2-amino-6-purinethiol SAM [101, 102]. The suppression was supposed to arise from hydrogen-bonded base pairing between the SAM and thymidine (cf. Fig. 8.20, right), which can prevent the approach of cyt. *c* to the electrode surface and suppress its redox reaction. The hydrogen-bonded pairing of complementary bases [adenine(A)-thiamine (T) and guanine (G)-cytosine (C)] is the ultimate molecular recognition system and plays important roles in biological systems. The interaction between the 6-amino-8-purinethiol SAM and thymidine in the solution was examined with SEIRAS from the viewpoint of molecular recognition [97].

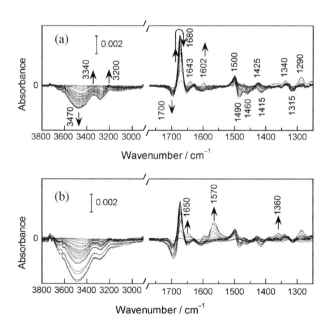

Fig. 8.18 Potential difference SEIRA spectra of a gold electrode modified with 6-amino-8-purinethiol measured in 0.1 M NaClO$_4$ solution without (a) and with 1 mM thymidine (b). The spectra were acquired during a potential sweep from −0.6 to +0.5 V (sweep rate, 5 mV s^{-1}) with respect to the reference spectrum acquired at −0.6 V. Arrows indicate the increase and decrease of band intensities associated with a positive potential sweep [97].

IR spectroscopy is one of the useful methods to study the hydrogen-bonded pairing between the bases. The measurements are generally carried out in aprotic solvents and the hydrogen-bonded pairing is discussed from the shift of v(NH) and v(C=O). Unfortunately, this approach cannot be used for the studies at the solid/water interface, because both adenine moiety of the SAM and thymidine are hydrated in water and the distinction between hydrogen bonds between the complementary base pair and those between the bases and water is practically impossible from the v(NH) and v(C=O) modes only. The interaction between the surface-confined 6-amino-8-purinethiol and thymidine was discussed from the potential dependence of the band intensities of the v(OH) mode of water and some modes of adenine moiety in this study.

Figure 8.18 shows the potential difference SEIRA spectra of a 6-amino-8-purinethiol SAM on Au measured in 0.1 M NaClO$_4$ solution without (a) and with 1 mM thymidine (b) acquired during a potential sweep from –0.6 V to +0.5 V (sweep rate of 5 mV s^{-1}) with respect to the reference spectrum acquired at –0.6 V. The intensities of some bands are plotted in Fig. 8.19 as a function of applied potential (open and closed circles correspond to the measurements without and with thymidine in the solution, respectively).

In the absence of thymidine in the solution (Fig. 8.18a), the 1700- and 1680-cm^{-1} bands are assigned to the δ(NH$_2$) modes of the protonated and unprotonated adenine moieties (at N1), respectively. The decrease and increase in the intensities of the 1700 and 1680 cm^{-1} bands, respectively, on the positive potential scan indicate the shift of the acid-base equilibrium towards the unprotonated form. The pK_a value for the adenine moiety at open circuit potential was evaluated to be ~6 from the SEIRA spectra, which is much larger than those for adenine

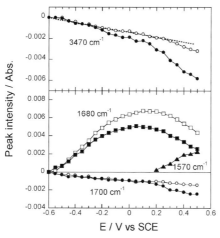

Fig. 8.19 Peak intensities of the 3470-, 1700-, 1670-, and 1570-cm^{-1} bands in the potential-dependent spectra measured in 0.1 M NaClO$_4$ solution without (open symbols) and with (closed symbols) 1 mM thymidine [97].

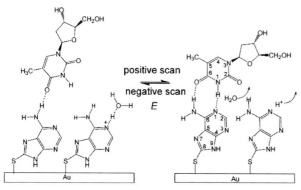

Fig. 8.20 Schematic drawing of the potential-dependent reorientation of 6-amino-8-purinethiol adsorbed on gold and hydrogen bonding with thymidine in solution [97].

($\sim$4.2) and adenosine ($\sim$3.9) in bulk water. The decrease in the 1700-cm^{-1} band intensity is nearly proportional to the potential, while the 1680-cm^{-1} band intensity increases at low potentials and then decreases at potentials more positive than 0 V (Fig. 8.19). If the spectral changes were caused simply by the shift of the acid-base equilibrium, the changes of the two bands should be proportional. The data suggest that some changes occur in the SAM with potential. On the basis of the surface selection rule, the decrease in the 1680-cm^{-1} band intensity at high potentials was ascribed to the reorientation of the adenine moiety from a largely tilted orientation at lower potential to a more upright model at higher potential, as schematically shown Fig. 8.20. Since the reorientation of the adenine moiety occurs around the pzc of polycrystalline Au ($\sim$0 V), the change in electrostatic interaction of the adenine moiety with the interfacial electric field triggered by the deprotonation are likely the driving force of the reorientation.

By the addition of thymidine into the solution, new bands appear at 1650, 1570, and 1360 cm^{-1}, as shown in Fig. 8.18 b. The potential dependences of other bands are also affected by the thymidine addition as shown in Fig. 8.19 (closed symbols). Quartz crystal microbalance measurements revealed that thymidine is adsorbed on the adenine SAM over the double-layer potential region of Au [102]. The noticeable decrease in the band intensity at 1680 cm^{-1} assigned to $\delta(NH_2)$ by the addition of thymidine suggests hydrogen bonding of thymidine to the NH$_2$ group of the adenine moiety. On the other hand, the effect of the thymidine addition for other bands is observed at potentials more positive of 0.1 V. The new bands at 1650, 1570, and 1360 cm^{-1} appear at the high potential range. The results suggest that thymidine interacts with the SAM differently at positive and negative potentials. The 1570-cm^{-1} band is a fundamental mode of the adenine moiety to which ring modes and $\delta(NH_2)$ are coupled and is known to increase its intensity by forming an A-T type hydrogen bond pair.

From these observations, it was concluded that an A-T type hydrogen bond pair is formed between the surface-confined adenine moiety and thymidine in solution as shown in Fig. 8.20 at positive potentials only. Such hydrogen bond pairing is difficult at negative potentials where the adenine moiety is largely tilted from the surface normal because of steric hindrance and also the partial protonation at N1 (Fig. 8.20, left).

8.5.4
Hydrogen Adsorption and Evolution on Pt

The electrocatalytic hydrogen evolution reaction (HER) on Pt is one of the most fundamental electrochemical reactions and has been studied extensively [103–107]. It is well established that HER occurs via two successive elementary steps: the initial discharge of protons to give adsorbed hydrogen atoms ($H^+ + e^- \rightarrow H_{ads}$; Volmer step) followed by the combination of two adsorbed hydrogen atoms ($2 H_{ads} \rightarrow H_2$; Tafel step) or the combination of an adsorbed hydrogen atom with a proton in the solution and an electron to give H_2 ($H^+ + e^- + H_{ads} \rightarrow H_2$; Heyrovsky step). Hydrogen adsorption on Pt at potentials positive to the thermodynamic reversible potential of HER ($0.05 < E < 0.4$ V vs RHE), so-called underpotential deposition (UPD), has been well studied [106, 107]. However, HER occurs at less positive potentials where the electrode surface has been fully covered by UPD hydrogen (H_{UPD}), suggesting that H_{UPD} is not the intermediate of HER. Despite the extensive studies, the nature of the intermediate (commonly called overpotentially deposited hydrogen, H_{OPD}) and its relation to the reaction mechanism have not yet been fully understood.

The first IRAS study of hydrogen adsorption on single- and poly-crystalline Pt electrodes [4, 108] showed that a vibrational band appears around 2090 cm^{-1} at ~ 0.1 V (vs RHE) and increases in intensity at less positive potentials. This band was assigned to hydrogen atoms adsorbed at atop sites from its frequency, and the hydrogen species was suggested to be the intermediate of HER from the potential dependence of the band intensity. However, it should be noted that these measurements in the external reflection mode were limited to narrow potential regions where HER is not so significant because evolved H_2 gas is trapped in the thin-layer cavity between the working electrode and the cell window and disturbs both spectral and electrochemical measurements.

ATR-SEIRAS is more convenient than IRAS for this purpose because of its higher sensitivity and its ability to probe the electrode surface without the disturbance by H_2 gas evolution. Figure 8.21 A shows a set of SEIRA spectra on a Pt thin-film electrode measured in 0.5 M H_2SO_4 at the various potentials indicated with respect to the reference potential of 0.6 V [109]. All of the potentials shown are the values after correcting *ohmic* drops. A band appears at 2079–2096 cm^{-1} and grows as the potential is made less positive. Since this band was shifted to around 1500 cm^{-1} in D_2O, it is undoubtedly ascribed to adsorbed hydrogen species. The potential dependence of the band intensity is shown in Fig. 8.21 B (closed circles). The open circles were taken from the IRAS study [108] and multiplied by 5. The SEIRAS measurement is consistent with the IRAS study, but it should be noted that the S/N of the SEIRA spectrum is much

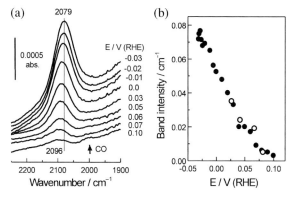

Fig. 8.21 Potential dependence of the SEIRA spectrum of hydrogen adsorbed on a Pt electrode in 0.5 M H$_2$SO$_4$ (A) and the integrated band intensity (B, closed circles). Open circles denote the IRAS data [108] multiplied by 5 [109].

higher than the IRA spectra, and, further, the measurements were extended to the potentials where vigorous H$_2$ evolution takes place. The better S/N and the expansion of the potential range revealed the potential-dependent peak shift of the band ($\sim$130 cm^{-1} V^{-1}). On the other hand, no bands corresponding to H$_{UPD}$ were observed in the spectral range above 1000 cm^{-1}, probably because of the adsorption at hollow sites on Pt [106, 107]; the Pt-H vibrational frequency for hydrogen atoms adsorbed at hollow sites is known to be located around 550 cm^{-1} [110].

The potential dependence of the band intensity suggests that the adsorbed hydrogen giving the band around 2090 cm^{-1} is the intermediate of HER. To confirm this, the SEIRAS data was quantitatively compared with the kinetics of HER (Tafel plot). The Tafel plot measured with the same electrode under the same conditions is shown in Fig. 8.22 A (open circles). The slope of 30 mV/decade can be explained in terms of the Volmer-Tafel mechanism, the latter process being rate-determining [103–105]. The slope of 60 mV/decade at $E > 0.02$ V in the semi-log plot of the intensity of the Pt-H band against potential (closed circles) is consistent with this mechanism (i.e., $\ln(\theta/\theta_0) = -FE/RT$, where θ_0 is the coverage at the equilibrium potential), although this relation holds also for the Volmer-Heyrovsky mechanism [103, 104].

A further conclusive support of the Volmer-Tafel mechanism is given by the log-log plot of the current density of HER (i) versus the band intensity shown in Fig. 8.22 B. The slope of 2 indicates $i \propto \theta^2$, which is the well-known rate equation of the Tafel step at limiting low θ (i.e., in the framework of the Langmuir isotherm of adsorption) [103–105].

The Tafel plot suggests that the Volmer-Tafel mechanism holds over the potential range examined; however, the slope in the semi-log plot of the band intensity deviates from 60 mV/decade at $E < 0.02$ V, where the intensity is proportional to E (Fig. 8.21 B). The reason for the inconsistency between the SEIRAS

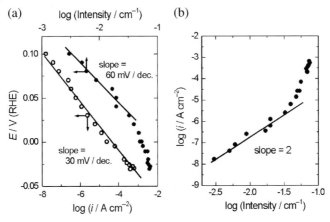

Fig. 8.22 (A) Semi-log plots of the current density (open circles) and band intensity of the Pt-H vibration (closed circles) versus applied potential, and (B) log-log plot of the band intensity versus the current density at the Pt electrode in 0.5 M H_2SO_4 [109].

and the kinetics data has not yet been clarified, but a possible explanation is the effect of the supersaturation of H_2 in the diffusion boundary layer region, which can give an apparent Tafel slope around 30 mV/decade [105, 106, 111].

8.5.5
Oxidation of C1 Molecules on Pt

Because of the great potential of methanol as a fuel for low-temperature fuel cells, the electro-oxidation of methanol on Pt or Pt-based alloy electrodes has been studied extensively in the past decades [112–115]. It is generally accepted that methanol is oxidized to CO_2 by the so-called dual-path mechanism [112]; via adsorbed CO (poison) and non-CO reactive intermediates. The formation of CO by dehydrogenation of methanol has been well confirmed, but no consensus has been reached so far on the nature of the reactive intermediates in the non-CO pathway. Various adsorbates such as CH_xOH [116], -COH [116], formyl (-HCO), [117] carboxy (-COOH) [117], a dimer of formic acid [35], and COO^- [38] have been claimed to be the reactive intermediates from IRAS and other physicochemical measurements. However, the spectra of the "reaction intermediates" are not well reproduced by other groups.

ATR-SEIRAS has been employed successfully to identify the reactive intermediate in methanol oxidation [118]. Figure 8.23a shows a series of SEIRA spectra of a Pt thin-film electrode in 0.1 M $HClO_4$ + 0.5 M methanol acquired during a potential sweep from 0.05 to 1.3 and back to 0.05 V (vs RHE) at a sweep rate of 5 mV s^{-1}. Two strong bands attributed to CO molecules linearly and bridge-bonded to the Pt electrode surface are observed at 2060 and

(a)
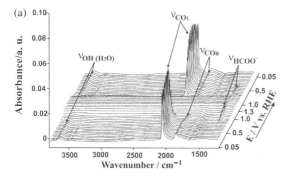

(b)
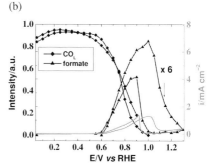

Fig. 8.23 (a) Series of SEIRA spectra of a Pt electrode in 0.1 M HClO$_4$ + 0.5 M methanol solution collected during a potential sweep from 0.05 to 1.3 and back to 0.05 V (vs RHE) at a rate of 5 mV s^{-1}. (b) Potential dependence of the band intensities of linear CO around 2060 cm^{-1} and formate at 1320 cm^{-1} taken from (a). The dotted trace shows the cyclic voltammograms recorded simultaneously with the spectra [118].

1860 cm^{-1}. The intensities of these bands are almost constant in the potential range of 0.05–0.5 V and decrease rapidly at more positive potentials. Accompanying the decrease in the CO band intensities, a new band appears around 1320 cm^{-1}. This band was assigned to the OCO symmetric stretching mode of formate, ν_s(OCO), adsorbed on the surface via two oxygen atoms at two adjacent Pt surface atoms. In addition, a sharp ν(OH) mode of water free from hydrogen bonding is seen around 3600 cm^{-1}. Since this band is observed on the CO-saturated Pt surface and disappears by CO oxidation, it was ascribed to water molecules embedded in the CO layer [32].

The intensity of the formate band is maximal at about 1 V on the positive-going sweep and about 0.9 V on the negative-going sweep as shown in Fig. 8.23b (triangles). The potential dependence of the formate band intensity can be well correlated with the CV recorded simultaneously with the SEIRA spectra (thin smooth trace in the figure), suggesting that formate is acting as a reactive intermediate of methanol oxidation. No bands corresponding to other intermediates proposed previously from IRAS studies were detected despite the very high sensitivity of SEIRAS.

It is well known that formic acid and formaldehyde are formed and dissolved into the solution as by-products in the methanol electro-oxidation [119–121]. SEIRAS studies revealed that both formaldehyde and formic acid also provide adsorbed formate on Pt electrodes in their oxidation [32, 33, 122, 123]. To identify the route of formate adsorption, the same experiment was carried out under flow conditions aiming to wash away produced formaldehyde and formic acid from the electrode surface, but no spectral changes were observed. The result suggests that formate is formed directly from methanol on the Pt surface.

The reaction scheme of methanol electro-oxidation proposed from the ATR-SEIRAS study is totally different from those proposed earlier, but is similar to that in methanol oxidation on Pt(111) covered with molecular oxygen in ambient gas phase [124, 125]. The adsorption of formate was not observed in earlier IRAS studies. The reason for this can be ascribed to the limited mass transport in the IRAS set-up: a simple calculation shows that the methanol concentration in the thin layer between the working electrode and a cell window is greatly reduced during the spectral acquisition [33]. Therefore, the detection of actual reaction intermediate is difficult with IRAS.

Although the processes involved in the formate formation are not yet understood, the decomposition (oxidation) of adsorbed formate to CO_2 is likely the rate-determining step common to all C1 molecules investigated. For further understanding, the kinetics of the formate decomposition to CO_2 was examined by using isotope-labeled formic acids ($H^{12}COOH$ and $H^{13}COOH$) [122]. Figure 8.24 shows a series of 80-ms time-resolved SEIRA spectra recorded during a solution exchange from 0.5 M H_2SO_4+0.1 M $H^{13}COOH$ to 0.5 M H_2SO_4+0.1 M $H^{12}COOH$ at a constant potential of 0.6 V (vs RHE). The $v_s(OCO)$ mode of ^{13}C-formate at 1305 cm^{-1} is replaced by that of ^{12}C-formate at 1323 cm^{-1} within 1 s (panel a), while the isotopic substitution for adsorbed CO was negligibly slow. The intensities of the ^{12}C- and ^{13}C-formate bands are plotted versus time after the solution exchange in Fig. 8.20 (triangles and circles, respectively). The sum of the two bands (squares) is nearly constant during the isotopic substitution, implying that the adsorption of formate is fast enough and in equilibrium with the desorption. The decay of the ^{13}C-formate band can be approximated by the first-order reaction kinetics, as is shown by the semi-logarithmic plot of the intensity against time (inset). From the slope of the plot, the rate constant of ~ 5 s^{-1} is evaluated.

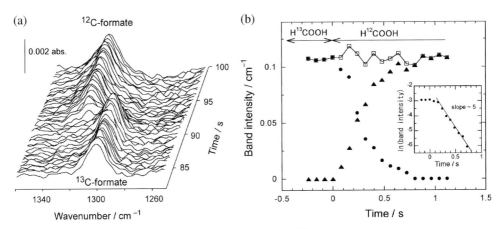

Fig. 8.24 (a) Series of 80-ms time-resolved SEIRA spectra of a Pt electrode recorded during a solution substitution of the solution from 0.5 M H_2SO_4+0.1 M $H^{13}COOH$ to 0.5 M H_2SO_4+0.1 M $H^{12}COOH$ at a constant potential of 0.6 V (vs RHE). (b) Plots of the band intensities of the ^{13}C-formate at 1300 cm^{-1} (circles) and ^{12}C-formate at 1325 cm^{-1} (triangles) as a function of time after the substitution. Squares are the sums of the two bands. Inset is a semi-log plot of the band intensities against time [123].

Two explanations are possible for the isotopic substitution of adsorbed formate; a simple adsorption/desorption of formate (HCOOH $\leftrightarrow$ HCOO$_{ads}$ +H$^+$+e$^-$) and the decomposition of formate (HCOO$_{ads}$ $\rightarrow$ CO$_2$+H$^+$+e$^-$). No net current flows in the former case, whereas current can flow in the latter case. The current is represented as

$$i = 2FSk_{decom}\theta_{formate}\Gamma_{Pt} \tag{7}$$

by tentatively assuming first-order reaction kinetics (inset in Fig. 8.24 b), where F is the Faraday constant, S is the real surface area of the electrode (10.8 cm^2), and Γ_{Pt} is the density of surface Pt atoms (1.3×10^{15} atoms cm^2). Assuming $\theta_{formate}$=0.05–0.5 (note 0.5 being the highest limit for formate in the bridging conformation), i=1–10 mA is calculated from Eq. (7). The current of 3 mA observed during the isotopic substitution experiment is well in the expected range. This result strongly argues that formate is a reactive intermediate in the formic acid oxidation (and also in methanol and formaldehyde oxidation) and the decomposition of formate to CO$_2$ is rate-determining. Further detailed analysis suggested that the oxidation of formic acid can be represented by a non-linear rate equation such as [122, 123]

$$i \propto -\frac{d\theta_{formate}}{dt} = k_{decom}\theta_{formate}(1 - 2\theta_{formate} - \theta_{CO\_L} - 2\theta_{CO\_B}) \tag{8}$$

where k_{decom} stands for the rate constant of the formate decomposition. The parenthesis represents the coverage of vacant sites. This equation can be reduced to the first-order rate equation in the analysis of the isotopic substitution experiment (Fig. 8.24) because θ_v is constant.

ATR-SEIRAS was also employed to examine methanol and CO oxidation on Pt-Ru and Pt-Fe alloy electrodes that have higher catalytic activity for methanol oxidation [36–38]. The enhanced catalytic activity is explained by a bifunctional mechanism [126] or a ligand (electronic) effect [127]. The former mechanism assumes that the second alloy element facilitates the adsorption of an oxygen source required for the oxidation CO (the poisoning species), while the ligand effect assumes the weakening of the Pt-CO bond. On the Pt-Ru electrode surface, three bands were observed at 3600–3620, ∼1615, and 1325 cm^{-1} in addition to linear and bridge CO at 1950–2060 and ∼1800 cm^{-1}, respectively [37, 38]. The first and second bands were assigned to the OH stretching and HOH bending modes of water adsorbed on Ru sites of the alloy electrode on the basis of the findings that the corresponding bands were observed on Ru electrodes but not on Pt electrodes. The frequency of the OH stretching band is much higher and sharper compared with the corresponding band for bulk water, suggesting that very weak hydrogen bonding occurs [28, 56]. The water and CO bands decreased in intensity together as the potential was made positive of 0.4 V (vs RHE) where oxidation current appears. The results were explained in terms of a bifunctional mechanism. That is, water adsorbed at Ru sites is discharged to form Ru-OH at 0.4 V, and CO is readily oxidized to CO_2 by the reaction with Ru-OH. The third band at 1325 cm^{-1} was assigned to Pt-COO$^-$, an intermediate for the CO oxidation [37, 38]. The frequency of this band is the same as that observed in methanol oxidation on a pure Pt electrode (Fig. 8.23). Since this band is shifted to 1300 cm^{-1} when CD_3OD is used instead of CH_3OH [118], the band is assigned more reasonably to formate rather than to COO$^-$.

8.6
Advanced Techniques for Studying Electrode Dynamics

8.6.1
Rapid-scan Millisecond Time-resolved FT-IR Measurements

Both dispersive monochrometers and Fourier Transform IR (FT-IR) spectrometers can be used in recording SEIRA spectra, but FT-IR spectrometers are more convenient owing to Fellget's, Jacquino's, and Connees's advantages [128]. The time-resolution of FT-IR spectroscopy is determined by the scan rate of the interferometer. Tens to hundreds of milliseconds time-resolution can easily be achieved by the use of modern rapid-scan FT-IR spectrometers at spectral resolutions of 4 or 8 cm^{-1}.

The combination of the high sensitivity of SEIRAS and a rapid-scan FT-IR spectrometer enables the spectral collection simultaneously with electrochemical measurements such as cyclic voltammetry and potential-step chronoamperometry. The time-resolved measurement can give some information on electrode kinetics and dynamics, as has been shown in Fig. 8.24. Figures 8.25 and 8.26 represent another example of millisecond time-resolved ATR-SEIRAS study of current oscillations during potentiostatic formic acid oxidation on a Pt electrode [123]. At a constant applied potential E of 1.1 V, the current oscillates as shown in Fig. 8.25a. Synchronizing with the current oscillations, the band intensities of linear CO and formate oscillate as shown in Fig. 8.26 (and also in Fig. 8.25c).

It is well established that ohmic drop plays an important role in the oscillation. Since the resistance of the system employed was 40 Ω (= solution resistance of 10 Ω + an external resistance of 30 Ω), the actual potential at the interface ($\phi = E - iR$) is ~0.5 V at the high current state (~15 mA) and ~0.7 V at the low current state (~10 mA). The periodic change in the actual potential at the interface is supported by the periodic frequency shift of the linear CO band (Fig. 8.25b). The adsorbed CO is oxidized at the low current state (i.e., at the high potential), which accelerates the direct oxidation path via formate and results in the decrease in potential through the increase in current. At the low potential, CO is accumulated and the direct path via formate is suppressed, which decreases the current yet again. This cycle repeats itself to give the sustained periodic current oscillations [123]. A similar scenario can be used to explain potential oscillations in galvanostatic formic acid oxidation [122].

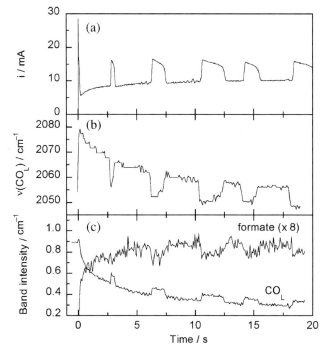

Fig. 8.25 (a) Current oscillations observed in the oxidation of formic acid at a constant potential of 1.1 V (RHE) in 0.5M $H_2SO_4 + 1M$ HCOOH with a 30-Ω external resistor. (b) Peak position of CO adsorbed at an atop site (CO_L). (c) Integrated band intensities of CO_L and formate taken from a series of SEIRA spectra acquired simultaneously with the current-time curve [122].

A higher time resolution of 28 ms was used in a rapid-scan time-resolved SEIRA study of adsorption kinetics of uracil on a quasi-Au(111) electrode surface [15]. Since the adsorption/desorption of uracil is reversible, the S/N of the spectra were enhanced by averaging 64 consecutive repetitions. This study showed that the kinetics is represented by the first-order rate equation, and that the rate constants evaluated from the time-resolved SEIRAS are much smaller than those for overall transition determined by chronoamperometry. The result suggests that the processes observed by the two measurements are not the same.

8.6.2
Step-scan Microsecond Time-resolved FT-IR Measurements

If the system to be examined is reversible or repeatable, much faster time resolutions are achieved by using a step-scan FT-IR spectrometer [129]. The interferometer is moved stepwise in the step-scan mode as shown in Fig. 8.27 (left). When the moving mirror of the interferometer is fixed at a position (i.e., at a

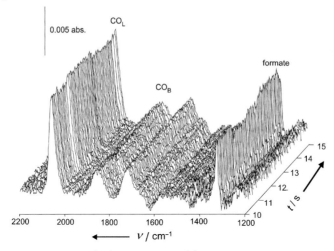

Fig. 8.26 Time-resolved SEIRA spectra of the Pt electrode surface acquired during the current oscillations with a time resolution of 80 ms, which corresponds to the 10–15 s region in Fig. 8.25 a. A spectrum of the bare Pt surface collected at 0.05 V (vs RHE) in the pure supporting electrolyte without formic acid was used as the reference [122].

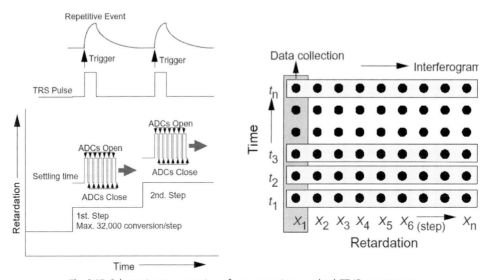

Fig. 8.27 Schematic representation of step-scan time-resolved FT-IR spectrometry.

fixed retardation), an event (reaction) is initiated by a trigger pulse from the controlling computer through a potentiostat, and the transient of the IR signal caused by the potential change is stored into the computer through a A/D converter. This cycle is repeated until the desired spectral range is covered. By rearranging the set of transient data at every retardation to time-resolved interfero-

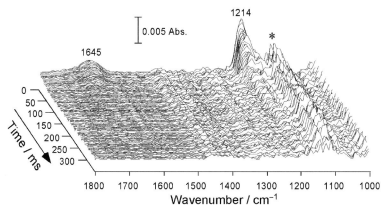

Fig. 8.28 Series of time-resolved infrared spectra of sulfate adsorbed on the Au(111) electrode in 0.1 M H_2SO_4 + 1 mM $CuSO_4$ recorded for the potential step from 0.1 to 0.3 V (vs SCE). The time resolution used was 1 ms. The 1224- and 1645-cm^{-1} bands are assigned to sulfate and water, respectively. The structure around 1100 cm^{-1} denoted by an asterisk arises from the silicon prism used [130].

grams at each time (Fig. 8.27, right), a set of time-resolved spectra can be obtained after Fourier transformation of the time-resolved interferograms. The time-resolution of the step-scan mode is limited by either the time-resolution of the A/D converter or the response of the detector (50–100 ns). The highest time resolutions of commercially available spectrometers are typically 5–10 µs, which is high enough for electrochemistry, because the response of most electrochemical systems is limited by the double-layer charging with the time-constant given by $\tau_{dl} = R_s C_{dl}$, where R_s and C_{dl} denote the solution resistance and double-layer capacitance, respectively. The typical value of τ_{dl} is 0.1–1 ms for ordinary electrodes (mm–cm in size) and 0.01–0.1 ms for microelectrodes.

As an example of step-scan time-resolved FT-IR measurement, the study of the dissolution dynamics of an underpotentially deposited (UPD) Cu layer on an Au(111) electrode is shown in Fig. 8.28 [130]. The set of 1-ms time-resolved spectra for a potential step from 0.1 to 0.3 V (vs SCE) represents the desorption of sulfate species coadsorbed with UPD Cu (the band at 1214 cm^{-1}). The transient of the intensity of the v(S-O) band of the sulfate species is plotted in Fig. 8.29b as a function of time. The corresponding current transient data is shown in Fig. 8.29a. The intensity data were fitted with a general rate equation

$$I = I_\infty[1 - \exp(-kt^n)] \tag{9}$$

where $n = 1$, 2, and 3 correspond to Langmuir kinetics and instantaneous and progressive nucleation-and-growth kinetics, respectively. A good fit to the experimental data was obtained for $n = 1$. On the other hand, the analysis of the chronoamperometry data suggests that the dissolution of the UPD Cu layer proceeds via an initial Langmuir-type process and a subsequent nucleation-and-

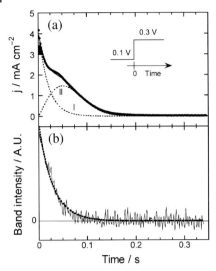

Fig. 8.29 Transients of current (a, open circles) and the sulfate band intensity (b) for a potential step from 0.1 to 0.3 V (vs SCE). The solid curve in (a) represents a numerical fit with Langmuir kinetics and spontaneous nucleation-and-growth kinetics. The contributions of the first and second kinetics in this equation are represented by dashed curves I and II, respectively. The infrared data in (b) were taken from Fig. 8.28. Numerical fit of the intensity transient using Langmuir kinetics is shown by the dashed curve [130].

growth process, as shown by dotted traces I and II, respectively. The rate constant of the initial current decay I was identical to that for the desorption of sulfate species, indicating that the initial process includes the desorption of sulfate species.

It is well established that the UPD Cu layer has the so-called honeycomb $(\sqrt{3} \times \sqrt{3})$ structure at the potential range of 0.1–0.2 V, in which Cu adatoms are arranged in honeycombs and sulfate species are adsorbed in the centers of the honeycombs [131, 132]. The coverages of Cu and sulfate species are 2/3 and 1/3 monolayers, respectively. By integrating the *I–t* curve, the charges for the initial and subsequent processes are calculated to be 85 and 125 μC cm^{-2}. The latter value corresponding to charge for the dissolution of 1/3 monolayer of Cu, while the former value corresponds to charge for the dissolution of 1/3 monolayer of Cu minus the charge for the desorption of sulfate species (40 μC cm^{-2} [132]). From these results, it was concluded that all sulfate species and 1/3 monolayer of Cu atoms are desorbed randomly in the initial process and then the remaining 1/3 monolayer of Cu atoms is desorbed in the subsequent nucleation-and-growth process. The conclusion is nearly consistent with a Dynamic Monte Carlo simulation [133]. This study demonstrates a fruitful combination of time-resolved IR and electrochemical measurements.

Another example demonstrates the importance of the choice of the time regions in time-resolved studies. Figure 8.30a shows the transient of the intensity

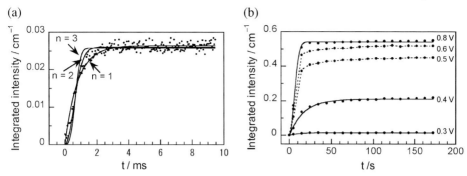

Fig. 8.30 Transient of the integrated band intensity of the $v_s(NO_2)$ mode of p-nitrobenzoate adsorbed on an Au electrode from 0.1 M $HClO_4$ + 1 mM p-nitrobenzoate for a potential step from 0.2 to 0.8 V (vs SCE) for (a) and from 0.2 V to the potentials shown in the figure for (b). Time resolution: (a) 50 µs and (b) 10 s. Solid curves in (a) represent the fittings with Langmuir kinetics (n=1) and spontaneous and progressive nucleation-and-growth kinetics (n=2 and 3, respectively) [14].

of the symmetric NO_2 stretching mode, $v_s(NO_2)$, of p-nitrobenzoic acid (PNBA) adsorbed on an Au electrode after stepping the potential from 0.2 to 0.8 V (vs SCE) (dots) measured with a time resolution of 50 µs [14]. The band intensity increases with time and saturates within about 3 ms. The rapid scan measurements with a time resolution of 10 s shown in Fig. 8.30b, however, shows that the band intensity further increases. The growth rate and saturated intensity are potential dependent. The results suggest that the adsorption of PNBA occurs via the fast and slow processes. Since the saturated intensity in the ms time-region is very small, PNBA is considered to adsorb in a flat orientation in the initial fast process and then to reorient to more vertical orientations via the carboxylate group of the molecule. The reorientation can increase both the band intensity and the coverage, which results in the second growth of the band intensity.

The transient intensity data were analyzed with Eq. (9). Although the very initial time region of the potential step is severely affected by the double-layer charging and the data points scatter, the best fit to the 50-µs time-resolved data was obtained for n=1 (i.e., random adsorption). On the other hand, reasonable fits to the 10-s time-resolved data were obtained for n=2. The change in kinetics can be explained by an initial random adsorption of flat-lying molecules followed by a gradual reorientaion via a spontaneous nucleation-and-growth mechanism. This explanation was supported by *in situ* STM observations.

Sub-millisecond time-resolved SEIRAS has also been applied for other dynamic processes such as the redox reactions of heptylviologen on Ag [12], adsorption/desorption of fumaric acid [13] on Au and sulfate on Ag [16], and phase transitions in uracil adlayers on Au [15]. In the latter two studies, discrepancies between the time-resolved IR and chronoamperometry were suggested.

8.6.3
Potential-modulated FT-IR Spectroscopy

Potential modulation techniques are used frequently in electrochemistry. The most well-known potential modulation electrochemical technique is a.c. impedance spectroscopy, in which current modulation caused by a potential modulation is analyzed. The potential modulation technique has also been used for *in-situ* IR spectroscopy (EMIRS and SNIFTIRS), but its use was aimed only to subtract the solution background and to enhance the S/N ratio of the spectrum. If the IR signal caused by a potential modulation is analyzed, some information on electrode dynamics could be obtained as in a.c. impedance spectroscopy.

To carry out such measurements with a conventional FT-IR spectrometer, the interferometer scan must be very slow or the potential modulation frequency must be very small or very large to avoid the cross-talk between the interferogram frequency and potential modulation frequency [134–136]. The slowest end of potential modulation corresponds to SNIFTIRS. A possible way to remove the cross-talk is with the use of the step-scan mode [135, 136]. A set-up for the potential-modulated FT-IR spectroscopy is schematically shown in Fig. 8.31 [137]. The electrode potential is modulated at a desired frequency and the infrared signal is demodulated with a lock-in amplifier (LIA) referenced to the potential modulation to give in-phase and quadrature (90° out-of-phase) components as in a.c. impedance spectroscopy. The d.c. component of the signal being passed through a low-pass filter can be used as the reference interferogram (i.e., the reference spectrum). Since the interferometer is moved step-wise and both a.c. and d.c. signals are collected at every optical retardation, the potential modulation frequency can be selected independently of the interferogram frequency. Very recently, a new technique that replaces the LIA with digital signal processing (DSP) technology has been developed [138].

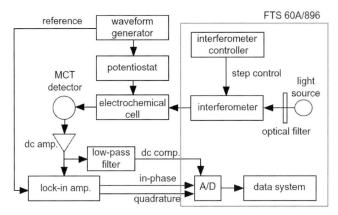

Fig. 8.31 Block diagram of the experimental set-up
for potential modulated FT-IR measurements [137].

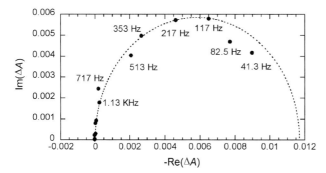

Fig. 8.32 Cole-Cole plot of the band intensity at 1618 cm^{-1} from a 4-PySH SAM on an Au electrode in in-phase and quadrature spectra [Re(ΔA) and Im(ΔA), respectively] measured for potential modulations between –0.1 and 0.3 V (vs SCE) at frequencies shown in the figure. Solution used was 0.1 M HClO$_4$ [137].

An example of potential-modulated FT-IR measurements is shown in Fig. 8.32. The figure shows a Cole-Cole plot of the in-phase [Re(ΔA)] and quadrature [Im(ΔA)] components of the band intensity at 1618 cm^{-1} from a 4-PySH SAM on an Au electrode. The data at several potential-modulation frequencies form a semicircle as in a.c. impedance, from which kinetic parameters can be extracted [137]. Kinetic parameters can also be obtained from the in-phase and quadrature components of the modulated IR signal measured at a fixed modulation frequency [138].

8.7
Summary and Future Prospects

In summary, SEIRAS in the ATR mode is superior to IRAS in surface sensitivity and interfacial selectivity, and is free from the mass transport limitation that is encountered in IRAS. These advantages of ATR-SEIRAS enable time-resolved monitoring of electrochemical reactions and have shifted the targets of *in-situ* IR spectroscopy from simple adsorption/desorption systems to more complicated reaction systems including bioelectrochemical reactions. The development of several new techniques for preparing thin-film electrodes of a variety of metals, especially Pt-group metals, has also significantly contributed to the widespread use of ATR-SEIRAS. Its application to electrocatalysis on Pt-group metal electrodes is giving fruitful results. Nevertheless, there still remains the drawback that well-defined single-crystal surfaces are not available. To study reactions that are sensitive to the crystallography of the electrode surface, the development of techniques to prepare single-crystal-like surfaces as for quasi-Au(111) is desirable.

As has been discussed in the preceding section, most of the recent ATR-SEIRAS measurements are carried out simultaneously with electrochemical measurements, from which some new kinetic and dynamic information can be obtained. The µs time resolutions achieved so far with the use of a step-scan FT-IR spectrometer are comparable to those of electrochemical measurements and are good enough for most electrochemical reaction systems. Since recent step-scan FT-IR spectrometers have nanosecond time-resolutions, faster reactions on microelectrodes that have smaller τ_{dl} could be examined with IR microscopy. Preliminary IR-microscopy measurements on microelectrodes have already been reported, although the time resolution is still limited to milliseconds [139, 140]. Further faster time-resolved measurements in the picosecond range are now becoming possible by coupling SEIRAS, a laser spectroscopy, and the so-called laser-induced temperature jump method [141]. Heating of the electrode surface with visible or near-IR pulses rapidly changes the electrode potential, and the reactions caused by the potential jump are monitored with a picosecond IR pulse in a pump-probe manner.

In addition to the expansion of time resolution, the expansion of spectral range to far IR is also desirable, since metal-adsorbate vibrations and surface vibrational modes occur in this range. The available spectral range is now limited by the prism (> 900 cm^{-1} for Si and > 700 cm^{-1} for Ge), the low intensity of the light source, and the low sensitivity of detectors. Since the absorption coefficient of Si fortunately decreases in the far-IR region, ATR-SEIRAS in the far-IR will be possible with the use of a synchrotron far-IR facility and a helium-cooled bolometer [142].

Acknowledgments

This work was supported partly by the Ministry of Education, Culture, Sports, Science and Technology of Japan (Grant-in-Aid for Basic Research No. 14205121 and for Scientific Research on Priority Areas 417) and the Japan Science and Technology Agency.

References

1 *Adsorption of Molecules at Metal Electrodes*; Lipkowski, J.; Ross, P. N., Eds.; VCH, New York, 1992.

2 *Interfacial Electrochemistry*; Wiekowski, A., Ed.; Mercel Dekker, New York, 1999.

3 Beden, B.; Lamy, C. In *Spectroelectrochemistry: Theory and Practice*; Gale, R. J., Ed.; Plenum, New York, 1988; pp 189.

4 Nichols, R. J. In *Adsorption of Molecules at Metal Electrodes*; Lipkowski, J., Ross, P. N., Eds.; VCH, New York, 1992; pp 347.

5 Iwasita, T.; Nart, F. C. *Prog. Surf. Sci.* **1997**, *55*, 271.

6 Bewick, A.; Kunimatsu, K.; Pons, B. S.; Russell, J. W. *J. Electroanal. Chem.* **1984**, *160*, 47.

7 Pons, B. S.; Davidson, T.; Bewick, A. *J. Electroanal. Chem.* **1984**, *169*, 63.

8 Osawa, M. *Bull. Chem. Soc. Jpn.* **1997**, *70*, 2861.

9 Osawa, M. *Top. Appl. Phys.* **2001**, *81*, 163.

10 Osawa, M. In *Handbook of Vibrational Spectroscopy*; Chalmers, J. M., Griffiths,

P. R., Eds.; Wiley: Chichester, 2002; Vol. 1; pp 785.

11 Osawa, M.; Ataka, K.; Yoshii, K.; Yotsuyanagi, T. *J. Electron Spectrosc. Relat. Phenom.* **1993**, *64/65*, 371.

12 Osawa, M.; Yoshii, K.; Ataka, K.; Yotsuyanagi, T. *Langmuir* **1994**, *10*, 640.

13 Noda, H.; Ataka, K.; Wan, L.-J.; Osawa, M. *Surf. Sci.* **1999**, *427/428*, 190.

14 Noda, H.; Wan, L.-J.; Osawa, M. *Phys. Chem. Chem. Phys.* **2001**, *3*, 3336.

15 Pronkin, S.; Wandlowski, T. *J. Electroanal. Chem.* **2003**, *550/551*, 131.

16 Rodes, A.; Orts, J.M.; Rerez, J.M.; Feliu, J.M.; Aldaz, A. *Electrochem. Commun.* **2003**, *5*, 56.

17 Bjerke, A.E.; Griffiths, P.R.; Theiss, W. *Anal. Chem.* **1999**, *71*, 1967.

18 *SurfaceEnhanced Raman Scattering*; Chang, R.K.; Furtak, T.E., Eds.; Plenum Press, New York, 1982.

19 Osawa, M.; Ataka, K.; Yoshii, K.; Nishikawa, Y. *Appl. Spectrosc.* **1993**, *47*, 1497.

20 Osawa, M.; Ikeda, M. *J. Phys. Chem.* **1991**, *95*, 9914.

21 Merklin, G.T.; Griffths, P.R. *Langmuir* **1997**, *13*, 6159.

22 Miyake, H.; Ye, S.; Osawa, M. *Electrochem. Commun.* **2002**, *4*, 973.

23 Osawa, M.; Kuramitsu, M.; Hatta, A.; Suëtaka, W.; Seki, H. *Surf. Sci. Lett.* **1986**, *175*, L787.

24 Osawa, M.; Ataka, K. *Surf. Sci. Lett.* **1992**, *262*, L118.

25 Kittel, C. *Introduction to Solid State Physics*, 7th ed.; Wiley, New York, 1996.

26 Pearce, H.A.; Sheppard, N. *Surf. Sci.* **1976**, *59*, 205.

27 Hoffmann, F.M. *Surf. Sci. Rep.* **1983**, *3*, 107.

28 Ataka, K.; Yotsuyanagi, T.; Osawa, M. *J. Phys. Chem.* **1996**, *100*, 10664.

29 Wandlowski, T.; Ataka, K.; Pronkin, S.; Diesing, D. *Electrochem. Acta* **2004**, *49*, 1233.

30 Miyake, H.; Osawa, M. *Chem. Lett.* **2004**, *33*, 278.

31 Loster, M.; Friedrich, K.A. *Surf. Sci.* **2003**, *523*, 287.

32 Miki, A.; Ye, S.; Osawa, M. *Chem. Commun.* **2002**, 1500.

33 Miki, A.; Ye, S.; Senzaki, T.; Osawa, M. *J. Electroanal. Chem.* **2004**, *563*, 23.

34 Zhu, Y.; Uchida, H.; Watanabe, M. *Langmuir* **1999**, *15*, 8757.

35 Zhu, Y.; Uchida, H.; Yajima, T.; Watanabe, M. *Langmuir* **2001**, *17*, 146.

36 Watanabe, M.; Zhu, Y.; Uchida, H. *J. Phys. Chem. B* **2000**, *104*, 1762.

37 Yajima, T.; Wakabayashi, N.; Uchida, H.; Watanabe, M. *Chem. Commun.* **2003**, 828.

38 Yajima, T.; Uchida, H.; Watanabe, M. *J. Phys. Chem. B* **2004**, *108*, 2654.

39 Nishikawa, Y.; Nagasawa, T.; Fujiwara, K.; Osawa, M. *Vib. Spectrosc.* **1993**, *6*, 43.

40 Yan, Y.-G.; Li, Q.-X.; Huo, S.-J.; Ma, M.; Cai, W.-B.; Osawa, M. *J. Phys. Chem. B* **2005**, *109*, 7900.

41 Zou, S.; Williams, C.T.; Chen, E.K.-Y.; Weaver, M.J. *J. Am. Chem. Soc.* **1998**, *120*, 3811.

42 Zou, S.; Weaver, M.J. *Anal. Chem.* **1998**, *70*, 2387.

43 Pronkin, S.; Wandlowski, T. *Surf. Sci.* **2004**, *573*, 109.

44 Sun, S.-G.; Cai, W.-B.; Wan, L.-J.; Osawa, M. *J. Phys. Chem.* **1999**, *103*, 2460.

45 Chang, S.-C.; Hamelin, A.; Weaver, M.J. *J. Phys. Chem.* **1991**, *95*, 5560.

46 Lu, G.-Q.; Sun, S.-G.; Chen, S.P.; Cai, L.-R. *J. Electroanal. Chem.* **1997**, *421*, 19.

47 Lu, G.-Q.; Sun, S.-G.; Cai, L.-R.; Chen, S.-P.; Tian, Z.-W. *Langmuir* **2000**, *16*, 778.

48 Zheng, M.-S.; Sun, S.-G. *J. Electroanal. Chem.* **2001**, *500*, 223.

49 Oritz, R.; Cuesta, A.; Marquez, O.P.; Marquez, J.; Mendez, J.A.; Gutierrez, C. *J. Electroanal. Chem.* **1999**, *465*, 234.

50 Orozco, G.; Gutierrez, C. *J. Electroanal. Chem.* **2000**, *484*, 64.

51 Pecharroman, C.; Cuesta, A.; Gutierrez, C. *J. Electroanal. Chem.* **2002**, *529*, 145.

52 Sun, S.-G. *J. Electroanal. Chem.* **2002**, *529*, 155.

53 Pecharroman, C.; Cuesta, A.; Gutierrez, C. *J. Electroanal. Chem.* **2004**, *563*, 91.

54 Hoon-Khosla, M.; Fawcett, W.R.; Chen, A.; Lipkowski, J.; Pettinger, B. *Electrochem. Acta* **1999**, *45*, 611.

55 Cai, W.-B.; Wan, L.-J.; Noda, H.; Hibino, Y.; Ataka, K.; Osawa, M. *Langmuir* **1998**, *14*, 6992.

56 Ataka, K.; Osawa, M. *Langmuir* **1998**, *14*, 951.

57 Li, N.; Zamlynny, V.; Lipkowski, J.; Hen-glein, F.; Pettinger, B. *J. Electroanal. Chem.* **2002**, *524/525*, 43.

58 Wilson, E. B. J.; Decius, J. C.; Cross, P. C. *Molecular Vibrations: The Theory of Infra-red and Raman Vibrational Spectra*; McGraw-Hill, New York, 1955.

59 Stöhr, J.; Outka, D. A. *Phys. Rev. B* **1987**, *36*, 7891.

60 Robert, J. T.; Friend, C. M. *J. Chem. Phys.* **1988**, *88*, 7172.

61 Carron, K. T.; Hurley, G. *J. Phys. Chem.* **1991**, *95*, 9979.

62 Stolberg, L.; Morin, S.; Lipkowski, J.; Irish, D. E. *J. Electroanal. Chem.* **1991**, *307*, 241.

63 Pfnür, H.; Menzel, D.; Hoffmann, F. M.; Ortega, A.; Bradshow, A. M. *Surf. Sci.* **1980**, *93*, 431.

64 Kunimatsu, K. *J. Electroanal. Chem.* **1986**, *213*, 149.

65 Leung, L.-W. H.; Wieckwski, A.; Weaver, M. J. *J. Phy. Chem.* **1988**, *92*, 6985.

66 Devllin, J. P.; Consani, K. *J. Phys. Chem.* **1981**, *85*, 2597.

67 Ferguson, E. E.; Matsen, F. A. *J. Am. Chem. Soc.* **1960**, *82*, 3268.

68 Diao, Y.-U.; Han, M.-J.; Wan, L.-J.; Itaya, K.; Uchida, T.; Miyake, H.; Yamakata, A.; Osawa, M. *Langmuir* **2006**, *22*, 3640.

69 Osawa, M.; Yoshii, K. *Appl. Spectrosc.* **1997**, *51*, 512.

70 Osawa, M.; Matsuda, N.; Yoshii, K.; Uchida, I. *J. Phys. Chem.* **1994**, *98*, 12702.

71 Cai, W.-B.; Amano, T.; Osawa, M. *J. Electroanal. Chem.* **2001**, *500*, 147.

72 Dornhaus, R.; Lang, M. B.; Benner, R. E. *Surf. Sci.* **1980**, *93*, 240.

73 Erdheim, G. R.; Birke, R. L.; Lombardi, J. R. *Chem. Phys. Lett.* **1980**, *69*, 495.

74 Hallmark, V. M.; Campion, A. *J. Chem. Phys.* **1986**, *84*, 2933.

75 Moskovits, M. *Rev. Mod. Phys.* **1985**, *57*, 783.

76 Muniz-Miranda, M.; Neto, N.; Sbrana, G. *J. Phys. Chem.* **1988**, *92*, 954.

77 Huang, Q. J.; Yao, J. L.; Mao, R. B.; Gu, R. A.; Tian, Z. Q. *Chem. Phys. Lett.* **1997**, *271*, 101.

78 Brolo, A. G.; Irish, E. D.; Szymanski, G.; Lipkowski, J. *Langmuir* **1998**, *14*, 517.

79 Arenas, J. F.; Woolley, M. S.; Tocon, I. L.; Otero, J. C.; Marcos, J. I. *J. Chem. Phys.* **2000**, *112*, 7669.

80 Osawa, M.; Matsuda, N.; Yoshii, K.; Uchida, I. *J. Phys. Chem.* **1994**, *98*, 12702.

81 Bewick, A.; Russell, J. W. *J. Electroanal. Chem.* **1982**, *132*, 329.

82 Bewick, A.; Russell, J. W. *J. Electroanal. Chem.* **1982**, *142*, 337.

83 Ataka, K.; Osawa, M. *J. Electroanal. Chem.* **1999**, *460*, 188.

84 Wan, L.-J.; Terashima, M.; Noda, H.; Osawa, M. *J. Phys. Chem. B* **2000**, *104*, 3563.

85 Wandlowski, T.; Lampner, D.; Lindsay, S. M. *J. Electroanal. Chem.* **1996**, *404*, 215.

86 Bewick, A.; Tuxford, A. M. *J. Electroanal. Chem.* **1973**, *47*, 255.

87 McIntyre, J. D. E. In *Advances in Electro-chemistry and Electrochemical Engineering*; Muller, R. H., Ed.; Wiley: New York, 1973; Vol. 9; pp 61.

88 Takamura, K.; Kusu, F. In *Methods of Biochemical Analysis*; Glick, D., Ed.; Wiley, New York, 1987; Vol. 32; pp 155.

89 Takamura, K.; Mori, A.; Watanabe, F. *Bioenerg.* **1981**, *8*, 125.

90 Futamata, M.; Diesing, D. *Vib. Spectrosc.* **1999**, *19*, 187.

91 Futamata, M. *Surf. Sci.* **1999**, *427/428*, 179.

92 Futamata, M.; Luo, L.; Nishihara, C. *Surf. Sci.* **2005**, *590*, 196.

93 Ye, S.; Zhou, W.; Abe, M.; Nishida, T.; Cui, L.; Uosaki, K.; Osawa, M.; Sakai, Y. *J. Am. Chem. Soc.* **2004**, *126*, 7343.

94 Zhou, W.; Ye, S.; Abe, M.; Nishida, T.; Uosaki, K.; Osawa, M.; Sasaki, Y. *Chem. Eur. J.* **2005**, *11*, 5040.

95 Taniguchi, I.; Yoshimoto, S.; Sunatsuki, Y.; Nishiyama, K. *Electrochemistry* **1999**, *67*, 1197.

96 Taniguchi, I.; Yoshimoto, S.; Yoshida, M.; Kobayashi, S.; Miyawaki, T.; Aono, Y.; Sunatsuki, Y.; Taira, H. *Electrochim. Acta* **2000**, *45*, 2843.

97 Sato, Y.; Noda, H.; Mizutani, F.; Yama-kata, A.; Osawa, M. *Anal. Chem.* **2004**, *76*, 5564.

98 Ataka, K.; Heberle, J. *J. Am. Chem. Soc.* **2003**, *125*, 4986.

99 Ataka, K.; Heberle, J. *J. Am. Chem. Soc.* **2004**, *126*, 9445.

100 Wan, L.-J.; Noda, H.; Hara, Y.; Osawa, M. *J. Electroanal. Chem.* **2000**, *489*, 68.

101 Sato, Y.; Mizutani, F. *J. Electroanal. Chem.* **1999**, *473*, 99.

102 Sato, Y.; Mizutani, F. *Electrochim. Acta* **2000**, *45*, 2869.

103 Parsons, R. *Trans. Faraday Soc.* **1958**, *34*, 1053.

104 Delahay, P. *Double Layer and Electrode Kinetics*; Interscience: New York, 1965.

105 Appleby, A. J.; Kita, H.; Chemila, M.; Bronoel, G. In *Encyclopedia of Electrochemistry of the Elements*; Bard, A. J., Ed.; Marcel Dekker: New York, 1982; Vol. IX-A; pp Chapter IX.

106 Conway, B. E. In *Interfacial Electrochemistry: Theory, experiment, and applications*; Wieckowski, A., Ed.; Marcel Dekker: New York, 1999; pp 131.

107 Jerkiewicz, G. *Prog. Surf. Sci.* **1998**, *57*, 137.

108 Nichols, R. J.; Bewick, A. *J. Electroanal. Chem.* **1988**, *243*, 445.

109 Kunimatsu, K.; Senzaki, T.; Tsushima, M.; Osawa, M. *Chem. Phys. Lett.* **2005**, *401*, 451.

110 Baro, A. M.; Ibach, H.; Bruchman, H. D. *Surf. Sci.* **1979**, *88*, 384.

111 Briter, M. In *Transactions of the Symposium on Electrode Processes*; Yeager, E., Ed.; Wiley: New York, 1961; pp 307.

112 Parsons, R.; VanderNoot, T. *J. Electroanal. Chem.* **1988**, *257*, 9.

113 Jarvi, T. D.; Stuve, E. M. In *Electrocatalysis*; Lipkowski, J., Ross, P. N., Eds.; Wiley-VCH, New York, 1998; pp 75.

114 Sun, S.-G. In *Electrocatalysis*; Lipkowski, J., Ross, P. N., Eds.; Wiley-VCH, 1998; pp 243.

115 Markovic, N. M.; Ross, P. N. *Surf. Sci. Rep.* **2002**, *45*, 117.

116 Xia, X. H.; Iwasita, T.; Ge, F.; Vielstich, W. *Electrochim. Acta* **1996**, *41*, 711.

117 Beden, B.; Leger, J. M.; Lamy, C. In *Modern Aspects of Electrochemistry*; Bockris, J. O. M., Conway, B. E., Eds.; Plenum Press: New York, 1992; Vol. 22; pp 97.

118 Chen, Y. X.; Miki, A.; Ye, S.; Sakai, H.; Osawa, M. *J. Am. Chem. Soc.* **2003**, *125*, 3680.

119 Korzeniewski, C.; Childers, C. L. *J. Phys. Chem. B* **1998**, *103*, 489.

120 Wang, H.; Loffler, T.; Baltruschat, H. *J. Appl. Electrochem.* **2001**, *31*, 759.

121 Jusys, Z.; Behm, R. J. *J. Phys. Chem. B* **2001**, *105*, 10874.

122 Sameské, G.; Miki, A.; Ye, S.; Yamakata, A.; Mukouyama, Y.; Okamoto, H.; Osawa, M. *J. Phys. Chem. B* **2005**, *109*, 23509.

123 Sameské, G.; Osawa, M. *Angew. Chem. Int. Ed.* **2005**, *44*, 5694.

124 Endo, M.; Matsumoto, T.; Kubota, J.; Domen, K.; Hirose, C. *J. Phys. Chem. B* **2000**, *104*, 4916.

125 Endo, M.; Matsumoto, T.; Kubota, J.; Domen, K.; Hirose, C. *J. Phys. Chem. B* **2001**, *105*, 1573.

126 Watanabe, M.; Motoo, S. *J. Electroanal. Chem.* **1975**, *60*, 267.

127 Iwasita, T.; Nart, F. C.; Vielstich, W. *Ber. Bunsenges. Phys. Chem.* **94**, *94*, 1030.

128 Griffiths, P. R. *Chemical Infrared Fourier Transform Spectroscopy*; Wiley: New York, 1975.

129 Nakano, T.; Yokoyama, T.; Toriumi, H. *Appl. Spectrosc.* **1993**, *47*, 1354.

130 Ataka, K.; Nishina, G.; Cai, W.-B.; Sun, S.-G.; Osawa, M. *Electrochem. Commun.* **2000**, *2*, 417.

131 Hachiya, T.; Honbo, H.; Itaya, K. *J. Electroanal. Chem.* **1991**, *315*, 275.

132 Shi, Z.; Lipkowski, J. *J. Electroanal. Chem.* **1994**, *365*, 303.

133 Rikvold, P. A.; Brown, G.; Novotny, M. A.; Wieckowski, A. *Colloids and Surfaces A* **1998**, *134*, 3.

134 Chazalviel, J.-N.; Dubin, V. M.; Mandel, K. C.; Ozanam, F. *Appl. Spectrosc.* **1993**, *47*, 1411.

135 Budevska, B. O.; Griffiths, P. R. *Anal. Chem.* **1993**, *65*, 2963.

136 Budevska, B. O.; Manning, C. J.; Griffiths, P. R.; Roginski, R. T. *Appl. Spectrosc.* **1993**, *47*, 1843.

137 Ataka, K.; Hara, Y.; Osawa, M. *J. Electroanal. Chem.* **1999**, *473*, 34.

138 Brevnov, D. A.; Hutter, E.; Hendler, J. H. *Appl. Spectrosc.* **2004**, *58*, 184.

139 Gong, H.; Sun, S.-G.; Li, J.-T.; Chen, Y.-J.; Chen, S.-P. *Electrochim. Acta* **2003**, *48*, 2933.

140 Zhou, Z.-Y.; Tian, N.; Chen, Y.-J.; Chen, S.-P.; Sun, S.-G. *J. Electroanal. Chem.* **2004**, *573*, 111.

141 Yamakata, A.; Uchida, T.; Kubota, J.; Osawa, M. *J. Phys. Chem. B* **2006**, *110*, 6423.

142 Melendres, C. A.; Hahn, F. *J. Electroanal. Chem.* **1999**, *463*, 258.

9

Quantitative SNIFTIRS and PM IRRAS of Organic Molecules at Electrode Surfaces

Vlad Zamlynny and Jacek Lipkowski

9.1
Introduction

The history of infrared reflection absorption spectroscopy (IRRAS) at metal surfaces begins with the work of Greenler [1–3] in the late 1960s. Greenler gave a theoretical background of the technique and measured infrared spectra of organic films adsorbed at metal mirrors. During the last few decades, this technique has been significantly expanded, and it is at the present time used to study thin films at both metal/gas (MG) and metal/solution (MS) interfaces.

The main challenge associated with IRRAS studies of organic films at the MS interface stems from the strong absorbance of infrared radiation by aqueous electrolyte. Two experimental configurations are currently used to overcome this difficulty. The first employs attenuated total internal reflection (ATR) spectroscopy using the cell in the Kretschmann configuration [4]. This cell is equipped with an optical window (usually a hemisphere of Si or Ge) that works in the total reflection regime. The beam enters the window and is totally reflected at the window/ solution interface. Only an evanescent wave penetrates into the solution at a depth of $\sim 1\ \mu m$. A film of metal nanoparticles ($\sim 20\ nm$ diameter) deposited directly onto the optical window, serves as an electrode in ATR spectroscopy. The most significant advantage of this method is a considerable enhancement of the electric field of the infrared beam at the window/electrolyte interface, i.e. at the location of the adsorbate molecules. Therefore, the technique is called surface-enhanced infrared reflection absorption spectroscopy (SEIRAS). In addition, this technique is very surface-selective, because the electric field of the totally reflected infrared radiation rapidly decays on the solution side of the interface. This method is described in detail in Chapter 8 of this book. Despite its many merits, ATR cannot be used to study adsorption of organic molecules at single crystal surfaces. Further, the exact structure of the film is not known, and hence it is difficult to model the interface in order to determine the mean square electric field strength (MSEFS) at the metal surface. Consequently it is difficult to determine the values of the tilt angles of molecules adsorbed at the electrode.

Advances in Electrochemical Science and Engineering Vol. 9.
Edited by Richard C. Alkire, Dieter M. Kolb, Jacek Lipkowski and Philip N. Ross
Copyright © 2006 WILEY-VCH Verlag GmbH & Co. KGaA, Weinheim
ISBN: 3-527-31317-6

The second approach applies IRRAS using a thin-layer cell in which the electrode is pressed against the optical window until it is separated by a several micrometers-thick layer of electrolyte [5, 6]. The beam is transmitted through the window/solution interface, travels through the thin layer of the electrolyte, and is reflected from the electrode surface. Such a configuration relies on the external reflection of the beam. Single-crystal electrodes may be used in this configuration, allowing studies of the effect of surface crystallography on the structure of the adsorbed layer. However, the spectrum of the analyzed film is superimposed on a spectrum of the electrolyte. It is difficult (even if it is possible) to acquire the background spectrum (the spectrum without the film) at exactly the same cavity thickness as that of the thin-layer cell during the acquisition of the film spectrum. Therefore, the background correction is made by either (a) potential modulation or (b) polarization modulation. The first technique is called subtractively normalized interfacial Fourier transform infrared spectroscopy (SNIFTIRS) [7, 8]. The second technique is called polarization modulation (Fourier transform) infrared reflection absorption spectroscopy (PM IRRAS) [9–15].

The object of this chapter is to describe recent methodological developments in SNIFTIRS and PM IRRAS that have made these techniques powerful quantitative tools to measure tilt angles of molecules in thin organic films adsorbed at metal electrode surfaces. This chapter complements a general description of recent advances in IR spectroscopy in Chapter 7 and a review of SEIRAS in Chapter 8.

9.2
Reflection of Light from Stratified Media

Quantitative analysis of IRRAS spectra is possible because the physics of reflection and refraction of electromagnetic radiation is well understood. Reflection of light in a thin-layer cell can be described theoretically using Fresnel equations. Analytical solutions of Fresnel equations are relatively simple for reflection at the interface between two phases. However, the complexity of these equations increases significantly when the interface consists of more than three layers [16]. In this case, it is more convenient to describe reflection using matrix algebra, first derived by Abelés [17, 18]. Both methods are described in detail in a seminal paper by Hansen [19]. A very good description of the reflection of light at the electrified interface is also given in a recent review by Lipert et al. [21].

Since this chapter is devoted to quantitative analysis of the IRRAS spectra, for completeness of the presentation, we provide fundamentals of reflection of light at interfaces, although this material is available elsewhere [19–21]. The next section consists of a brief introduction to the reflection, refraction, and absorption of electromagnetic radiation at a simple two-phase boundary. The section that follows presents Fresnel equations in the matrix form that describes the reflection and refraction of electromagnetic radiation at a multilayer interface [19].

9.2.1
Reflection and Refraction of Electromagnetic Radiation at a Two-phase Boundary

Infrared radiation is a transverse, electromagnetic wave that is characterized by the orthogonal electric and magnetic fields oscillating in the directions perpendicular to the direction of the wave propagation. The energy of the magnetic field is exactly equal to the energy of the electric field of an electromagnetic wave. Hence, it is convenient to consider only the electric field.

A planar electromagnetic wave (i.e., a wave with a flat wavefront) that propagates along the x axis in vacuum can be described by a wave equation that expresses the electric field strength E at an arbitrary moment of time t and at a distance x from the origin:

$$E = E_0 \cos\left(\frac{2\pi}{\lambda}x + \delta - \omega t\right) = \mathrm{Re}\left(E_0 \exp\left[-i\left(\frac{2\pi}{\lambda}x + \delta - \omega t\right)\right]\right) \tag{1}$$

where E_0 denotes the amplitude of the electric field strength and λ, ω, and δ are the wavelength, the angular frequency, and the phase shift, respectively. The last two quantities are related to the frequency ν and the displacement Δx of the electric field strength amplitude with respect to the coordinate origin as described by the following equations:

$$\omega = 2\pi\nu \quad \text{and} \quad \delta = \Delta x \cdot 2\pi/\lambda. \tag{2}$$

The propagation of the electromagnetic wave in a condensed medium is described by

$$E = \mathrm{Re}\left\{E_0 \exp\left[-i\left(\frac{2\pi n}{\lambda}x + \delta - \omega t\right)\right] \exp\left(-\frac{2\pi k}{\lambda}x\right)\right\} \tag{3}$$

where $n = c/v$ is the refractive index that takes into account the fact that the rate of propagation v in a condensed medium is slower than the speed in vacuum (c). The attenuation coefficient k takes into account absorption of the electromagnetic wave by the medium. It is convenient to define a complex refractive index $\hat{n}$ of a condensed phase:

$$\hat{n} = n + ik. \tag{4}$$

The electromagnetic wave equation may then be written in a shorter form:

$$E = \mathrm{Re}\left\{E_0 \exp\left[-i\left(\frac{2\pi\hat{n}}{\lambda}x + \delta - \omega t\right)\right]\right\}. \tag{5}$$

It is important to note that the convention that regards k as a positive number is used throughout this chapter.

A detector of infrared radiation measures absorbed radiation, and the signal at the detector terminal is proportional to the energy dissipated per unit volume of the detector element per unit time. The time-averaged value of this energy is called the intensity I, irradiance, or radiation power. For non-magnetic materials it is related to the amplitude of the electric field of the incident radiation according to the formula [19]:

$$I = \varepsilon_0 c E_0^2 / 2 = \varepsilon_0 c \langle E_0^2 \rangle \tag{6}$$

where ε_0 is the permittivity of vacuum.

When the electromagnetic radiation is incident at a boundary between two phases, part of the beam is reflected and part of it is transmitted into the second medium. Figure 9.1 shows the plane of incidence at the boundary formed by a transparent and an absorbing phase. The plane of incidence is the plane that contains the incident and reflected beams. Figure 9.1 shows that the incident beam can be envisaged as composed of two linearly polarized orthogonal components. These components, called p- and s-polarized components, are characterized by the directions of the electric field vectors in the plane of incidence and perpendicular to it, respectively. When the electric field of the radiation is located only in the plane of incidence, the beam is referred to as p-polarized. On the other hand, when the electric field of the radiation is located only in the plane perpendicular to the plane of incidence the beam is said to be s-polarized.

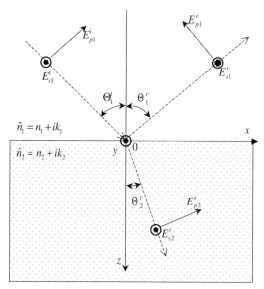

Fig. 9.1 Reflection and refraction of electromagnetic radiation at the boundary between a transparent and an absorbing medium. Dashed lines denote the direction of propagation of radiation. Arrows indicate the electric field vector of p-polarized light, which lies in the plane of incidence. Symbols ⊙ and ⊕ indicate the advancing and retreating electric field vectors of s-polarized radiation (which is perpendicular to the plane of incidence). In addition, symbol ⊙ shows the (advancing) direction of the y coordinate axis.

The coordinate system has the x and z axes in the plane of incidence and the y axis normal to the plane of incidence, as shown in the figure. The propagation directions of the electromagnetic radiation are indicated by dashed lines.

In accordance with the law of reflection

$$\Theta_1^i = \Theta_1^r \tag{7}$$

and following Snell's law of refraction

$$n_2 \sin \Theta_2^t = n_1 \sin \Theta_1^i \tag{8}$$

where n_j denotes the refractive index of a phase j. Equations (7) and (8) were discovered empirically; however, they can be derived from Maxwell's equations for isotropic phases and using the conditions of continuity of the electric and magnetic field components tangential to the phase boundary. It follows from Eq. (8) that if $n_1 > n_2$ there is an angle Θ_1^i at which $\Theta_2^t = 90°$. This is the critical angle Θ_{1c}^i, and its value can be determined from the expression

$$\sin \Theta_{1c}^i = \frac{n_2}{n_1} . \tag{9}$$

At $\Theta_1^i > \Theta_{1c}^i$ a total reflection takes place at the interface between phase 1 and phase 2 and the beam is no longer transmitted into phase 2.

Fresnel equations relate the electric field strength amplitudes of the incident, reflected, and transmitted waves. They are solutions of Maxwell's equations by applying the above-mentioned boundary conditions. It can be shown that for a plane boundary between two non-magnetic isotropic phases of infinite thickness, schematically depicted in Fig. 9.1, the Fresnel reflection (r) and transmission (t) coefficients for s- and p-polarized light are given by the following equations:

$$r_s = \frac{E_{s1}^r}{E_{s1}^i} = \frac{\xi_1 - \xi_2}{\xi_1 + \xi_2} \tag{10}$$

$$r_p = \frac{E_{p1}^r}{E_{p1}^i} = \frac{\hat{n}_2^2 \xi_1 - \hat{n}_1^2 \xi_2}{\hat{n}_2^2 \xi_1 + \hat{n}_1^2 \xi_2}$$

$$t_s = \frac{E_{s2}^t}{E_{s1}^i} = \frac{2\xi_1}{\xi_1 + \xi_2} \tag{11}$$

$$t_p = \frac{E_{p2}^t}{E_{p1}^i} = \frac{2\hat{n}_2^2 \xi_1}{\hat{n}_2^2 \xi_1 + \hat{n}_1^2 \xi_2} \left(\frac{\hat{n}_1}{\hat{n}_2}\right)$$

where

$$\xi_j = \hat{n}_j \cos \Theta_j^t = \sqrt{\hat{n}_j^2 - n_1^2 \sin^2 \Theta_1^t} = \mathrm{Re}\,\xi_j + i\,\mathrm{Im}\,\xi_j \tag{12}$$

and the symbols E_y denote the amplitudes of the electric fields, oriented along the y axis, for s- and p-polarized radiation, respectively. Note that ξ_j can be calculated from experimentally available optical constants of the corresponding phases $\hat{n}_j$ and the angle of incidence Θ_1^t. Both $\mathrm{Re}\,\xi_j$ and $\mathrm{Im}\,\xi_j$ must always be taken as positive values when calculating the square root of the complex number in Eq. (12), because, consistently with convention, k must be a positive number.

The reflectance (also called reflectivity) (R) and transmittance (T) are, respectively, given by the intensities of the reflected and transmitted radiation normalized by the intensity of the incident radiation. They are related to the Fresnel reflection and transmission coefficients in the following way:

$$R_{s,p} = \frac{I_{s,p1}^r}{I_{s,p1}^i} = |r_{s,p}|^2 = \hat{r}_{s,p} \cdot \hat{r}_{s,p}^* \tag{13}$$

$$T_s = \frac{I_{s2}^t}{I_{s1}^i} = \frac{\mathrm{Re}(\xi_2)}{\xi_1}|t_s|^2 = \frac{\mathrm{Re}(\xi_2)}{\xi_1}\hat{t}_s \cdot \hat{t}_s^* \tag{14}$$

and

$$T_p = \frac{\mathrm{Re}(\xi_2/\hat{n}_2^2)}{\xi_1/\hat{n}_2^2}|t_p|^2 = \frac{\mathrm{Re}(\xi_2/\hat{n}_2^2)}{\xi_1/\hat{n}_2^2}\hat{t}_p \cdot \hat{t}_p^* . \tag{15}$$

Finally, the phase usually changes on reflection or refraction. The phase shift δ can be determined from the amplitude coefficients using the following relationships:

$$\delta_{s,p}^r = \mathrm{Arg}(r_{s,p}) = \tan^{-1}\left[\frac{\mathrm{Im}(r_{s,p})}{\mathrm{Re}(r_{s,p})}\right] . \tag{16}$$

The electric field of the electromagnetic wave at the surface is a vectorial sum of the fields of the incident and reflected waves. Depending on the phase shift of the reflected wave, the sum gives enhancement or attenuation of the electric field. The components of the MSEFS of the IR photon at the surface in directions perpendicular and parallel to the surface ($z = 0$) can be calculated with the help of the following formulas [19]:

$$\frac{\langle E_s^2(z=0)\rangle}{\langle E_{s1}^2\rangle} = (1 + R_s + 2\sqrt{R_s}\cos\delta_s)/2$$

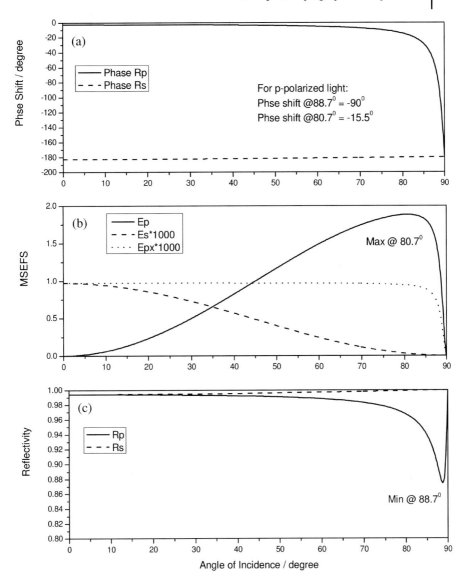

Fig. 9.2 Angular dependences of the phase shift (a), mean square electric field strength (MSEFS) (b), and reflectivity (c) for an air/Au interface at 1600 cm^{-1}. Optical constants for gold at this wavenumber are $n=3.04$, $k=45.26$.

$$\frac{\langle E_{px}^2(z=0)\rangle}{\langle E_{p1x}^2\rangle} = (\cos^2\theta_1[1 + R_p - 2\sqrt{R_p}\cos\delta_p])/2$$

$$\frac{\langle E_{pz}^2(z=0)\rangle}{\langle E_{p1z}^2\rangle} = (\sin^2\theta_1[1 + R_p + 2\sqrt{R_p}\cos\delta_p])/2 \, . \tag{17}$$

Equations (16) and (17) can now be used to calculate the phase shift and MSEFS of p- and s-polarized electromagnetic waves as a function of the angle of incidence. Figure 9.2a and b show results of such calculations for waves with $1600 \, \text{cm}^{-1}$ wavenumber reflected at the air/gold interface. The optical constants for gold were taken from Ref. [22]. For an s-polarized wave, the phase shift is ca. $-180°$ at all angles. Consequently, the incident and reflected waves cancel each other, and MSEFS is always negligibly small for this polarization. For a p-polarized beam, the phase shift is small for angles of incidence less than $80°$. The MSEFS increases with the angle of incidence because of an increase in the magnitude of the components of the electric field vectors of the incident and reflected beams in the z direction. The vectorial addition of the two vectors in Figure 9.3 shows that the sum increases in the case of a p-polarized wave and cancels in the case of an s-polarized wave. It is important to note that the x-component of the p-polarized wave is also cancelled at the metal surface. The maximum enhancement of the p-polarized wave approaches a factor of two at the angle of incidence $80.7°$ (grazing angle of incidence). Figures 9.2a and 9.3 illustrate the so-called surface selection rules [23], which state that the electric field of p-polarized radiation is enhanced and the electric field of s-polarized beam is attenuated in the vicinity of a metal surface.

Panel c plots reflectance of the gold surface. The reflectance has a minimum at the angle of $88.7°$, which is different from the angle of maximum MSEFS of $80.7°$. Hence, the best performance of spectroscopic measurements is achieved at angles that are different from the position of reflection minimum (maximum absorption by a metal surface).

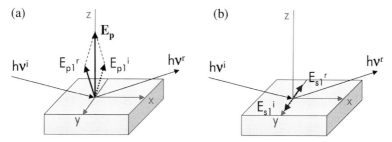

Fig. 9.3 Directions of the electric field vectors of the incident (dotted arrows) and the reflected (solid arrows) IR beams at the air/gold interface for (a) p-polarized and (b) s-polarized radiation.

9.2.2
Reflection and Refraction of Electromagnetic Radiation at a Multiple-phase Boundary

When electrochemical *in-situ* IRRAS experiments are performed using an electrode separated from an IR window by a thin layer of electrolyte, the reflection takes place in a four-phase system. The first medium is an infrared transparent window (non-absorbing with $k_1 = 0$), while the other three phases that represent electrolyte, an adsorbed film and a metal, may absorb light. The optical properties of a phase j are defined by its complex refractive index $\hat{n}_j$. It is convenient to employ matrix algebra to derive Fresnel equations for such a complex interface. The equations given below were derived assuming that the phases are non-magnetic, isotropic, homogeneous, and infinitely parallel, with perfectly sharp boundaries [19]. The first and the last phases are considered to be infinitely thick, and all the intermediate phases j are characterized by the thicknesses h_j. The incident beam was regarded to be a planar wave of wavelength λ directed at an angle Θ_1^t with respect to the z axis, which coincides with the normal to the planes of the phase boundaries.

For N phases with $N-1$ boundaries, one can define $N-1$ matrices M_j that fully characterize the electromagnetic radiation at each phase boundary. These matrices are called characteristic matrices and are given by Eqs. (18) and (19) for s- and for p-polarized electromagnetic radiation, respectively.

$$M_j^s = \begin{bmatrix} \cos\beta_j & \frac{-i}{\xi_j}\sin\beta_j \\ -i\xi_j\sin\beta_j & \cos\beta_j \end{bmatrix} \tag{18}$$

$$M_j^p = \begin{bmatrix} \cos\beta_j & \frac{-i\hat{n}_j^2}{\xi_j}\sin\beta_j \\ -i\frac{\xi_j}{\hat{n}_j^2}\sin\beta_j & \cos\beta_j \end{bmatrix} \tag{19}$$

where ξ_j is defined by Eq. (12) and

$$\beta_j = \frac{2\pi\xi_j h_j}{\lambda} \tag{20}$$

and λ is the wavelength of the incident radiation and h_j is the thickness of a layer j. The characteristic matrix of the whole stratified medium is obtained by multiplication of individual characteristic matrices of all its constituents that are sandwiched between the first and the last phase

$$M = \prod_{j=2}^{j=N-1} M_j . \tag{21}$$

The reflection and transmission coefficients for the whole stack can be extracted from the characteristic matrix of the whole stratified medium using the following relationships:

$$\hat{r}_s = \frac{(m_{11} + m_{12}\xi_N)\xi_1 - (m_{21} + m_{22}\xi_N)}{(m_{11} + m_{12}\xi_N)\xi_1 + (m_{21} + m_{22}\xi_N)} \tag{22}$$

$$\hat{r}_p = \frac{(m_{11} + m_{12}\xi_N/\hat{n}_N^2)\xi_1/\hat{n}_1^2 - (m_{21} + m_{22}\xi_N/\hat{n}_N^2)}{(m_{11} + m_{12}\xi_N/\hat{n}_N^2)\xi_1/\hat{n}_1^2 + (m_{21} + m_{22}\xi_N/\hat{n}_N^2)} \tag{23}$$

$$\hat{t}_s = \frac{2\xi_1}{(m_{11} + m_{12}\xi_N)\xi_1 + (m_{21} + m_{22}\xi_N)} \tag{24}$$

$$\hat{t}_p = \frac{2\xi_1/\hat{n}_1^2}{(m_{11} + m_{12}\xi_N/\hat{n}_N^2)\xi_1/\hat{n}_1^2 + (m_{21} + m_{22}\xi_N/\hat{n}_N^2)} \left(\frac{n_1}{\hat{n}_N}\right) \tag{25}$$

where m_{jk} are the elements of the 2×2 characteristic matrix M of the whole stack. Reflectance of the whole assembly for s- and p-polarized light can be calculated from the corresponding reflection coefficients using Eq. (13), while transmittance of the whole stack for s- and p-polarized light can be determined from the corresponding transmission coefficients using Eqs. (14) and (15), respectively.

Quantitative analysis of the orientation of organic molecules at the metal electrode requires precise knowledge of the mean square electric field strength (MSEFS) at the metal surface and in the bulk of the thin-layer cavity. The tangential (with respect to the propagation direction) fields $U_k(z)$ and $V_k(z)$ at an arbitrary point within the stratified medium are related to the fields U_1 and V_1 at the first interface $(z=z_1=0)$ by the following matrix:

$$\begin{bmatrix} U_k(z) \\ V_k(z) \end{bmatrix} = N_k(z) \prod_{j=k-1}^{2} N_j \begin{bmatrix} U_1 \\ V_1 \end{bmatrix} \tag{26}$$

where for s-polarized radiation $U_1 = E_y$ and $V_1 = H_x$, while for p-polarized radiation $U_1 = H_y$ and $V_1 = E_x$. E and H symbolize the amplitudes of electric and magnetic fields, respectively.

$U_k(z)$ and $V_k(z)$ denote the tangential fields at an arbitrary point within the stratified medium. Matrix N_j is a reciprocal to matrix M_j described by Eqs. (18) and (19). The other terms of Eq. (26) are described by the following expressions for s-polarized light:

$$\begin{bmatrix} U_1 \\ V_1 \end{bmatrix} = \begin{bmatrix} 1 + r_s \\ \xi_1(1 - r_s) \end{bmatrix} E_{y1}^i \tag{27}$$

$$N_k^s(z) = \begin{bmatrix} \cos\left(\dfrac{2\pi\xi_k(z - z_{k-1})}{\lambda}\right) & \dfrac{i}{\xi_k}\sin\left(\dfrac{2\pi\xi_k(z - z_{k-1})}{\lambda}\right) \\ i\xi_k\sin\left(\dfrac{2\pi\xi_k(z - z_{k-1})}{\lambda}\right) & \cos\left(\dfrac{2\pi\xi_k(z - z_{k-1})}{\lambda}\right) \end{bmatrix} \tag{28}$$

and for p-polarized light:

$$
\begin{bmatrix} U_1 \\ V_1 \end{bmatrix} = \begin{bmatrix} n_1(1 + r_{\mathrm{p}}) \\ \frac{\xi_1}{n_1}(1 - r_{\mathrm{p}}) \end{bmatrix} E_{x1}^{i}
\tag{29}
$$

$$
N_k^{\mathrm{p}}(z) = \begin{bmatrix} \cos\left(\dfrac{2\pi\xi_k(z - z_{k-1})}{\lambda}\right) & \dfrac{i\hat{n}_k^2}{\xi_k}\sin\left(\dfrac{2\pi\xi_k(z - z_{k-1})}{\lambda}\right) \\[3mm] \dfrac{i\xi_k}{\hat{n}_k^2}\sin\left(\dfrac{2\pi\xi_k(z - z_{k-1})}{\lambda}\right) & \cos\left(\dfrac{2\pi\xi_k(z - z_{k-1})}{\lambda}\right) \end{bmatrix}.
\tag{30}
$$

Finally, the expressions for the MSEFS at k interface are as follows. For s-polarized radiation:

$$
\langle E_{yk}^{s\,2} \rangle = \frac{1}{2}|U_k(z)|^2 .
\tag{31}
$$

For p-polarized radiation there are two relationships – for each component of the electric field vector oriented along the x and z coordinate axes:

$$
\langle E_{xk}^{p\,2} \rangle = \frac{1}{2}|V_k(z)|^2
\tag{32}
$$

$$
\langle E_{zk}^{p\,2} \rangle = \frac{1}{2}\left|\frac{n_1 \sin\Theta_1^{i}}{\hat{n}_k^2} U_k(z)\right|^2 .
\tag{33}
$$

Combined, they yield the total MSEFS for p-polarized light:

$$
\langle E_k^{p\,2} \rangle = \langle E_{xk}^{p\,2} \rangle + \langle E_{zk}^{p\,2} \rangle .
\tag{34}
$$

Although the above relationships are not very transparent, they are readily handled by a computer. They allow one to calculate the MSEFS using measurable quantities such as the wavelength and the angle of incidence of incoming radiation, thicknesses, and optical constants of the materials that constitute strata in a multilayer system. These equations offer a tremendous help when the optimization of experimental conditions is undertaken. This point will be illustrated by the material presented in the next section.

9.3
Optimization of Experimental Conditions

The absorbance of an adsorbed film is proportional to the MSEFS at the electrode surface [Eq. (3)] [23]. Therefore, calculation of the MSEFS helps one to determine experimental conditions that ensure the best S/N for the IR spectra of adsorbed species. The first calculations of MSEFS were carried out by Greenler

[1–3], who determined the optimal angles of incidence for IRRAS experiments at the air/metal interface of different metals. He also demonstrated how the optical constants of the organic film affect the resulting spectra. Dluhy expanded IRRAS to study thin films adsorbed at the air/water interface [24]. He calculated the reflectivity and the MSEFS of infrared radiation at the air/water interface and compared these values with the corresponding data for the air/metal interface. Seki et al. [14] have calculated the MSEFS distribution within a thin-layer electrochemical cell. They demonstrated that absorption of radiation by the electrolyte has a profound effect on the MSEFS and consequently on the S/N.

Roe et al. [25, 26] performed analysis of the influence of both the angle of incidence and the thin-cavity thickness on the reflectivity of a spectroelectrochemical cell. They showed that precise control of the thin-cavity thickness is necessary to fully optimize the experimental set-up. They also concluded that the best experimental conditions to perform IRRAS using p-polarized light are achieved when the reflectivity of the interface reaches a minimum. This conclusion was recently challenged by Blaudez et al. [27], who demonstrated that a significant reduction in S/N is observed at these conditions. Indeed, Fig. 9.2 shows that at the air/gold interface the maximum of MSEFS corresponds to the angle of incidence of 80.7°, while the minimum of the reflectivity is observed at 88.7°.

Popenoe et al. [28] performed calculations of MSEFS as a function of the angle of incidence for a thin-layer cavity with water and deuterium oxide as a solvent and two IR window materials: CaF_2 and Si. Faguy et al. [29–31] calculated MSEFS at the metal electrode surface for several window materials such as CaF_2, ZnSe, and Ge. They concluded that the best surface enhancement is achieved at angles of incidence just above the critical angle for the window/electrolyte interface. Under these conditions, only the evanescent wave penetrates into the electrolyte, and the spectra have to be recorded using a very small thin-cavity thickness of 1 μm or less [32, 33]. This configuration can be used to study films adsorbed on non-reflecting or porous electrodes [34, 35].

The geometry of the IR window has also an impact on the quality of recorded spectra. A flat CaF_2 window was used in the first IRRAS studies of the metal/solution interface by Bewick, Kunimatsu and Pons [5–8]. Golden, Kunimatsu, and Seki [14, 15] were the first to replace a flat CaF_2 window by a CaF_2 prism bevelled at 60° to optimize the optical throughput. Later, Faguy et al. [29–31] demonstrated that an improvement in the signal-to-noise ratio (S/N) is achieved in SNIFTIRS when a ZnSe hemisphere is used as the window. The information concerning optimization of the IRRAS experiment is scattered across the literature. Below, we describe calculations of MSEFS at a gold electrode surface for various experimental conditions. The calculations were performed using the matrix method and Fresnel 1 software, which was written in house. The software is available on demand [36].

9.3.1
Optimization of the Angle of Incidence and the Thin-cavity Thickness

Since absorbance by an organic molecule in front of a metal mirror is proportional to the mean square electric field strength (MSEFS) of the p-polarized radiation [23], optimization of experimental conditions is equivalent to finding a global maximum of the MSEFS of p-polarized light at the electrode surface. Figure 9.4 shows a model of the spectroelectrochemical cell that was used for calculations of the MSEFS. For a given wavelength, the MSEFS is a function of the angle of incidence, the angle of the incident beam convergence, the thin-cavity thickness, and the optical constants of all the constituents of the interface. First, we will discuss optimization of the angle of incidence and the thin-cavity thickness. The influence of the beam collimation and the choice of the optical window will be described next.

The calculations described in this section were performed for CaF_2, D_2O, and Au as the optical window, electrolyte solvent, and working electrode, respectively. The optical constants of these materials are available in the literature [22, 37–39]. We show optimization of experimental conditions for the wavenumber of $1600\ cm^{-1}$ (mid-IR spectral region) and for $2850\ cm^{-1}$ (C-H stretching region). The 3D graph in Fig. 9.5 plots the calculated MSEFS at the gold surface as a function of two variables: the angle of incidence and the thin-cavity thickness for the convergent ($\pm 6°$) beam and wavenumber of $1600\ cm^{-1}$. (To simulate the incident beam convergence, the calculations were done for several angles within the $\pm 6°$ of convergence and then the results were averaged.) The MSEFS surface displays a global maximum at the angle of $66°$ and the thin-cavity thickness of $3.4\ \mu m$. Hence, these values should be used for the acquisition of experimental spectra in the $1600\ cm^{-1}$ wavenumber region.

To analyze the MSEFS surface in a further detail, solid lines in Fig. 9.6 A and B plot the angular and spatial profiles of the MSEFS that cross the global maxi-

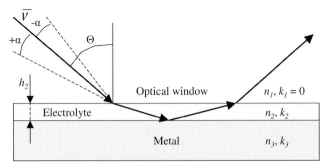

Fig. 9.4 Stratified medium that models the experimental thin-layer cell. $\bar{v}$, $\pm a$ and Θ are the wavenumber, convergence and the angle of incidence of the infrared beam, respectively. n, k and h are the refractive index, attenuation coefficient and thickness of the three phases that comprise the stratified medium, respectively.

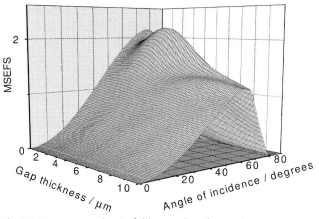

Fig. 9.5 Mean square electric field strength at the metal surface for a p-polarized beam as a function of the angle of incidence and the thin-cavity (gap) thickness. Calculate for the convergent ($\pm 6°$) radiation of 1600 cm^{-1}. For stratified medium: $CaF_2/D_2O/Au$.

mum. The value of the MSEFS at the maximum, equal to 2.2, indicates that enhancement of the electric field strength is observed as the result of the constructive interference of the incident and reflected p-polarized light at the metal surface. This value is slightly higher than 1.84, the maximum value of the MSEFS calculated for the air/gold interface and convergent ($\pm 6°$) radiation of 1600 cm^{-1}. The angular dependence of the MSEFS, shown in Fig. 9.6 a, demonstrates that the electric field is reduced by about 20% if the angle of incidence is set to 60°, the value commonly used when the equilateral prism is employed. The spatial dependence of the MSEFS, presented in Fig. 9.6 b, indicates that not only the angle of incidence but also the thin-cavity thickness should be optimized to achieve the maximum MSEFS at the electrode surface.

The position of the global maximum on the MSEFS surface depends also on the frequency of the incident radiation. Solid lines in Fig. 9.7 a and b plot the angular and spatial profiles intersecting the global maximum at the MSEFS surface, calculated for the wavenumber of 2850 cm^{-1}. These profiles differ significantly from those calculated for the wavenumber of 1600 cm^{-1} and presented in Figs. 9.6 a and b. The new maximum of MSEFS is found at the angle of incidence of 55° and the thin-cavity thickness of 2.3 μm. The maximum enhancement of the MSEFS is equal to 4.12 and is about a factor of 2 higher than for wavenumber 1600 cm^{-1}.

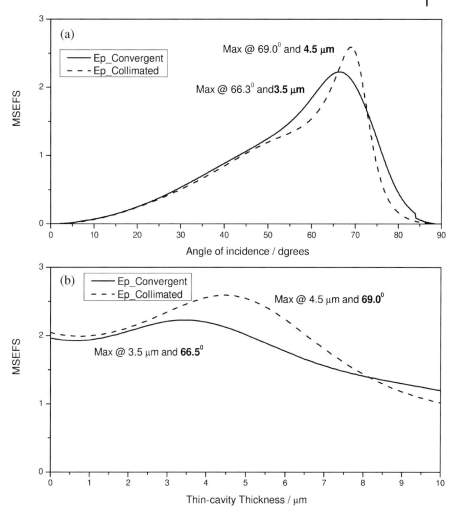

Fig. 9.6 Mean square electric field strength at the metal surface for a p-polarized beam as a function of (a) the angle of incidence at the thin cavity thickness shown, (b) the thin cavity thickness at the angle of incidence shown. Calculated at 1600 cm^{-1} for collimated (dashed line) and convergent (solid line) radiation. Angle of convergence was 6°, and stratified medium was CaF$_2$/ D$_2$O/Au.

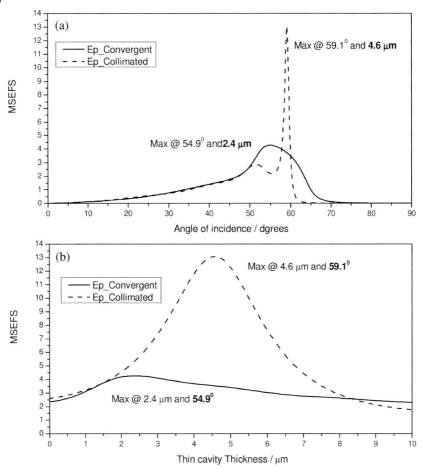

Fig. 9.7 Mean square electric field strength at the metal surface for a p-polarized beam as a function of (a) the angle of incidence at the thin cavity thickness shown, (b) the thin cavity thickness at the angle of incidence shown. Calculated at 2850 cm^{-1} for collimated (dashed line) and convergent (solid line) radiation. Angle of convergence was 6° and stratified medium was CaF$_2$/D$_2$O/Au.

9.3.2
The Effect of Incident Beam Collimation

The calculation of the MSEFS was repeated, for a collimated beam. Dashed lines in Figs. 9.6 and 9.7 plot the MSEFS surface profiles at the global maximum for the parallel beam. For the wavenumber of 1600 cm^{-1} the effect of the collimation is small. However, for the wavenumber of 2850 cm^{-1} the effect is huge. The maximum of the MSEFS is now equal to ~ 13 and is about three times larger than for the convergent radiation. This additional enhancement of MSEFS at the gold surface for the electrode within the thin-layer cavity is chiefly due to multiple reflections and constructive interference of the beam bouncing back and forth between the gold and IR window surfaces.

Further, the MSEFS plot for the collimated beam displays a sharper maximum than for the convergent beam. This behavior is observed when the angle of incidence is close to the critical angle at the window/solution interface. At the wavenumber of 2850 cm^{-1}, $n_{D_2O} = 1.22$ and $n_{CaF_2} = 1.41$. These numbers inserted into Eq. (9) give the critical angle equal to 60°. Thus, the maximum of the MSEFS is observed "just below" the critical angle. It is useful to note that commercially available CaF$_2$ prisms are beveled at 60° [14]. Since this is the critical angle at 2850 cm^{-1}, one may have difficulty in finding a spectrum of an adsorbed species if a beam perpendicular to the prism surface is used. The calculations presented in Figs. 9.6 and 9.7 indicate that in order to achieve an optimum S/N for the measured spectra the experimental set-up must be realigned for each spectral region of interest.

9.3.3
The Choice of the Optical Window Geometry and Material

Four types of optical windows are presently used for *in situ* IRRAS experiments: flat disks, equilateral prisms, hemispheres, and hemicylinders. Flat disks are disadvantageous because of significant losses in the throughput (20–40%) due to reflection from the front air/window interface at high values of the angle of incidence. Prisms are made of low refractive index materials, such as CaF$_2$ and BaF$_2$, and are cut to admit radiation at normal incidence. They have very small losses at the front air/window interface (under 10%), but they have no collimating properties. Hemispheres and hemicylinders act as lenses and can be utilized to collimate or refocus the incident beam. Hemispheres focus the beam onto a spot while hemicylinders focus it onto a line.

The differences between the properties of a prism and a hemispherical window are schematically shown in Fig. 9.8. In order to collimate the beam, the focal point of the convergent beam should be located at a distance d from the hemisphere surface equal to [32]

$$d = r/(n-1) \tag{35}$$

where r and n denote radius and refractive index of the window material. The diagram in Fig. 9.8 shows that the distance between the focal point and the hemisphere surface must be small in order to achieve a narrow collimated beam. Therefore, the refractive index of the hemisphere must be greater than 2. This restricts window materials to those having a high refractive index, e.g., ZnSe, Ge, or Si.

However, when the refractive index of the window is high, significant losses in throughput are observed due to the reflection at the window/air interface (up to 30–80% even for the normal angle of incidence). The exact values of the reflectivity at the air/window interface at normal incidence for several infrared

(a)

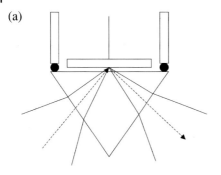

(b)

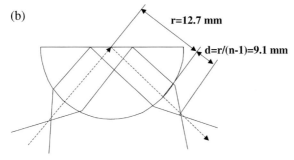

r=12.7 mm

d=r/(n-1)=9.1 mm

Fig. 9.8 Optical setup for *in-situ* IRRAS based on the following windows: (a) CaF$_2$ equilateral prism (transmission range 76900–1100 cm^{-1}; R_{Air/CaF_2}=0.03), (b) ZnSe hemisphere (transmission range 22200–700 cm^{-1}; $R_{Air/ZnSe}$=0.18).

window materials are shown in Table 9.1. The total losses of throughput due to reflection from the front air/window interface are approximately twice as high as the corresponding reflectivities, because the beam crosses the air/window interface twice. These losses can be significantly reduced with the help of anti-reflective coatings. Unfortunately, anti-reflective coatings are fragile and diminish the frequency range over which the window is transparent to the incident radiation.

To compare the performance of different optical windows, the global maxima of the MSEFS at the metal surface were calculated for different materials and a collimated (convergence of ±1°) radiation of 1600 cm^{-1}. The results are summarized in Table 9.1. It is apparent that the greater the difference between the refractive indices of D$_2$O and the window material, the higher are the values of the MSEFS at the metal surface. (The larger is the enhancement due to multiple reflections between the window and the gold surface.) However, Si and Ge windows give the maximum of the MSEFS at very low values of the angle of incidence, imposing geometrical constraints on the optical set-up. These constraints and significant losses of the total throughput due to reflection of the incident beam from the air/window interface make these materials rather difficult to work with. Nevertheless, using antireflective coatings and specially designed

Table 9.1 The refractive index, reflectance of the air/window interface (at normal incidence), the maximum MSEFS at the metal surface, the coordinates of the maximum (thin cavity thickness and the angle of incidence), the full width at half maximum (FWHM) of the MSEFS for different optical window materials and the low frequency cut off limit.

Material	n	R (%)	MSEFS	Gap (µm)	Angle (°)	FWHM (°)	Limit (cm^{-1})
CaF$_2$	1.38	2.5	2.6	4.4	69	19	1020
BaF$_2$	1.44	3.2	3.6	4.3	63	10	800
ZnSe	2.42	17	9.8	4.1	31.8	4	600
Si	3.42	30	12.3	4	21.8	4.5	700
Ge	4.01	36	13.5	4.9	18.2	3.8	600

FWHM is shown for the dependence of the MSEFS on the angle of incidence.
Data in the table were calculated for a collimated (convergence $=\pm1°$) p-polarized radiation of 1600 cm^{-1} and stratified medium: Window/D$_2$O/Au.

long-focal-length set-ups it is in principle possible to use these windows. A very large enhancement of the MSEFS at the metal electrode surface may be achieved in this case.

At present, the best window material for reflection absorption spectroscopy is the ZnSe hemisphere. It allows one to achieve high values of MSEFS because it acts as a collimating lens, and, because of the high value of the refractive index, gives a significant enhancement of the MSEFS at the electrode surface through multiple reflections [32]. The reflection losses at the air/window interface for ZnSe are moderate and can be significantly reduced if an anti-reflective coating is used. Figure 9.9 compares the MSEFS versus angle of incidence profiles at the gold surface for air/gold, CaF$_2$(equilateral prism)/D$_2$O/Au, and ZnSe(hemisphere)/D$_2$O/Au systems for incident-radiation wavenumbers of 1600 cm^{-1} and 2850 cm^{-1}. Clearly, the enhancement of the MSEFS increases by a factor of ~ 6 by moving from the air/gold to the ZnSe(hemisphere)/D$_2$O/Au system at 1600 cm^{-1} and by a factor of ~ 17 at a wavenumber of 2850 cm^{-1}. The maximum progressively shifts to lower angles of incidence and the profiles become sharper. The sharpness of the profile makes the alignment of the optical set-up rather a difficult task. Table 9.1 shows the values of the full width at half maximum of the MSEFS, which characterizes the sharpness of the MSEFS peak as a function of the angle of incidence.

Figure 9.10 illustrates the advantages of using a ZnSe hemisphere instead of a CaF$_2$ prism as the optical window. The figure shows SNIFTIRS spectra of pyridine adsorbed at gold electrode surfaces. The spectrum acquired using a cell equipped with the ZnSe hemispherical window is presented at the bottom of the figure, and the spectrum measured with the use of the CaF$_2$ equilateral prism is shown at the top. The potential was modulated between a value where

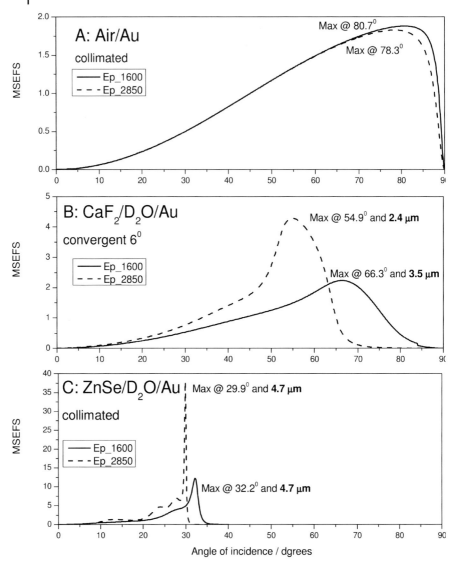

Fig. 9.9 Influence of the optical window on the enhancement of the MSEFS at a gold electrode surface: (A) no window, (B) CaF$_2$ window, (C) ZnSe window. Data show calculations for window/D$_2$O/Au interface for 1600 and 2850 cm^{-1}. Note that calculations were carried out for collimated beam for ZnSe and air, but for convergent beam for CaF$_2$ (angle of convergence was 6°). These conditions approximate typical experimental settings.

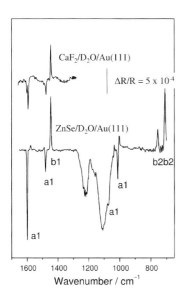

a total desorption of pyridine molecules from the electrode takes place and a value at which a compact monolayer is formed at the two gold surfaces. The results show that the use of a ZnSe window allows one to record the SNIFTIRS spectra over a much broader frequency range, so that many more bands can be seen in the spectrum recorded with this window. In addition, the bands are significantly stronger when the CaF$_2$ prism is replaced by the ZnSe hemisphere. Specifically, the band at ~ 1600 cm^{-1} is approximately 3 times stronger when the spectrum is acquired using the ZnSe window. This increase correlates very well with the enhancement of MSEFS at the gold surface calculated for the two windows [40].

Carbon monoxide has a large attenuation coefficient for absorption of IR radiation. Therefore, the SNIFTIRS spectra of CO adsorbed at a Pt electrode surface are a convenient standard to test the S/N of a spectroelectrochemical set-up. The left panel in Fig. 9.11 shows SNIFTIRS spectra of a monolayer of CO at Pt electrode surface recorded using a cell equipped with a CaF$_2$ prism [91]. The right panel shows similar spectra recorded using a ZnSe hemispherical window. The IR bands of CO adsorbed at Pt are significantly Stark-shifted when the electrode potential is modulated between -200 and $+200$ mV versus SCE. Consequently, the potential difference spectrum displays bipolar bands. Clearly a much better S/N for these bands is achieved when a ZnSe hemisphere is used as the window.

While ZnSe is a superior window material for potential modulation spectroscopy, this advantage is lost in photon polarization modulation spectroscopy because of significant differences between reflectivity of the s- and p-polarized light at the window/electrolyte interface at angles at which the enhancement of MSEFS for the p-polarized radiation is at maximum. One has to reduce the an-

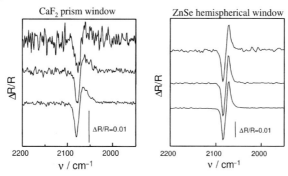

Fig. 9.11 Comparison of SNIFTIRS spectra for CO absorbed at a Pt electrode surface acquired using CaF_2 prism and ZnSe windows, 0.1 M $HClO_4$ solution, potential modulated between $E_1 = -0.2$ V and $E_2 = +0.2$ V versus SCE, taken with permission from Ref. [91].

gle of incidence significantly in order to achieve a comparable transmittance for the s- and p-polarized beams at the window/solution interface. Figure 9.9 shows that the enhancement of the p-polarized photon is then significantly reduced.

Typically CaF_2 and BaF_2 are used in polarization modulation reflection absorption spectroscopy. BaF_2 is the better of the two because higher values of the MSEFS are attained for this material and the maximum of the MSEFS is observed at lower values of the angle of incidence. It is important to have moderate values of this angle since a significant elongation of the footprint of the infrared beam at the electrode surface occurs if the angle of incidence exceeds $60°$. The cut-off of a BaF_2 prism is also shifted to lower wavenumbers (~ 900 cm^{-1}) relative to that of a CaF_2 prism (~ 1100 cm^{-1}). Unfortunately BaF_2 has about two orders of magnitude higher solubility than CaF_2 in aqueous media. However, if 0.1 M NaF is used as the electrolyte, the dissolution of the window can be suppressed.

9.4
Determination of the Angle of Incidence and the Thin-cavity Thickness

The previous section demonstrated that it is essential to control both the angle of incidence and the thin-cavity thickness in the IRRAS experiment. While it is usually not too difficult to control the angle of incidence, a measurement of the thin-cavity thickness with a precision of a fraction of a micrometer is not a trivial task. A new method [40] of determination of these important parameters is described in Fig. 9.12 for an experiment performed using BaF_2 as the window, D_2O as the solvent, and Au as the electrode.

First, one measures the reflectivity R_o of an empty cell. In the absence of the electrolyte, the beam is totally reflected from the inner surface of the BaF_2

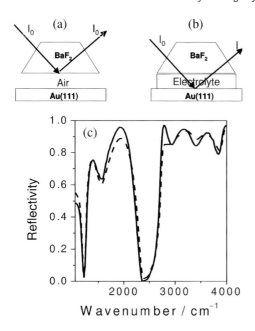

Fig. 9.12 (a) Total reflection from the empty cell; (b) reflection from the filled cell; (c) reflectivity spectrum of the setup: (dashed line) experimental values; (solid line) calculated using Fresnel formulae for $BaF_2/D_2O/Au$ interface and the angle of incidence of 60°. Thin-cavity thickness was determined to be 2.4 μm.

prism, because the angle at which the IRRAS experiment is performed is greater than the critical angle for the BaF_2/air interface. Total internal reflection is very efficient, and hence R_o is a good measure of the intensity of the incident beam. Next, the cell is filled with the electrolyte and the electrode is pressed against the window to form the thin-layer configuration. Now, one measures the reflectivity R of a beam that is transmitted through the BaF_2/solution interface, travels through the thin layer of the electrolyte, and is reflected from the electrode surface. The solid line in Fig. 9.12 plots the experimentally determined R/R_o ratio for the thin-layer cell.

Next, Fresnel equations in the matrix form are used to calculate the ratio R/R_o using optical constants of the three phases, $BaF_2/D_2O/Au$. The thickness of the thin-layer cavity (thickness of D_2O layer) and the angle of incidence are the two adjustable parameters. One uses initial approximate values of the thickness and the angle of incidence, and then with the help of iterative software the computer searches for the values of these parameters that give the best fit to the experimental R/R_o. The dashed line in Fig. 9.12 shows the curve calculated using the Fresnel equations and the optical constants of the three parallel phases giving the best fit to the experimental data. In the example shown in Fig. 9.12, the best fit was observed for the angle of incidence and the thin-cavity thickness equal to 60° and 2.4 μm, respectively. Although two parameters are allowed to vary in this procedure, the angle of incidence can be measured quite precisely at the beginning of the experiment, and the initial value is usually close to the final one. The thickness of the thin-layer cavity is the parameter that is difficult to measure, and one relies on the above iterative procedure to determine its value. If the angle of incidence and the thickness of the thin-layer cavity do not correspond to the values of the global maximum of the MSEFS, the set-up is realigned again and the procedure is repeated until the optimal parameters are found.

9.5
Determination of the Isotropic Optical Constants in Aqueous Solutions

One needs to know optical constants to calculate IRRAS spectra of molecules either adsorbed at the electrode surface or resident inside the thin-layer cavity. The isotropic optical constants of a given compound are usually determined from transmittance spectra. A pressed pellet, prepared by grinding the dispersion of the compound with a KBr or KCl powder, is typically used as a sample. Recently, Arnold et al. [41] have demonstrated that this method can yield nonreproducible results due to different histories of the sample preparation. In addition, the optical constants determined using the powder method can be quite different from those of the film at the metal/electrolyte interface because of the difference in the environment.

Below, we describe a procedure, adapted from the work of Allara et al. [42], which allows one to determine optical constants of a given compound in a solution. A thin-layer transmission optical cell, shown in Fig. 9.13 A, is employed to obtain the experimental data. Two ZnSe or, better, BaF$_2$ discs are used as windows, and a 10- or 25-μm thick Teflon gasket is used as a spacer. The cell is placed in a Teflon housing, clamped between two aluminum plates, and mounted inside the main compartment of an FTIR spectrometer. Teflon tubing is used to fill the cell with fluid samples. Two samples are necessary to acquire a set of experimental spectra needed for calculations: the pure solvent is used as a background sample and a solution of a given compound is used as the analyte sample.

The precise size of the gap between the optical windows is determined from the number of interference fringes in the spectrum recorded when the thin-layer cell is empty [43]. Panel 13 B shows the transmittance of an empty cell ($n_\infty \approx 1$) equipped with ZnSe windows and a $\sim$ 10-μm thick Teflon spacer. An interference pattern arises because of the multiple reflections from the internal walls of the cell. The number of interference fringes ΔN in the frequency range $\bar{\nu}_1 - \bar{\nu}_2$ is related to the cell thickness d in the following fashion [43]:

$$d = \frac{\Delta N}{2n_\infty (\bar{\nu}_1 - \bar{\nu}_2)} \tag{36}$$

where n_∞ is the average refractive index of the medium within the cell. The arrows point to an interval (from 4575 cm^{-1} to 3557 cm^{-1}) that includes 2 full interference fringes. Using Eq. (36), one can calculate that the exact thickness of the cell is equal to 9.82 μm.

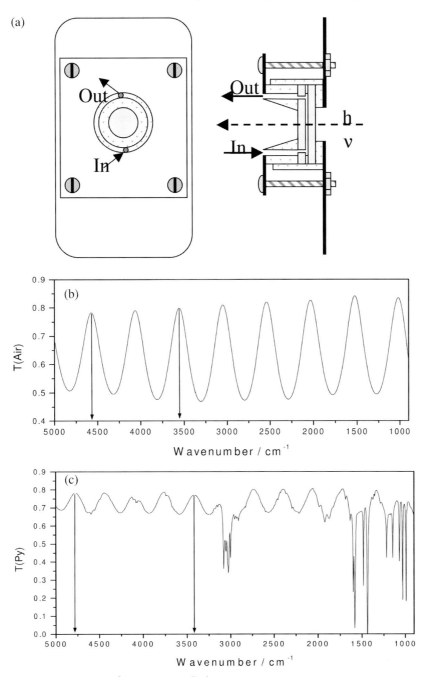

Fig. 9.13 (a) Diagram of transmission cell, (b) transmittance of an empty cell, (c) transmittance of the cell filled with neat pyridine. The cell: ZnSe windows separated by a 10-μm Teflon spacer.

Next, the cell is filled with a liquid sample, and Eq. (36) is utilized to determine the average refractive index of the sample. Panel 13C shows the transmittance of neat pyridine as an example. The arrows point to an interval between 4776 cm^{-1} and 3415 cm^{-1} that includes 4 fringes. Using the previously calculated value of d, the average refractive index of the sample can be determined ($n_{\infty} = 1.49$ for pyridine). This value is needed for calculation of the refractive index from the spectrum of the attenuation coefficient, with the help of the Kramers-Krönig transformation.

Finally, the transmittance of the analyzed solution is measured by filling the cell with a solution of a known concentration of the analyte, taking the sample spectrum, and then filling the cell with the pure solvent and taking the background spectrum. The isotropic optical constants of the analyte can now be calculated using the transmittance spectra of the sample and pure solvent and knowing the concentration, the average refractive index of the analyte in the sample, the thickness of the cell, and the optical constants of the optical windows and the solvent. The calculation procedure consists of the following steps:

Step 1. The approximate values of the attenuation coefficient of the analyte k_i are determined from the two transmittance spectra using the Lambert law [44] modified to take into account dilution of the analyte (the Beer-Lambert law):

$$\frac{I}{I_0} = \exp\left(\frac{-4\pi f_i k_i d}{\lambda}\right) \tag{37}$$

where I and I_0 are the intensities of radiation transmitted through the thin-layer cell filled with the analyte solution and the pure solvent, respectively, d is the thickness of the thin-layer cell, λ is the wavelength of the incident infrared radiation in vacuum, and f_i is the volume fraction of the analyte in the sample. The latter can be calculated using the following equation:

$$f_i = C_i \frac{M_i}{\rho_i} \tag{38}$$

where C_i is the molar concentration, M_i is the molar mass, and ρ_i is the density of the analyte in the solution. It is important to note that such an approach allows one to determine the attenuation coefficient k_i of the pure analyte, including the effects of solvent-analyte interactions taking place in the aqueous environment.

Step 2. Using values of k_i as an initial estimate, the approximate values of the refractive index of pure analyte n_i are determined using the Kramers-Krönig transformation [45]:

$$n_i(\bar{\nu}_0) = n_{\infty} + \frac{2}{\pi} P \int_{\bar{\nu}_1}^{\bar{\nu}_2} \frac{\bar{\nu} k_i(\bar{\nu})}{(\bar{\nu}^2 - \bar{\nu}_0)} d\bar{\nu} \tag{39}$$

where n_{∞} is the average refractive index in the mid-infrared that represents the contribution to the refractive index of regions removed from any absorption band. P indicates that the Cauchy principal value of the integral must be taken because of the singularity at $\bar{v} = \bar{v}_0$, and $\bar{v}_1$ and $\bar{v}_2$ are the lower and the upper limits of the spectrum of k, respectively.

Step 3. The refractive indices and attenuation coefficients of the solution are calculated using the following equations [46]:

$$\frac{\sum x_i M_i}{\rho} \frac{n^2 - 1}{n^2 + 2} = \sum x_i \frac{M_i}{\rho_i} \frac{n_i^2 - 1}{n_i^2 + 2} \tag{40}$$

$$\frac{\sum x_i M_i}{\rho} k = \sum x_i \frac{M_i}{\rho_i} k_i \tag{41}$$

where ρ and ρ_i denote the densities of the solution and the pure component i of the solution, n, k and n_i, k_i are the corresponding refractive indices and attenuation coefficients, x_i is the mole fraction, and M_i is the molar mass of component i.

Step 4. The approximate values of n and k of the solution are used as starting values in the exact Fresnel equations (see Section 9.2.2) to calculate the theoretical transmittance spectrum. This spectrum is computed as the ratio of the simulated spectrum, corresponding to the analyte solution, and the simulated background spectrum of the pure solvent. Both spectra are calculated using a thin layer of solution, situated between the two optical windows, as the model of the transmission cell. In one case, the layer is represented by the optical constants of the pure solvent and in the other case it has the optical constants of the solution, calculated using Eqs. (40) and (41).

Step 5. The attenuation coefficients of the analyte k_i are perturbed at each wavelength and used to calculate the refractive indices n_i using Eq. (39). Both optical constants are then used to calculate the optical constants of the solution using Eqs. (40) and (41). The attenuation coefficients of the solution are later used to calculate another theoretical transmittance spectrum using the approach described in Step 4.

Step 6. Both theoretical transmittance spectra (obtained in Step 4 and Step 5) are compared with the experimental transmittance and a refined spectrum of attenuation coefficients k_i is obtained (Steps 5 and 6 use the Newton-Raphson iterative method of numerical solutions of non-linear and transcendental equations [47]).

Step 7. Refined attenuation coefficients k_i are used to calculate the improved refractive indices n_i using Eq. (39), and Steps 4–7 are repeated, using the refined values of k_i and n_i as an initial estimate in Step 4, until the theoretical spectra deviate from the experimental transmittance by no more than a specified value of precision.

The procedure is highly convergent, and usually two iterations are sufficient to achieve the final set of optical constants, yielding the theoretical spectrum within a 1% deviation from the measured transmittance. Figure 9.14 plots the optical constants of pyridine in a solution of D_2O and Fig. 9.15 plots the optical

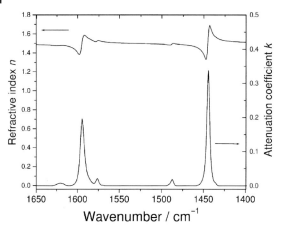

Fig. 9.14 Optical constants of pyridine, calculated from the transmittance spectrum of a 0.1 M solution in D_2O. n is the refractive index, k is the attenuation coefficient. Taken with permission from Refs. [36] and [40].

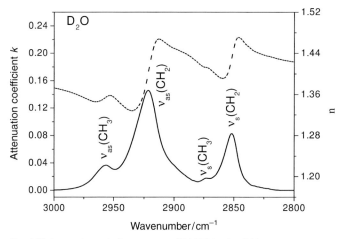

Fig. 9.15 Isotropic optical constants of DMPC in the C-H stretching region calculated from the transmittance of its 0.6286 (v/v)% solution in D_2O, n (- - - - -) refractive index, k (_____) attenuation coefficicients. Adapted from Refs. [79] and [87].

constants of 1,2-dimyristoyl-sn-glycero-3-phosphocholine (DMPC) determined with the help of this procedure. These constants will be used later in this chapter to calculate spectra of molecules in a thin-layer spectroelectrochemical cell.

9.6
Determination of the Orientation of Organic Molecules at the Electrode Surface

When linearly polarized light is absorbed by a sample, the integrated intensity of the absorption band is proportional to the square of the absolute value of the dot product of the transition dipole vector and the electric field vector of the incident radiation [48, 49]:

$$\int Ad\nu \propto \Gamma|\vec{\mu} \cdot \vec{E}|^2 = \Gamma|\mu|^2\langle E^2\rangle \cos^2\Theta. \tag{42}$$

At the metal surface, the direction of the electric field vector of the p-polarized radiation is normal to the surface. Therefore, the integrated band intensity of adsorbed molecules is proportional to $\cos^2\Theta$, where Θ is the angle between the average direction of the transition dipole moment and the surface normal. If the direction of the transition dipole with respect to the coordinates of the adsorbed molecule is known, Eq. (42) may be used to determine the orientation of that molecule at the metal surface. The direction of the transition dipole moment with respect to coordinates of a molecule can be determined by performing normal coordinate *ab initio* calculations using commercial software such as Gaussian 98 [50].

There are two methods to determine the molecular orientation from Eq. (42). These were recently discussed by Arnold et al. [41]. The first approach, called the relative method, relies on the availability of at least two intense absorption bands that correspond to differently oriented transition dipoles. An elegant application of this method is to determine the tilt angle of α-helical thio-peptides attached to a gold surface [51]. The method relies on measurements of intensities of amide I and amide II bands at 1660 cm^{-1} and 1545 cm^{-1}, respectively. The transition dipoles of these bands are oriented at angles $a_I = 39°$ and $a_{II} = 75°$ with respect to the helix axis. Using Eq. (42) and simple geometrical arguments, the ratio of integrated intensities of amide I to amide II bands can be related to the angle between the helix axis and the surface normal γ by the formula [51]:

$$\frac{\int A_I d\nu}{\int A_{II} d\nu} = \frac{|\mu_I|^2}{|\mu_{II}|^2} \cdot \frac{\cos^2\Theta_I}{\cos^2\Theta_{II}} = 1.5 \frac{(3\cos^2\gamma - 1)(3\cos^2 a_I - 1) + 2}{(3\cos^2\gamma - 1)(3\cos^2 a_{II} - 1) + 2} \tag{43}$$

The numerical value of the ratio $|\mu_I|^2/|\mu_{II}|^2 = 1.5$ was determined from the absorbance ratio in an independently measured transmission spectrum. The relative method is very useful when either the surface concentration is not available or the MSEFS is difficult to calculate.

The second method, developed by Allara et al. [42, 48, 49], relies on calculation of the theoretical reflection absorption spectrum for the same angle of incidence and the thin-cavity thickness as the values used during the collection of the experimental data. The optical constants of the window, electrolyte, and metal can be taken from the literature [22, 37–39], while the isotropic optical constants of the investigated sample have to be determined using the method described in the preceding section. The matrix method described in Section 9.2 is used to perform these calculations. In order to apply this method, one also has to determine the surface concentration of the adsorbed species Γ or the thickness of the film, using an independent method. Isotropic optical constants correspond to randomly oriented molecules. When a linearly polarized beam is absorbed by randomly oriented molecules, $\cos^2 \Theta = 1/3$. Using Eq. (42), the ratio of the integrated band intensity in the experimental spectrum of adsorbed molecules to the band intensity in the calculated spectrum is then given by the formula

$$\frac{\int A_{exp} dv}{\int A_{cal} dv} = \frac{\cos^2 \Theta}{1/3}. \tag{44}$$

The main advantage of this method is that the absolute value of Θ can be determined directly. However, supplementary information such as the surface concentration, the angle of incidence, the thin-cavity thickness, and the optical constants of the film have to be determined from independent measurements. Below, we show several examples of how to apply this method.

9.7
Development of Quantitative SNIFTIRS

9.7.1
Description of the Experimental Set-up

We recommend the optical set-up described in Fig. 9.16. A photograph showing this accessory and the spectroelectrochemical cell is shown in Fig. 9.17. It is a modified version of the set-up described by Faguy and Marinkovic [32]. The set-up employs a hemispherical ZnSe window, for which a ray diagram is shown at the bottom of Fig. 9.16. The window is mounted on a platform equipped with three micrometer screws that are used for alignment of the flat surface of the hemispherical window parallel to the optical bench of the FTIR instrument. These screws are also used to adjust the height (H) of the window (and hence the height of the cell) relative to the focal point of the beam F. Two flat mirrors are mounted on posts attached to riders that can easily be moved back and forth along the straight path of the beam to adjust their position (b) relative to the focal point F. The tilt angle can be adjusted precisely with a set of micrometer screws that press levers attached to the mirrors. The location of the focal point of the unobstructed beam is marked F. After deflection of the beam by the mirrors, the new focal point must be at a distance d from the window surface described by Eq. (35) in Section 9.3.3. For a given angle of incidence θ, the distance between the center of the mirror and the focal point F must be equal to [32]

(a)

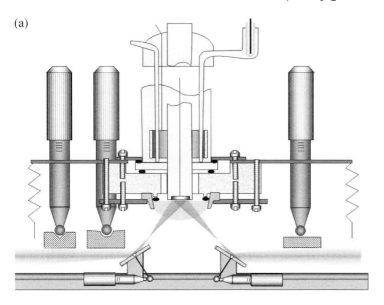

(b)

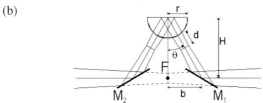

Fig. 9.16 (a) Schematic diagram of the accessory for
SNIFTIRS experiment, courtesy of D. Malevich,
(b) Ray diagram of the optical path of the beam, courtesy
T. Doneux [52].

$$b = R\left(\frac{n}{n-1}\right)\left(\frac{\sin\theta}{1-\sin\theta}\right) \tag{45}$$

The distance between the flat surface of the hemispherical window and the focal point is then given by [52]

$$H = (r + d + b)\cos\theta. \tag{46}$$

Therefore, adjustment of a proper angle of incidence requires not only selection
of a proper tilt angle for the mirrors but also adjustment of distances b and H.
Figure 9.18 plots values of distances b and H as a function of the angle of incidence for the beam wavenumber 1600 cm^{-1}. These calibration curves are essentially independent of the photon frequency within the region where ZnSe is
transparent.

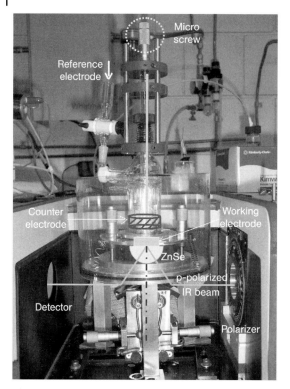

Fig. 9.17 Photograph of the accessory for SNIFTIRS with the spectroelectrochemical cell, courtesy of Jie Li.

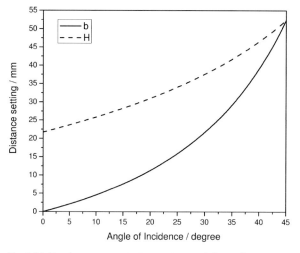

Fig. 9.18 For the accessory in Figure 9.14 and the mid IR beam with 1600 cm^{-1}, plots of distance settings H and b versus the angle of incidence.

9.7.2
Fundamentals of SNIFTIRS

In SNIFTIRS, the electrode potential is modulated between the base value (E_1) and the sample value (E_2), and the spectra of the reflected infrared radiation are measured. The reflection absorption spectrum ($\Delta R/R$) is obtained by plotting

$$\frac{\Delta R}{R} = \frac{R(E_2) - R(E_1)}{R(E_1)} \tag{47}$$

as a function of the wavenumber of the incident radiation, where $R(E_1)$ and $R(E_2)$ are the reflectivities at the base and the sample potentials, respectively. In the presence of the absorbing film, the reflectivity $R(E_i)$ may be written as

$$R(E_i) = R_o e^{-2.3\Gamma\varepsilon(E_i)} \tag{48}$$

where R_o is the reflectivity of the spectroelectrochemical cell in the absence of the film. The term $e^{-2.3\Gamma\varepsilon(E_i)}$ represents attenuation of the beam by the film; $A = \Gamma\varepsilon(E_i)$ is the absorbance of the film and $\varepsilon(E_i)$ is the molar absorptivity coefficient of the film at a potential E_i. For a very thin film, $2.3\Gamma\varepsilon(E_i) \ll 1$ and the exponential term $e^{-2.3\Gamma\varepsilon(E_i)} \approx 1 - 2.3\Gamma\varepsilon(E_i)$. Consequently, Eq. (47) may be written as:

$$\frac{\Delta R}{R} = 2.3\{\Gamma\varepsilon(E_1) - \Gamma\varepsilon(E_2)\}. \tag{49}$$

Alternatively, one may plot the SNIFTIRS spectrum as $\log\frac{R(E_2)}{R(E_1)}$, and in this case:

$$\log\frac{R(E_2)}{R(E_1)} = \Gamma\varepsilon(E_1) - \Gamma\varepsilon(E_2). \tag{50}$$

Because of the subtraction, SNIFTIRS spectra are devoid of the common background signal from the aqueous electrolyte. However, they represent the difference between absorbances of organic molecules at potentials E_1 and E_2 (or precisely the difference multiplied by 2.3). In order to facilitate interpretation of such spectra, it is convenient to choose a value of the base potential E_1 at which the molecules are totally desorbed into the bulk of the thin-layer cavity. The sample potential E_2 corresponds then to the adsorbed state of the film. Consequently, SNIFTIRS spectra plot a difference between the absorbances of molecules desorbed into the thin-layer cavity at potential E_1 and those adsorbed at the electrode surface at potential E_2. Thus, positive bands (or positive lobes in bipolar bands) are due to absorbance by desorbed molecules and negative bands (or negative lobes in bipolar bands) are the result of absorbance by adsorbed molecules.

9.7.3
Calculation of the Tilt Angle from SNIFTIRS Spectra

SNIFTIRS spectra do not allow direct determination of the tilt angles of ad-sorbed molecules because of the superposition of the bands of molecules in the adsorbed state onto the bands of molecules desorbed into the thin-layer cavity. In order to determine the tilt angle, one has to calculate the absorbance for the adsorbed film:

$$2.3\Gamma\varepsilon(E_2) = 2.3\Gamma\varepsilon(E_1) - \left(\frac{\Delta R}{R}\right)_{\text{exp}}. \tag{51}$$

For this purpose, the absorbance of desorbed molecules $2.3\Gamma\varepsilon(E_1)$ has to be de-termined independently. The subscript "exp" in $\left(\frac{\Delta R}{R}\right)_{\text{exp}}$ emphasizes that this is the experimentally measured SNIFTIRS spectrum. We have recently demon-strated that $2.3\Gamma\varepsilon(E_1)$ can be calculated from the isotropic optical constants of the investigated compounds and optical constants of the window, solvent, and electrode [40], knowing the number of adsorbed molecules per unit area Γ, the thickness of the thin layer of electrolyte, and the angle of incidence.

The spectrum of molecules in the desorbed state can be calculated using the following equation:

$$2.3\Gamma\varepsilon(E_1) = 1 - \left(\frac{R(\Gamma_S)}{R_o}\right)_{\text{cal}} \tag{52}$$

where $R(\Gamma_S)$ denotes the reflectivity of the stratified medium (window/elec-trolyte/metal) that contains a homogeneous solution of Γ molecules in the thin layer of electrolyte, and R_o is the reflectivity of the same stratified medium in the absence of the investigated molecules, the subscript "cal" in $\left(\frac{R(\Gamma_S)}{R_o}\right)_{\text{cal}}$ indicates that this is a calculated quantity. The reflectivities $R(\Gamma_S)$ and R_o can be calculated using the Fresnel equations in the matrix form de-scribed in Section 9.2.2.

Once the quantity $2.3\Gamma\varepsilon(E_2)$ is known, the angle between the transition di-pole and the surface normal can be determined using Eq. (44). However, this requires additional calculation of the hypothetical spectrum of randomly orient-ed adsorbed molecules $2.3\Gamma\varepsilon(E_2)_{\text{ran}}$ given by the equation

$$2.3\Gamma\varepsilon(E_2)_{\text{ran}} = 1 - \left(\frac{R(\Gamma_A)_{\text{ran}}}{R_o}\right)_{\text{cal}} \tag{53}$$

where $R(\Gamma_A)_{\text{ran}}$ corresponds to the reflectivity of the four-phase interface consist-ing of the window, thin layer of the electrolyte, an ultrathin film of Γ randomly oriented molecules adsorbed at the metal surface, and the metal. Such calculation can also be performed using the matrix method. The thickness of the ad-sorbed film can be calculated using the value of independently determined Γ and the molecular volume of the investigated compound.

$$l = \Gamma \frac{M}{\rho} \tag{54}$$

where M is the molar mass and ρ is the density.

Finally, combining Eqs. (44) and (47)–(53), the angle between the transition dipole of a given vibration and the surface normal may be expressed by

$$\cos^2 \theta = \frac{\int \varepsilon(E_2) dv}{3 \int \varepsilon(E_2)_{\text{ran}} dv} = \frac{\int \left\{ 1 - \left(R(\Gamma_S)_{\text{ocal}} - \left(\frac{\Delta R}{R} \right)_{\text{exp}} \right) \right\} dv}{3 \int \left\{ 1 - \left(\frac{R(\Gamma_A)_{\text{ran}}}{R_o} \right) \right\} dv}. \tag{55}$$

Below, we apply this equation to calculating the tilt angle of several molecules adsorbed at a gold electrode surface.

9.7.4
Applications of Quantitative SNIFTIRS

Quantitative SNIFTIRS was introduced in 2002 and so far has been applied to the investigation of the orientation of adsorbed molecules in three systems: pyridine at Au(110) [40], citrate at Au(111) [53], and 2-mercaptobenzimidazole at Au(111) [54, 55]. Pyridine adsorption at gold single-crystal surfaces has long been used as a model system to study the coordination of organic molecules to metal electrode surfaces. The thermodynamics of pyridine adsorption has been thoroughly investigated with the help of the chronocoulometric technique [56–64]. The availability of the thermodynamic data made this an ideal system to test the performance of the quantitative SNIFTIRS.

Figure 9.19 explains why the Au(110) surface was selected for these investigations. Pyridine adsorption starts at $E \approx -0.6$ V (SCE), and at $E = -0.35$ V it attains a limiting surface concentration of 6.2×10^{-10} mol cm^{-2}, which corresponds to the packing density of a closely packed monolayer of vertically oriented molecules. The limiting surface concentration maintains a constant value over a very broad range of electrode potentials, from $E = 0.35$ V to $E = 0.55$ V (SCE). The charge density at the electrode changes in this region from a value of $\sigma_M \approx 8$ μC cm^{-2} to a value of $\sigma_M \approx 35$ μC cm^{-2}. This corresponds to a change in the electric field strength at the interface from 4.5×10^9 V m^{-1} to 2×10^{10} V m^{-1}, estimated using Gauss's theorem $F = \sigma_M / \varepsilon$ and assuming that $\varepsilon = n^2 \varepsilon_0$, where $n = 1.5$ is the refractive index of pyridine and $\varepsilon_0 = 8.85 \times 10^{-12}$ C V^{-1} m^{-1} (the permittivity of vacuum).

In the plateau region, the coverage of the electrode surface by adsorbed molecules is constant. However, the static electric field at the interface changes sig-

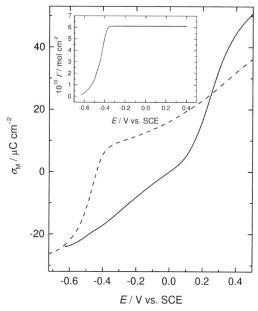

Fig. 9.19 Charge density at the Au(110) surface plotted against the electrode potential for 0.1 M KClO$_4$ solution. Solid line: bare electrode. Dashed line: with addition of 0.001 mM pyridine. Inset: (solid line) Gibbs excess of pyridine versus electrode potential plot. Electrochemical data were taken from Ref. [58].

nificantly. The pyridine molecule has a rigid ring-like structure and a large dipole moment (~ 2 D) parallel to its C$_2$ axis. Therefore, the energy of the dipole-field interaction varies from 0.17 eV to 0.8 eV. This very significant change should have a pronounced effect on the orientation of molecules coordinated to the electrode surface.

The transmission spectrum of neat pyridine is shown in Fig. 9.20. In the SNIFTIRS spectrum, the two broad bands located at ~ 1230 and ~ 1100 cm^{-1} correspond to absorption by D$_2$O and perchlorate ions, respectively. The remaining narrow bands are due to pyridine. An isolated pyridine molecule has C$_{2v}$ symmetry. Four bands at 1581, 1483, 1031, and 991 cm^{-1} have a_1 symmetry, the bands at 1438 and 1146 cm^{-1} have b_1 symmetry, and the two bands at 748 and 704 cm^{-1} have b_2 symmetry. The a_1 and b_1 bands correspond to in-plane ring deformations and b_2 bands to out-of-plane vibrations. The inset to Fig. 9.20 shows that the a_1 bands correspond to changes of the dipole moment along the C$_2$ axis of the molecule and the b_1 band is due to the change of the dipole moment in the direction perpendicular to the C$_2$ axis. The direction of the transition dipole of the b_2 band is normal to the plane of the molecule. Therefore, using the ZnSe window one can record SNIFTIRS spectra over a sufficiently broad range of frequencies such that the spectrum encompasses bands whose transition dipole moments are directed along all three coordinates.

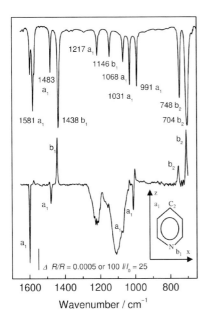

Fig. 9.20 Comparison of the IR spectrum of neat pyridine recorded in the transmission mode with the SNIFTIRS spectrum for the configuration Au(110) electrode/0.001 M pyridine solution in D_2O/ZnSe hemispherical window, recorded using the sample potential $E_2 = 0.4$ V and the reference potential $E_1 = -0.75$ V (SCE). Inset shows directions of transition dipole moments of a_1 and b_1 bands in the IR spectrum of pyridine. Taken with permission from Refs. [36] and [40].

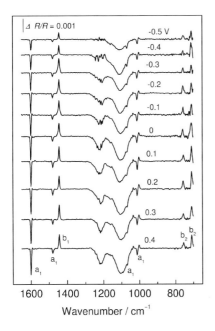

Fig. 9.21 A series of SNIFTIR spectra recorded for pyridine adsorption at the Au(110) surface using p-polarized light. The base potential E_1 was set to -0.75 V (SCE) and the sample potential E_2 varied. Its values are indicated at the corresponding spectrum. The solution was 0.1 M $KClO_4$ and 0.001 M pyridine in D_2O. Taken with permission from Refs. [36] and [40].

Figure 9.21 shows a series of SNIFTIRS spectra acquired for a variable sample potential and the same reference potential equal to -0.75 V (SCE). The a_1 bands are negative, while the b_1 and b_2 bands are positive for all investigated potentials. This behavior indicates that a_1 bands correspond to adsorbed pyridine while b_1 and b_2 bands belong to desorbed species. For adsorbed molecules, the transition dipole of the a_1 bands has a large component in the direction normal to the surface, while the component is small or close to zero for b_1 and b_2 bands. This is consistent with a vertical orientation of the adsorbed molecules. The integrated intensity of the a_1 band and the procedure described in the preceding section can be used to determine the angle between the transition dipole of the a_1 band and surface normal and at the same time the angle between the C_2 axis of the molecule and the normal.

Figure 9.22 illustrates the procedure. Plot A represents the calculated spectrum of 6.2×10^{-10} mol cm^{-2} of pyridine molecules homogeneously distributed within a 7.2 µm thick layer of D_2O between the gold electrode and the hemispherical ZnSe window. The spectrum was calculated from optical constants of pyridine shown in Fig. 9.14 using Eq. (52). This spectrum is equal to $1 - \left(\dfrac{R(\Gamma_s)}{R_0} \right)_{cal}$. Spectrum B in Fig. 9.22 is the experimental SNIFTIRS spec-

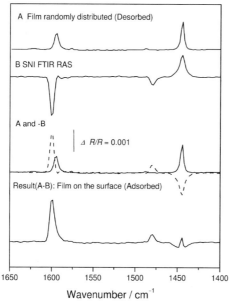

Wavenumber / cm^{-1}

Fig. 9.22 An illustration of the SNIFTIRS decoupling procedure. (A) Spectrum calculated for the slab of 6.2×10^{-10} mol cm^{-2} of pyridine molecules evenly distributed within the thin cavity between the optical window and the electrode. (B) SNIFTIRS spectrum of 6.2×10^{-10} mol cm^{-2} of pyridine molecules. A (solid) and B (dashed) are, respectively, spectrum A and inverted spectrum B, overlapped. The resulting spectrum at the bottom corresponds to the film of adsorbed pyridine molecules. Cell configuration: ZnSe/7.2 µm thick layer of 0.001 M solution of pyridine in D_2O/Au(110). SNIFTIR spectrum was obtained using the sample potential $E_2 = 0$ V and the background potential $E_1 = -0.75$ V (SCE). Taken with permission from Refs. [36] and [40].

trum. In order to compare the magnitude of the infrared bands in the SNIFTIRS spectrum and in the spectrum corresponding to pyridine molecules inside the thin-layer cavity, spectrum A and inverted spectrum B are overlaid in Fig. 9.22 (A and −B). Finally, the bottom plot in the figure shows the sum of A and −B. Consistent with Eq. (51), this is the spectrum of a monolayer of pyridine molecules adsorbed at the electrode surface. It represents

$$1 - \left(\frac{R(\Gamma_S)}{R_0}\right)_{cal} - \left(\frac{\Delta R}{R}\right)_{exp}.$$

At this point one performs a second calculation of a hypothetical film of randomly oriented pyridine molecules $1 - \left(\frac{R(\Gamma_A)_{ran}}{R_0}\right)$. The matrix method is used again and the calculation is performed for a four-phase system consisting of a window, thin-layer of electrolyte, thin film of adsorbed pyridine, and gold electrode. The thickness of the electrolyte layer is again taken as 7.2 μm of D_2O. The thickness of the film of adsorbed pyridine is calculated from Eq. (54). When $\Gamma = 6.2 \times 10^{-10}$ mol cm^{-2}, $M = 79.10$ g mol^{-1}, $\rho = 0.9814$ g cm^{-3}, and l is equal to 0.50 nm.

Finally, all terms in Eq. (55) are determined and one can calculate $\cos^2\theta$. The calculated values are plotted as a function of $1/\sigma_M$ in Fig. 9.23. The lowest value of $\cos^2\theta$ corresponds to $\theta = 34°$ and the highest to $\theta = 16°$. Therefore the average tilt angle of adsorbed pyridine changes by $\sim 18°$ when the charge density at the metal surface varies from $\sigma_M \approx 8$ μC cm^{-2} to a value of $\sigma_M \approx 35$ μC cm^{-2}. Figure 9.23 shows that $\cos^2\theta$ depends linearly on $1/\sigma_M$. Such behavior indicates that indeed the change of the tilt angle is driven by the interaction between the permanent dipole of the pyridine molecule and the static electric field at the electrode surface. Such interaction is described by the Langevin function, and $\cos^2\theta$ would be expected to depend on the charge at the metal according to the formula [40]

$$\cos^2\theta = \left\{1 - \frac{2\varepsilon kT}{p\sigma_M}\right\}, \tag{56}$$

where ε, k, T, p and σ_M are the permettivity Boltzmann's constant, absolute temperature, dipole moment and charge density at the metal surface, respectively. This example illustrates that by applying quantitative SNIFTIRS, one is able to provide direct evidence that pyridine molecules assume a vertical N-bonded state at the Au(110) surface over the whole range of electrode potentials investigated. The analysis revealed that the tilt angle of adsorbed pyridine molecules changes with the electrode potential. These changes are described by a simple electrostatic model that assumes interplay between an orienting force of the dipole-field interaction and a disorienting force associated with the thermal energy. The spectroscopic data indicate that, driven by the thermal energy, the adsorbed molecules may display a significant waving motion when the field at the interface is small. The energy of the dipole-field interaction restricts this movement, and when the field increases the amplitude of the waving motion de-

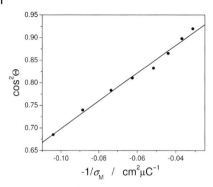

Fig. 9.23 Plot of $\cos^2 \theta = A_{exp}/3A_{cal}$ versus inverse of the charge density at the electrode surface, where A_{exp} corresponds to the integrated intensity of the 1599 cm^{-1} band for adsorbed pyridine (Fig. 9.20) and A_{cal} is the integrated intensity of this band calculated for the four parallel phases (ZnSe/D$_2$O/monolayer of pyridine/ZnSe) from the optical constants, using the matrix method and assuming that pyridine molecules are oriented with the C$_2$ axis perpendicular to the metal surface. Taken with permission from Refs. [36] and [40].

creases. At very large values of the static electric field, the monolayer becomes effectively frozen.

A similar analysis was performed to determine the character of citrate ion adsorption at an Au(111) electrode surface [53]. The symmetric (-COO$^-$) band was used to determine the orientation of carboxylic groups with respect to the electrode surface. The transition dipole of the symmetric (-COO$^-$) band is located in plane of the carboxylic group along its diagonal. Therefore the angle θ between this direction and the surface provides useful information concerning the coordination of the citrate ion to the electrode surface. Figure 9.24 shows a model of citrate ion deduced from the combined chronocoulometric and quantitative SNIFTIRS studies. The three carboxylic groups are oriented towards the gold surface such that the angle between the diagonal of each (-COO-) and the surface normal is 55°.

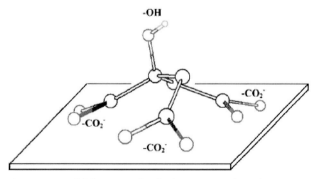

Fig. 9.24 A model for citrate adsorbed on Au(111) through 3 coordinating carboxylate groups. Taken with permission from Ref. [53].

The studies of 2-mercaptobenzimidazole (MBI) adsorption at an Au(111) electrode surface [54, 55] constitute the third example of quantitative analysis of
SNIFTIRS spectra being employed to reveal the orientation of an adsorbed molecule. The inset to Fig. 9.25 shows that the transition dipoles of the A_1 bands
are directed along the C_2 axis of the MBI molecule. Hence, quantitative analysis
of these bands allows one to determine the tilt of this axis with respect to the
surface normal. The angle θ, determined from quantitative analysis of SNIF-
TIRS spectra, is plotted as a function of the electrode potential in Fig. 9.25. This
angle is relatively large, but its value progressively decreases when the potential
is changed in the positive direction. The SNIFTIRS spectra also showed that
the intensity of B_2 bands for adsorbed molecules is very weak, and hence their
transition dipoles (in the plane of the molecule and perpendicular to the C_2
axis) are essentially perpendicular to the surface normal.

In conclusion, quantitative SNIFTIRS is a powerful technique for studying
the orientation of organic molecules at electrode surfaces. It works best when

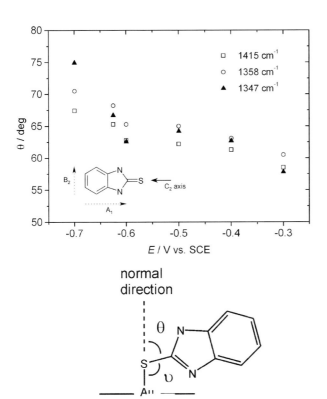

Fig. 9.25 The tilt angle of the C_2 axis plotted against the
electrode potential. The angle was calculated from the
intensity of A_1 bands at 1347 (full triangles), 1358 (open
circles) and 1415 cm^{-1} (open squares), adopted from
Ref. [54].

applied to investigate the adsorption of small soluble molecules that are reversibly adsorbed at the metal surface, such as Gibbs films. Water-soluble molecules can be periodically switched between the adsorbed and the desorbed state by modulating the potential. The application of this technique to the case of irreversible adsorption is more challenging. The technique is certainly not very helpful for the investigation of Langmuir films formed by insoluble surfactants. In this case, PM IRRAS should be the technique of choice. It is described in the next section.

9.8
Development of Quantitative *in-situ* PM IRRAS

9.8.1
Introduction

Hipps and Crosby [10] introduced the photoelastic modulator to infrared spectroscopy in 1979. They recorded the first PM IR spectra in the transmission mode. The first polarization modulation IRRAS experiments using a metal mirror were performed by Golden et al. [9] in 1981. An ultrahigh vacuum chamber and a dispersive instrument were utilized in that work. The first experiments that employed polarization modulation IRRAS to study an electrochemical interface were carried out by Russell et al. [11] in 1982. Two years later, Golden, Kunimatsu, and Seki [12, 13] used a Fourier transform spectrometer to perform the first PM IRRAS experiments. They described the layout of their experimental set-up and calculated the mean square electric field strength (MSEFS) within the thin-layer cell [14]. The performance of dispersive instruments was compared with that of FTIR instruments by Golden et al. [15] using an ultrahigh vacuum-based PM IRRAS set-up. The group concluded that the use of FTIR instruments allows one to achieve better resolution and S/N ratio. Since then, FT instruments began to dominate, and we use the name PM IRRAS with the understanding that it involves application of a Fourier transform instrument. The advantages of PM IRRAS over SNIFTIRS were demonstrated by Faguy et al. [30, 31]. These researchers have shown that, as well as having superior S/N, the PM technique is almost insensitive to atmospheric H_2O and CO_2. An important advance in PM IRRAS was made by Corn and coworkers [65, 66], who introduced a novel method of experimental signal demodulation in 1991. Their device, a synchronous sampling demodulator (SSD), allowed one to enhance the signal almost twofold in comparison to the conventional lock-in technique. The PM IRRAS technique was employed recently to study the adsorption of ions at electrode surfaces by Gewirth and coworkers [67, 68].

The main difficulty of the PM IRRAS technique for quantitative analysis of ultrathin films at metal electrodes stems from the broad-band background that is introduced to the spectra by the photoelastic modulator (PEM), the so called PEM response functions. Recently, Buffeteau et al. described methods to remove the artefacts introduced by the PEM via normalization by the spectrum of the film-free substrate [69] or by utilization of calibration spectra, acquired in an independent experiment [70].

The exceptionally low sensitivity of PM IRRAS to atmospheric CO_2 and H_2O vapor prompted interest in the application of this technique to the study of insoluble surfactant monolayers at the air/water interface in a Langmuir trough [27, 71–78]. These studies significantly advanced the PM IRRAS technique.

9.8.2
Fundamentals of PM IRRAS and Experimental Set-up

In PM IRRAS, the incident beam polarization is modulated between two orthogonal directions: (a) perpendicular (s-polarization) and (b) parallel (p-polarization) to the plane of incidence of the infrared beam. At the metal surface, the electric field of the s-polarized infrared beam almost vanishes, while the electric field of the p-polarized beam is enhanced (see Fig. 9.3). Thus, the first component of the modulated signal is insensitive to the presence of the film of adsorbed molecules and can be used to obtain the background spectrum, while the other component can be utilized to obtain the spectrum of the film. Modulation between s- and p-polarizations is carried out at a high frequency (50 kHz), and the reflection absorption spectrum $\Delta R / \langle R \rangle$ is obtained as

$$\frac{\Delta R}{\langle R \rangle} = \frac{|R_s - R_p|}{(R_s + R_p)/2} \tag{57}$$

where R_s and R_p represent the reflectivities of the s- and p-components of the modulated infrared beam from the thin-layer cell. When throughputs of the set-up are the same for s- and p-polarized beams, any common-mode signal, such as that due to absorption by atmospheric CO_2 and H_2O, is removed from the spectrum. Since there is no need to modulate the electrode potential, this method is particularly useful to study films of biological molecules that are often irreversibly desorbed by a high-amplitude potential modulation. It is important to note that PM IRRAS is sensitive not only to the surface species but also to the molecules that are in the vicinity of the metal surface at a distance which is less than 1 μm (ca. 1/4 of the wavelength of the incident radiation). Hence, the background due to the aqueous electrolyte cannot be completely removed from the $\Delta R / \langle R \rangle$ spectrum.

A diagram of an experimental set-up used for PM IRRAS experiments is shown in Fig. 9.26 and a photograph of the spectroelectrochemical cell in Fig. 9.27. The set-up is built from components purchased from Newport and custom-machined parts in an external tabletop optical mount (TOM) box. A convergent infrared beam from the spectrometer enters a port of the TOM box where it is deflected by a flat mirror and focused onto the working electrode by a parabolic mirror (f=6 in, Nicolet). Before entering the cell, the beam passes through a static polarizer (diameter 1 in, with an anti-reflective coating –

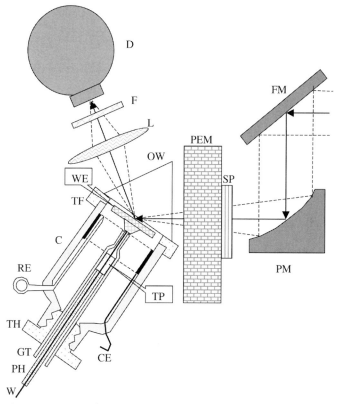

Fig. 9.26 Experimental PM IRRAS setup (top view). FM, flat mirror; PM, parabolic mirror; SP, static polarizer; PEM, photoelastic modulator; OW, optical window; L, lens; F, optical filter; D, detector; WE, working electrode; TF, Teflon flange; C, cell; RE, reference electrode; TH, Teflon holder, GT, glass tube; PH, piston handle; W, gold wire; CE, counter electrode, TP, Teflon piston.

Specac) and the photoelastic modulator (PEM) (Hinds Instruments PM-90 with II/ZS50 (ZnSe 50 kHz) optical head). The beam, exiting the cell, is refocused by a ZnSe lens (f=1 in, diameter 1.5 in, with an anti-reflective coating – Janos Technology) onto an MCT-A detector (TRS 50 MHz bandwidth, Nicolet). A bandpass optical filter (3–4 μm pass, Specac) is placed at the entrance window of the detector when the spectra of the hydrocarbon stretching modes are acquired.

The spectroelectrochemical cell (Fig. 9.27) is custom-made of glass and connected to the optical window (BaF$_2$ 1 in equilateral prism, Janos Technology) via a Teflon flange. The working electrode is a 15-mm disk-shaped Au(111) single crystal inserted into a glass tube, which is used to push the electrode against the optical window through a threaded Teflon holder, as indicated in Fig. 9.26. A piece of gold wire is used as a conductor between the potentiostat clip and the gold single crystal. It is inserted into a small orifice in a Teflon piston and

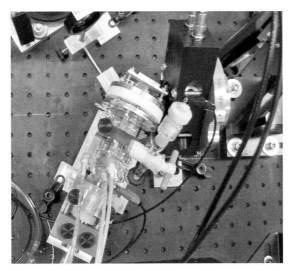

Fig. 9.27 Photograph of a spectroelectrochemical cell used in PM IRRAS experiments.

placed inside the glass tube as shown in the figure. To ensure tight sealing of the cell around the gold wire, the Teflon piston is pushed against the conical end of the glass tube by a hollow brass handle attached to the piston. A Pt foil cylinder was used as a counter electrode, and a glass capillary was used to connect a saturated Ag/AgCl reference electrode as shown in Fig. 9.27.

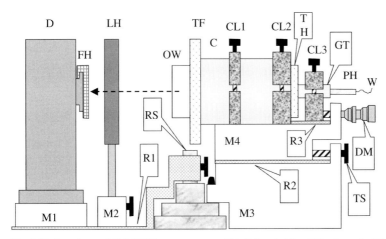

Fig. 9.28 Experimental PM IRRAS setup (side view). D, detector; FH, optical filter holder; M1, mount 1; LH, lens holder; M2, mount 2; R1, rail 1; RS, rotation stage; OW, optical window; TF, Teflon flange; C, cell; CL1, clamp 1; CL2, clamp 2; TH, Teflon holder; CL3, clamp 3; GT, glass tube; PH, piston handle; W, gold wire; M4, mount 4; R3, rail 3; DM, differential micrometer; R2 rail 2; TS, translation screw; M3, mount 3.

The components of the set-up are assembled on 4 mounts as shown in Fig. 9.28. The detector is placed on mount 1. The lens is inserted in a lens holder (Newport), mount 2. Both mounts could be translated along rail 1, attached to a rotation stage (Newport). The rotation stage is fixed to the optical table such that its pivoting axis coincides with the focal point of the parabolic mirror. The rotation stage is used to independently set both the angle of incidence and the angle of reflection. The angle of reflection is adjusted by rotation of rail 1 and the angle of incidence is changed by rotation of mount 3, which serves as the base for the cell holder. The cell holder is assembled on mount 4, which could be translated with respect to mount 3 along rail 2 using the translation screw. The cell holder consists of clamp 1 and clamp 2, which are used to attach the cell to the holder, and clamp 3, which is used to secure the glass tube of the working electrode assembly. Clamp 3 could be translated with respect to mount 4 along rail 3 using the differential micrometer (Newport). This translation allows one to adjust the thickness of the thin layer between the electrode and the optical window.

9.8.3
Principles of Operation of a Photoelastic Modulator

Figure 9.29a shows a simplified diagram of a set-up for polarization modulation spectroscopy. The set-up consists of a static polarizer P and a photoelastic modulator PEM. The incoming radiation can be represented by two orthogonal components I_{0V} and I_{0H}, which denote intensities of the vertical and horizontal components of the incident beam with respect to the plane of the optical bench. When the beam crosses the static polarizer P, the vertical component is rejected and the radiation becomes linearly polarized in the direction parallel to the optical bench, indicated as I_1. The modulator is mounted at an angle of 45° with respect to the plane of the optical bench. Hence, the beam of linearly polarized light I_1 can be regarded as composed of two orthogonal components, I_{1x} and I_{1y}, oriented along the main axis of the PEM and perpendicular to it, respectively.

The PEM is made of a piezoelectric transducer that is glued to a ZnSe crystal. The piezoelement converts a periodic voltage to a periodic mechanical (acoustic) wave, which compresses or expands the crystal. This movement changes the refractive index in the x direction and imposes a periodic retardation (or acceleration) of the I_{1x} component of the incident linearly polarized wave. The I_{1y} component remains unchanged. The PEM is operated at its resonant frequency (50 kHz). If the optical element is at rest, the polarization of the radiation remains unchanged. If the optical element undergoes compression or expansion, the component I_{1x} has a positive (retardation) or negative (acceleration) phase shift relative to the phase component of the component I_{1y}.

Figure 9.29b shows the two components when the amplitude of the applied voltage is set such that the PEM imposes the maximum relative phase shift of 90° (half-wave retardation). Figure 9.29c shows that in this case the incident

(a)

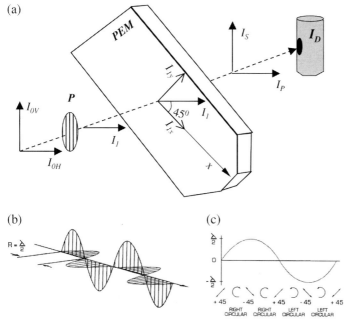

(b)

(c)

Fig. 9.29 (a) Simplified diagram of the experimental setup and orientations of the electric field vectors of the electromagnetic radiation. (b) Relative orientation of the electric components at half-wave retardation. (c) Effect of retardation on the polarization of the beam that exits the PEM. The arrows mark directions of linearly polarized light. The angle +45° indicates that the electric field of the photon is parallel to x axis and angle −45° indicates that the electric field is parallel to y axis. Adapted from Ref. [92].

beam changes its polarization every 90°. During one cycle of the applied voltage the half-wave retardation occurs twice: once when the optical element becomes compressed (the beam is then linearly polarized at an angle of +45° with respect to the incident beam) and once when it becomes stretched (the beam is then polarized at an angle of −45° with respect to the incident beam). Thus, the PEM can be used to alter the polarization of the incident linearly polarized beam at a frequency that is twice the frequency of the voltage applied to the piezoelement.

Hipps and Crosby [10] derived an analytical expression that describes the signal at the detector measured using such a set-up:

$$I_D(\varphi) = \frac{I_s + I_p}{2} + \frac{I_p - I_s}{2} \cos \varphi. \tag{58}$$

Here I_D denotes the intensity of the radiation at the detector placed at the location shown in Fig. 9.29a. I_p and I_s represent the components of the beam in front of the detector that are oriented parallel and perpendicular to the optical bench, respectively. The notations were chosen to be consistent with the orientations of the s- and p-polarized beam at the metal electrode, which is not shown. The phase shift φ depends on the frequency of the PEM ω_m and time t in the following way:

$$\varphi = \varphi_0 \cos(\omega_m t) . \tag{59}$$

Here, φ_0 is the maximum phase shift introduced by the PEM, which depends on the magnitude of the voltage (V_m) that is applied to the PEM and the wavelength of the polarized radiation λ given by:

$$\varphi_0 = \frac{G V_m}{\lambda} \tag{60}$$

where G is the proportionality factor. The amplitude of the voltage V_m is usually optimized such that $\varphi_0 = \pi$ for a selected wavelength λ_0, and hence

$$G V_m = \pi \lambda_0 . \tag{61}$$

The phase shift φ introduced by the PEM to a beam of monochromatic light of wavelength λ other than λ_0 is therefore given by

$$\varphi = \frac{\lambda_0}{\lambda} \pi \cos(\omega_m t) . \tag{62}$$

Combining Eqs. (58) and (62), one can finally write

$$I_D(t) = \frac{I_s + I_p}{2} + \frac{I_p - I_s}{2} \cos\left(\frac{\lambda_0}{\lambda} \pi \cos(\omega_m t)\right) . \tag{63}$$

It follows from Eq. (63) that

$$I_D(t) = I_p \tag{64}$$

when $\omega_m t = \frac{\pi}{2} \pm N\pi$ for $N = 0, 1, 2\ldots$ and

$$I_D(t) = I_s + \frac{I_p - I_s}{2} \left\{1 + \cos\left(\frac{\lambda_0}{\lambda} \pi\right)\right\} \tag{65}$$

when $\omega_m t = 0 \pm N\pi$ for $N = 0, 1, 2\ldots$

It is worth noting that, when $\lambda = \lambda_0$, the second term in Eq. (65) disappears and

$$I_D(t) = I_s . \tag{66}$$

Thus, at λ_0 (for which the PEM is optimized), the polarization of the incident beam is periodically changed between p and s, and every 90° of a period either

(I_s) and (I_p) is incident at the detector. If the wavelength of the incident beam departs from λ_0, the polarization of the incident beam becomes circular (i.e. light that has both s- and p-components). Consequently, one can fully optimize an experiment for a single wavelength only.

Therefore, when using polarization modulation FTIR spectroscopy one has to distinguish between a theoretical spectrum equal to

$$\left(\frac{\Delta I}{\langle I \rangle}\right)_{\text{theor.}} = \frac{I_p - I_s}{(I_s + I_p)/2} = \frac{|R_s - R_p|}{(R_s + R_p)/2} = 2.3\Gamma\varepsilon \tag{67}$$

and the experimental quantity which is the ratio of the intensity difference signal $I_D^{\text{Diff}}(2\omega)$ at the second harmonic frequency to the average intensity signal I_D^{Ave} given by the formula

$$\frac{I_D^{\text{Diff}}(2\omega)}{I_D^{\text{Ave}}} = \frac{\Delta I \cdot J_2(\varphi_0)}{\langle I \rangle + \frac{1}{2}\Delta I \cdot J_0(\varphi_0)} . \tag{68}$$

There are two demodulation techniques that are used to measure the signal arriving at the detector into I_D^{Ave} and $I_D^{\text{Diff}}(2\omega)$. The first uses a lock-in amplifier and a low-pass filter [10, 69], while the second relies on the synchronous sampling demodulator (SSD) to obtain the intensity difference and average signals [65, 66]. In the first case, J_2 and J_0 are the second-order and zero-order Bessel functions. In the second case they are described by the following expressions:

$$J_2(\varphi_0) = \left\{\frac{1 + \cos\varphi_0}{2}\right\} \tag{69}$$

and

$$J_0(\varphi_0) = \left\{\frac{1 + \cos\varphi_0}{2}\right\} . \tag{70}$$

The polarization modulation reflection absorption spectrum acquired using the SSD can therefore be written as

$$\left(\frac{I_D^{\text{Diff}}(2\omega)}{I_D^{\text{Ave}}}\right)_{\text{Exp.}} = \frac{\Delta I\{(1 - \cos\varphi_0)/2\}}{\langle I \rangle + \frac{1}{2}\Delta I\{(1 + \cos\varphi_0)/2\}} . \tag{71}$$

Equations (59)–(66) show that $\dfrac{I_D^{\text{Diff}}(2\omega)}{I_D^{\text{Ave}}} = \dfrac{\Delta I}{I} = \dfrac{\Delta R}{R}$ only when $\lambda = \lambda_0$. For other wavelengths, the absorbance by the investigated sample is superimposed on the background caused by the J_2 and J_0 functions.

9.8.4

Correction of PM IRRAS Spectra for the PEM Response Functions

In order to calculate $\Delta I / \langle I \rangle$ from the measured PM IRRAS spectra, one has to determine functions J_2 and J_0 in an independent experiment. A reliable method to measure the PEM response functions was described by Buffeteau et al. [69]. Below we describe a similar method that we adapted with minor changes to use for electrochemical systems [81]. The spectroelectrochemical cell is replaced by the dielectric total external reflection mirror (a CaF_2 equilateral prism can be used for this purpose). The second polarizer is inserted just after the PEM and set to admit p-polarized light (identical setting to that of the first polarizer). The PEM is turned off and the reference spectrum is acquired. This spectrum gives the intensity of the p-polarized light $I_p(cal)$, which passes through the whole optical bench.

Next, the PEM is turned on and polarization modulation spectra are acquired. (Unity gain on the difference channel has to be used during the acquisition of these calibration spectra.) The two calibration signals $I_D^{Ave}(cal)$ and $I_D^{diff}(cal)$ are measured independently. Since only p-polarized light is passed through the second static polarizer, the I_s component is equal to zero for the calibration signal, and, following Eq. (68), one can write

$$I_D^{Ave}(cal) = \frac{I_p(cal)}{2} + \frac{I_p(cal)}{2} J_0(\varphi_0) \tag{72}$$

and hence

$$J_0(\varphi_0) = 2 \frac{I_D^{Ave}(cal)}{I_p(cal)} - 1 . \tag{73}$$

The measurement on the second channel gives

$$I_D^{Diff}(cal) = I_p(cal) J_2(\varphi_0) \tag{74}$$

from which $J_2(\varphi_0)$ can easily be calculated.

Solid lines in Fig. 9.30 show the experimental $J_0(\varphi_0)$ and $J_2(\varphi_0)$ determined for the set-up in Figs. 9.26 and 9.27 using the above calibration procedure and the SSD. In addition, the theoretical Bessel functions (dotted lines) and the cosine functions defined by Eqs. (69) and (70) (dashed lines) are also plotted in Fig. 9.30. The experimental response functions deviate significantly from both the Bessel functions and the cosine terms. This result shows that it is important to determine the response function using the calibration procedure, because the performance of the PEM device may deviate from the ideal behavior assumed when theoretical functions are used.

Knowing the PEM response functions, the measured signals I_D^{Ave} and $I_D^{Diff}(2\omega)$ can be corrected, and the average $\langle I \rangle$ and difference ΔI signals may be calculated using equations:

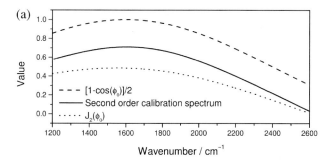

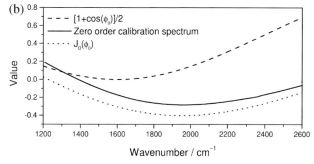

Fig. 9.30 The theoretical response functions of the PEM and empirical calibration spectra. (a) Second order and (b) zero order. Dotted line: theoretical Bessel functions, solid line: empirical calibration spectra, dashed line: theoretical cosine terms. The PEM was optimized for 1600 cm^{-1}.

$$\langle I \rangle = \frac{I_s + I_p}{2} = I_D^{Ave} - \frac{I_D^{Diff}(2\omega)}{2} \frac{I_p(cal)}{I_D^{Diff}(cal)} \left(2 \frac{I_D^{Ave}(cal)}{I_p(cal)} - 1 \right) \tag{75}$$

and

$$\Delta I = I_p - I_s = I_D^{Diff}(2\omega) \frac{I_p(cal)}{I_D^{Diff}(cal)} . \tag{76}$$

Buffeteau and coworkers [69] demonstrated that signals $\langle I \rangle$ and ΔI have to be corrected further to take into account the fact that the ratio of the optical throughputs of the experimental set-up for p- and s-polarized light γ is not equal to unity. When $\gamma \neq 1$, the corrected PM IRRAS spectrum can be calculated with the help of the formula [36]

$$\frac{\Delta I}{\langle I \rangle} = 2 \frac{(\gamma + 1)\Delta I(\omega) + 2(\gamma - 1)\langle I \rangle(\omega)}{(\gamma - 1)\Delta I(\omega) + 2(\gamma + 1)\langle I \rangle(\omega)} . \tag{77}$$

For the experimental set-up built in our laboratory, γ was found to be 1.06.

9.8.5
Background Subtraction

The last step in processing the experimental PM IRRAS spectra involves the subtraction of the background. This background arises from the slowly varying broad-band absorbance of infrared radiation by the aqueous electrolyte. In order to remove the background, a procedure similar to that published by Barner et al. [66] was developed by Zamlynny [36]. The baseline is created from the experimental data points using spline interpolation. Successful interpolation requires knowledge of the exact positions of the absorption bands and a little experience.

Figure 9.31 illustrates the procedure of background subtraction from the PM IRRAS spectrum of DMPC in the C-H stretching region. Panel 31a shows a typical PM IRRAS spectrum after corrections for the PEM response functions and the difference between the throughputs of s- and p-polarized radiation. Open circles in panel 31b denote points that were used to build the spline, shown as the dotted line. Panel 31c shows that an error is introduced by this spline interpolation. Since point d was used to build the spline, the value of ΔS_{cor} at this point was artificially set to zero. The spectrum simulated from transmittance measurement (panel 31d) shows that the descending branch of the Fermi resonance and the ascending branch of the symmetric stretch of the methyl group are present in this spectral region.

The following approach [79] can be used to eliminate the error introduced by this spline. In plots 31c and 31d, points a and c are at the centers of the symmetric ($v_s(CH_2)$) stretching bands in the simulated spectrum and the PM IRRAS spectrum, respectively. Point b in panel 31d is located at exactly the same wavenumber as the point d which was used to build the spline in panel 31c. We assume that the following relationship should apply to the PM IRRAS spectrum:

$$\left(\frac{\Delta S_a}{\Delta S_b}\right)_{trans} = \left(\frac{\Delta S_c}{\Delta S_d}\right)_{PM\ IRRAS}, \tag{78}$$

where ΔS denotes a basseline-corrected PM IRRAS signal. The value of ΔS_d at point d can be calculated from the above relationship. This value is subtracted from the original spectrum (plot 31a) and the spline interpolation is repeated using the new $\Delta S - \Delta S_d$ value at the wavenumber of point d. The new spline is shown in plot 31e. Finally, plot 31f shows the baseline-corrected spectrum after background subtraction. This spectrum is in absorbance units and is equal to

$$\Delta S_{corr} = \left(2\frac{|R_s - R_p|}{R_s + R_p}\right)_{exp} \approx 2.3\Gamma\varepsilon. \tag{79}$$

The subscript exp is used to emphasize that this is the experimental spectrum of the film of adsorbed molecules.

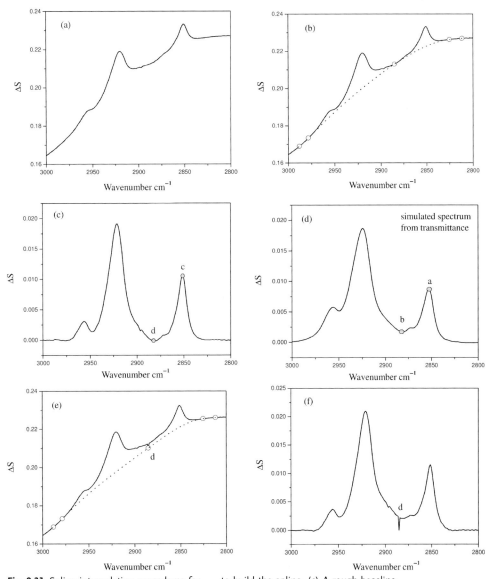

Fig. 9.31 Spline interpolation procedures for background corrections of PM IRRAS spectra. (a) A raw PM IRRAS spectrum of a DMPC bilayer on a Au(111) surface at the electrode potential E = −0.2 V. (b) A spline (dashed line) for the first rough baseline correction, open circles are data points used to build the spline. (c) A rough baseline-corrected PM IRRAS spectrum. (d) A simulated spectrum from transmittance measurements. (e) Modified spline interpolation. (f) A spectrum obtained after subtraction of the background. Taken with permission from Ref. [79].

9.8.6
Applications of Quantitative PM IRRAS

Horswell et al. [80] made the first attempt to apply PM IRRAS to determine tilt angles of acyl chains of DMPC in a bilayer of this phospholipid deposited at an Au(111) electrode surface in 2002. No correction of the measured spectra for the PEM response function was made at that time, and the tilt angle was determined using the relative method. It was assumed that the transition dipole of the asymmetric $\nu_{as}(CH_3)$ stretch was randomly oriented. The tilt angle was then calculated from the ratio of the integrated intensity of the asymmetric $\nu_{as}(CH_2)$ and symmetric $\nu_s(CH_2)$ methylene stretches to the integrated intensity of the $\nu_{as}(CH_3)$ band using Eq. (43).

The correction of the measured PM IRRAS signal for the PEM response functions was introduced in the studies of the potential-induced reorientation of a film formed by 4-pentadecyl pyridine [81], one year later. For the first time, the absorbance of a Langmuir film adsorbed at the electrode surface was determined in that paper. The theoretical spectrum of a film of randomly oriented molecules was calculated from independently measured optical constants, and tilt angles of the pentadecyl chain and the pyridine moiety were determined using the absolute method and Eq. (44). This work laid the methodological foundations for future PM IRRAS studies on monolayers and bilayers formed by amphiphilic molecules at electrode surfaces.

Thus, quantitative PM IRRAS has been employed to investigate the orientation and the field-induced transformations of a monolayer [82] and a bilayer [83] of *n*-octadecanol at a gold electrode surface. The *n*-octadecanol forms 2D solid-like films. These studies illustrate how to use PM IRRAS to determine not only the tilt angle of adsorbed molecules but also the packing and orientation of molecules into a unit cell of the 2D lattice of a solid film.

Recently, Zawisza et al. [84–86] and Bin et al. [87, 88] demonstrated that quantitative PM IRRAS has a very important application in biomimetic research. This technique provides unique information concerning potential-induced changes in the orientation and conformation of molecules in a model biological membrane supported at an electrode surface. This point is illustrated by the application of PM IRRAS to study the structure of a bilayer of DMPC formed at the Au(111) electrode surface by fusion of unilamellar vesicles [87].

Figure 9.32 shows a model of the DMPC molecule and the directions of the transition dipoles of the major bands. This information is either available in the literature or could be determined from *ab initio* normal coordinate calculations. A DMPC molecule contains 24 methylene groups in the two acyl chains and four methylene groups in its head group. The CH_2 groups of the chains dominate the methylene bands. Consequently, these bands provide information concerning conformation, orientation, and physical state of the chains.

Figure 9.33a shows the IR spectra in the CH stretching region. The top, thicker line plots $\left(2\dfrac{|R_s - R_p|}{R_s + R_p}\right)_{cal}$ which shows bands for the bilayer of ran-

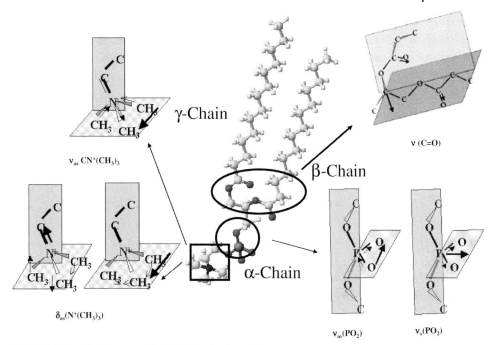

γ-Chain

ν (C=O)

ν$_{as}$ CN$^+$(CH$_3$)$_3$

β-Chain

α-Chain

δ$_{as}$(N$^+$(CH$_3$)$_3$)

ν$_{as}$(PO$_2$)   ν$_s$(PO$_2$)

Fig. 9.32 Schematic diagram of the DMPC molecule and the directions of the transition dipole moments of the major IR bands. Taken with permission from Refs. [79] and [87].

domly oriented molecules, calculated from the optical constants shown in Fig. 9.15 and taking the thickness of the bilayer as equal to 5.4 nm [87]. The four bottom curves plot the PM IRRAS spectra for the bilayer of DMPC at the electrode surface at selected electrode potentials. This spectral region consists of four overlapping bands corresponding to $\nu_{as}(CH_3)$, $\nu_{as}(CH_2)$, $\nu_s(CH_3)$, and $\nu_s(CH_2)$, and two Fermi resonances between the overtones of the symmetric bending mode and symmetric methyl and methylene stretches [87]. In order to extract quantitative information concerning the methylene bands, it was necessary to de-convolute this spectral region. In this region, a Lorentzian function was used as the de-convolution band shape. The band de-convolution is shown in Fig. 9.33 b.

Figure 9.33 a shows that the intensities of the methylene stretching bands change with potential. The intensity is low at negative potentials, where the bilayer is detached from the metal, and increases at more positive E, where the bilayer is adsorbed on the gold surface. This behavior indicates that orientation of DMPC molecules and consequently the angle θ between the direction of the transition dipole of a given vibration and the surface normal change with potential.

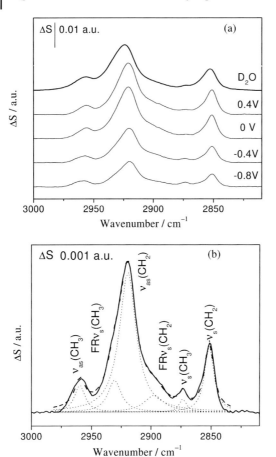

Fig. 9.33 (a) CH stretching region of PM IRRAS spectra of a DMPC bilayer on an Au(111) electrode in 0.1 M NaF/D$_2$O solution at potentials indicated in the figure. The top trace plots spectrum calculated for 5.5-nm thick film of randomly oriented DMPC molecules using optical constants of DMPC for vesicles dispersion in D$_2$O.

(b) Example of a de-convolution of the overlapping ν_{as} (CH$_3$), ν_{as} (CH$_2$), ν_s (CH$_3$), ν_s (CH$_2$) bands and two Fermi resonances between the overtones of the symmetric bending mode and symmetric methyl FRν_s (CH$_2$) and methylene stretch FRν_s (CH$_3$) modes for the DMPC bilayer at $E=-0.5$ V. Adapted from Refs. [79] and [87].

The angle θ can be calculated from the ratio of the integrated intensity of a given band in the $\left(2\dfrac{|R_s - R_p|}{R_s + R_p}\right)_{exp}$ spectrum to the integrated intensity of the same band in the calculated spectrum $\left(2\dfrac{|R_s - R_p|}{R_s + R_p}\right)_{cal}$ with the help of Eq. (44). Figure 34a and b plot θ_s(CH$_2$) and θ_{as}(CH$_2$) as functions of the electrode potential determined using this procedure.

For a fully stretched all-trans conformation of acyl chains, θ_s(CH$_2$), θ_{as}(CH$_2$) and the chain tilt angle θ_{tilt} are related by the formula [89]

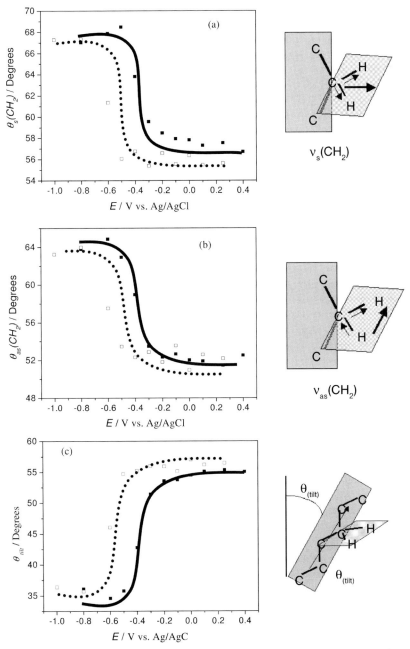

Fig. 9.34 The dependence on potential of the angle (θ) between the directions of the transition dipole moment and the electric field of the photon (normal to the surface) for the DMPC bilayer on the Au(111) electrode in 0.1 M NaF/D$_2$O solution for (a) v_s(CH$_2$), (b) v_{as}(CH$_2$), and (c) the direction of the acyl chain; (solid line) positive, (dashed line) negative potential steps. Taken with permission from Refs. [79] and [87].

$$\cos^2 \theta_{as} + \cos^2 \theta_s + \cos^2 \theta_{tilt} = 1 . \tag{80}$$

Therefore, Eq. (80) allows one to calculate the chain tilt angle, which is plotted as a function of the electrode potential in Fig. 9.34c. The tilt angle θ_{tilt} of the hydrocarbon chains in a DMPC bilayer supported at the gold electrode surface changes from $\sim 35°$ at potential range $-1.0 < E < -0.5$ V to $\sim 55°$ at $E > -0.4$ V. Independent neutron reflectivity experiments have shown that at $E < -0.5$ V the bilayer is detached from the electrode but remains in its proximity, separated from the gold surface by a ~ 1-nm thick layer of the electrolyte [90]. When $E > -0.5$ V the bilayer is directly adsorbed at the metal surface. The PM IRRAS studies have demonstrated that the adsorption and detachment of the bilayer involve a significant change in the tilt of the acyl chains.

A similar analysis has been performed on the bands in the polar head region of the DMPC molecule. The molecular models in Fig. 9.35 show the chain orientation and the polar head conformation for the two states of the adsorbed and detached states of the bilayer. For the bilayer deposited at a gold electrode surface, the differences between the chain tilt angles and conformations of the polar head region in the adsorbed and detached states are dramatic. The result shows that structural changes between the detached and adsorbed state of the film allow maximum contact of the polar head and specifically the phosphate group with the metal. This transformation causes the chains to tilt and the heads to be less densely packed. It opens a space for water to penetrate the polar head region. The PM IRRAS data show that the transition from the detached to the adsorbed state of the bilayer involves a significant increase in the hydration of the ester and phosphate groups [87].

This example demonstrates that quantitative PM IRRAS spectroscopy is a powerful tool to study the potential-driven transformation of a phospholipid bi-

(a)   (b)

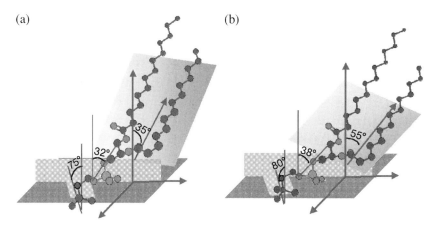

Fig. 9.35 Scheme of orientation of the DMPC molecule in the bilayer at (a) $E = -0.8$ V (detached film) and (b) $E = 0.2$ V (adsorbed film). Taken with permission from Refs. [79] and [87].

layer supported at an electrode surface. It should find a broad application in biomimetic research and in general in the studies of Langmuir films deposited at metal surfaces.

9.9
Summary and Future Directions

The quantitative SNIFTIRS and PM IRRAS were developed three years ago. They allow one to measure the angle between the direction of the transition dipole of an IR band and the normal to the electrode surface. Since the direction of the transition dipole with respect to the coordinates of the adsorbed molecule could be determined with the help of *ab initio* normal coordinate calculations, these methods ultimately allow one to determine the orientation of molecules adsorbed at the electrode surface. In a short period of time, they have been successfully applied to provide information concerning the orientation of molecules in several organic films. Today, thanks to the availability of these methods, investigations of thin films at the metal/solution interface may be as detailed as they were at the metal/gas or metal/vacuum interfaces in the past.

The two techniques are complementary. SNIFTIRS is powerful in determining orientation of molecules in Gibbs films, formed by adsorption of soluble molecules. In contrast, PM IRRAS finds application in studies of Langmuir films, formed by insoluble surfactants. Two important future applications of PM IRRAS can be envisaged. The first is to study the structure of barrier films used for corrosion protection or packaging in microelectronics. The second is in biomimetic research to study model biological membranes supported at metal electrode surfaces. A combination of electrochemical methods and PM IRRAS spectroscopy opens up a unique opportunity to study voltage-gated phenomena such as opening or closing of ion channels, voltage-dependent host-guest interactions in bio-electrochemical sensors, and protein- and cholesterol-induced changes in the structure of the biological membrane. The biophysical applications of PM IRRAS constitute the future direction of our research.

Acknowledgments

The authors acknowledge many helpful discussions with Bruno Pettinger, who introduced them to the Fresnel equations calculations. This work was supported by grants from the Natural Sciences and Engineering Research Council of Canada. JL acknowledges the Canadian Foundation of Innovation for the Canada Research Chair Award. VZ acknowledges the Canadian Foundation for Innovation for the New Opportunities Award.

References

1 R. G. Greenler, J. Chem. Phys. 44 (1966) 310.

2 R. G. Greenler, J. Chem. Phys. 50 (1969) 1963.

3 R. G. Greenler, J. Vac. Sci. Technol. 12 (1975) 1410.

4 M. Osawa, M. Kuramitsu, A. Hatta, W. Sutaka and H. Seki, Surf. Sci. 175 (1986) L787.

5 A. Bewick, K. Kunimatsu and B. S. Pons, Electrochim. Acta 25 (1980) 465.

6 A. Bewick and K. Kunimatsu, Surf. Sci. 101 (1980) 131.

7 T. Davidson, B. S. Pons, A. Bewick and P. P. Schmidt, J. Electroanal. Chem. 125 (1981) 237.

8 S. Pons, T. Davidson and A. Bewick, J. Am. Chem. Soc. 105 (1983) 1802.

9 W. G. Golden, D. S. Dunn and J. Overend, J. Catal. 71 (1981) 395.

10 K. W. Hipps and G. A. Crosby, J. Phys. Chem. 83 (1979) 555.

11 J. W. Russel, J. Overend, K. Scanlon, M. Severson and A. Bewick, J. Phys. Chem. 86 (1982) 3066.

12 W. G. Golden, K. Kunimatsu and H. Seki, J. Phys. Chem. 88 (1984) 1275.

13 K. Kunimatsu, H. Seki and W. G. Golden, Chem. Phys. Lett. 108 (1984) 195.

14 H. Seki, K. Kunimatsu and W. G. Golden, Appl. Spectrosc. 39 (1985) 437.

15 W. G. Golden, D. D. Saperstein, M. W. Severson and J. Overend, J. Phys. Chem. 88 (1984) 574.

16 A. W. Crook, J. Opt. Soc. Am. 38 (1948) 954.

17 F. Abelés, Ann. Phys. (Paris) 5 (1950) 596.

18 F. Abelés, in "Advanced Optical Techniques" A. C. S. Van Heel (Ed.), John Wiley & Sons, New York (1967) 145.

19 W. N. Hansen, J. Opt. Soc. Am. 58 (1968) 380.

20 A. Hecht, A. Zajac, Optics, 2nd edn. Addison-Wesley, Toronto (1975).

21 R. J. Lipert, B. D. Lamp and M. D. Porter, "Specular Reflection Spectroscopy" in "Modern Techniques in Applied Molecular Spectroscopy", F. M. Mirabella (Ed.), Techniques in Analytical Chemistry Series, John Wiley & Sons, Inc., New York, 1998.

22 E. D. Palik (Ed.), Handbook of optical constants of solids. Academic Press, London (1998).

23 M. Moskovits, J. Chem. Phys. 77 (1982) 4408

24 R. A. Dluhy, J. Phys. Chem. 90 (1986) 1373.

25 D. K. Roe, J. K. Sass, D. S. Bethune and A. C. Luntz, J. Electroanal. Chem. 216 (1987) 293.

26 D. S. Bethune, A. C. Luntz, J. K. Sass and D. K. Roe, Surf. Sci. 197 (1988) 44.

27 D. Blaudez, T. Buffeteau, B. Desbat, P. Fournier, A.-M. Ritcey and M. Pézolet, J. Phys. Chem. B 102 (1998) 99.

28 D. D. Popenoe, S. M. Stole and M. D. Porter, Appl. Spectrosc. 46 (1992) 79.

29 P. W. Faguy and W. R. Fawcett, Appl. Spectrosc. 44 (1990) 1309.

30 P. W. Faguy and W. N. Richmond, J. Electroanal. Chem. 410 (1996) 101.

31 P. W. Faguy, W. N. Richmond, R. S. Jackson, S. C. Weibel, G. Ball and J. Payer, Appl. Spectrosc. 52 (1998) 557.

32 P. W. Faguy and N. S. Marinković, Appl. Spectrosc. 50 (1996) 394.

33 P. W. Faguy and N. S. Marinkovi, Anal. Chem. 67 (1995) 2791.

34 P. A. Brooksby and W. R. Fawcett, Anal. Chem. 73 (2001) 1155.

35 S.-M. Moon, C. Bock and B. MacDougall, J. Electroanal. Chem. 568 (2004) 225

36 V. Zamlynny, "Electrochemical and Spectroscopic Studies of Pyridine Surfactants at the Gold-Electrolyte Interface", PhD thesis, University of Guelph, 2002. Inquiries concerning Fresnel 1 software should be addressed to *Vlad.Zamlynny@acadiau.ca.*

37 M. A. Ordal, L. L. Long, R. J. Bell, S. E. Bell, R. R. Bell, R. W. Alexander, Jr. and C. A. Ward, Appl. Opt. 22 (1983) 1099.

38 A. N. Rusk, D. Williams and M. Querry, J. Opt. Soc. Am. 61 (1971) 895.

39 J. E. Bertie, M. K. Ahmed and H. H. Eysel, J. Phys. Chem. 93 (1989) 2210.

40 N. Li, V. Zamlynny, J. Lipkowski, F. Henglein and B. Pettinger, J. Electroanal. Chem. 524/525 (2002) 43.

41 R. Arnold, A. Terfort and C. Wöll, Langmuir 17 (2001) 4980.

42 D. L. Allara, A. Bacca and C. A. Pryde, Macromolecules 11 (1978) 1215.

43 D. A. Skoog, Principles of Instrumental Analysis, 3rd edn, CBS College Publishing, New York (1985) 341.

44 J. D. E. McIntyre, in P. Delahay, C. W. Tobias (Eds.), Advances in Electrochemistry and Electrochemical Engineering, Vol. 9, John Wiley & Sons, New York (1973) 61.

45 F. Stern, in F. Seitz and D. Turnbull (Eds.), Solid State Physics. Advances in Research and Application, Vol. 15. Academic Press, New York (1963) 299.

46 R. H. Muller, in P. Delahay, C. W. Tobias (Eds.), Advances in Electrochemistry and Electrochemical Engineering, Vol. 9. John Wiley & Sons, New York (1973) 288.

47 V. P. Diakonov, Handbook of algorithms and programs for personal computers in Basic language. Nauka, Moscow (1989) p. 86.

48 D. L. Allara and J. D. Swalen, J. Phys. Chem. 86 (1982) 2700.

49 D. L. Allara and R. G. Nuzzo, Langmuir 1 (1985) 52.

50 M. J. Frisch et al., Gaussian 98, Rev. A. 11, Gaussian Inc., Pittsburgh PA, 2001.

51 Y. Miura, S. Kimura, Y. Imanishi and J. Umemura, Langmuir, 14 (1998) 6935.

52 T. Doneux, PhD thesis, Université Libre de Bruxelles, 2005.

53 R. J. Nichols, I. Burgess, K. L. Young, V. Zamlynny and J. Lipkowski, J. Electroanal. Chem. 563 (2004) 33.

54 T. Doneux, C. Buess-Herman and J. Lipkowski, R. Guidelli special issue, J. Electroanal. Chem. 564 (2004) 65.

55 T. Doneux, C. Buess-Herman, M. G. Hosseini, M. R. Arshadi, R. J. Nichols and J. Lipkowski, Electrochim. Acta 50 (2005) 4275.

56 L. Stolberg, J. Richer, J. Lipkowski and D. E. Irish, J. Electroanal. Chem. 207 (1986) 213.

57 L. Stolberg, J. Lipkowski and D. E. Irish, J. Electroanal. Chem. 238 (1987) 333.

58 L. Stolberg, J. Lipkowski and D. E. Irish, J. Electroanal. Chem. 296 (1990) 171.

59 L. Stolberg, J. Lipkowski and D. E. Irish, J. Electroanal. Chem. 300 (1991) 563.

60 L. Stolberg, S. Morin, J. Lipkowski and D. E. Irish, J. Electroanal. Chem. 307 (1991) 241.

61 D. F. Yang, L. Stolberg and J. Lipkowski, J. Electroanal. Chem. 329 (1992) 259.

62 J. Lipkowski and L. Stolberg in Adsorption of molecules at metal electrodes, J. Lipkowski and P. N. Ross (Eds.), VCH Publishers, New York (1992).

63 F. Henglein, J. Lipkowski and D. M. Kolb, J. Electroanal. Chem. 303 (1991) 245.

64 J. Lipkowski, L. Stolberg, D.-F. Yang, B. Pettinger, S. Mirwald, F. Henglein and D. M. Kolb, Electrochim. Acta 39 (1994) 1057.

65 M. J. Green, B. J. Barner and R. M. Corn, Rev. Sci. Instrum. 62 (1991) 1426.

66 B. J. Barner, M. J. Green, E. I. Sáez and R. M. Corn, Anal. Chem. 63 (1991) 55.

67 M. E. Biggin and A. A. Gewirth, J. Electrochem. Soc. 148 (2001) C339.

68 C. M. Teague, X. Li, M. E. Biggin, L. Lee, J. Kim and A. A. Gewirth, J. Phys. Chem. B, 108 (2004) 1974.

69 T. Buffeteau, B. Desbat and J. M. Turlet, Appl. Spectrosc. 45 (1991) 380.

70 T. Buffeteau, B. Desbat, D. Blaudez and J. M. Turlet, Appl. Spectrosc. 54 (2000) 1646.

71 D. Blaudez, T. Buffeteau, J. C. Cornut, B. Desbat, N. Escafre, M. Pézolet and J. M. Turlet, Appl. Spectrosc. 47 (1993) 869.

72 D. Blaudez, J.-M. Turlet, J. Dufourcq, D. Bard, T. Buffeteau and B. Desbat, J. Chem. Soc. Faraday Trans. 92 (1996) 525.

73 A. Dicko, H. Bourque and M. Pézolet, Chem. Phys. Lipids 96 (1998) 125.

74 H. Bourque, I. Laurin, M. Pézolet, J. Klass, R. B. Lennox and G. R. Brown, Langmuir 17 (2001) 5842.

75 I. Pelletier, H. Bourque, T. Buffeteau, D. Blaudez, B. Desbat and M. Pezolet, J. Phys. Chem. B 106 (2002) 1968.

76 W.-P. Ulrich and H. Vogel, Biophys. J. 76 (1999) 1639.

77 J. Gallant, B. Desbat, D. Vaknin and C. Salesse, Biophys. J. 75 (1998) 2888.

78 J. Saccani, T. Buffeteau, B. Desbat and D. Blaudez, Appl. Spectrosc. 57 (2003) 1260.

79 X. Bin, "Electrochemical and polarization modulation Fourier transform infrared reflection absorption spectroscopic studies of phospholipids bilayers on a Au(111) electrode surface". PhD thesis, University of Guelph, 2005.

80 S. L. Horswell, V. Zamlynny, H.-Q. Li, A. R. Merrill and J. Lipkowski, Faraday Discuss. 121 (2002) 405.

81 V. Zamlynny, I. Zawisza and J. Lipkowski, Langmuir 19 (2003) 132.

82 I. Zawisza, I. Burgess, G. Szymanski, J. Lipkowski, J. Majewski and S. Satija, Electrochim. Acta 49 (2004) 3651.

83 I. Zawisza and J. Lipkowski, Langmuir 20 (2004) 4579.

84 I. Zawisza, A. Lachenwitzer, V. Zamlynny, S. L. Horswell, J. D. Goddard and J. Lipkowski, Biophys. J. 85 (2003) 4055.

85 I. Zawisza, X. Bin and J. Lipkowski, Bioelectrochemistry 63 (2004) 137.

86 I. Zawisza, X. Cai, V. Zamlynny, I. Burgess, J. Majewski, G. Szymanski and J. Lipkowski, Pol. J. Chem. 78 (2004) 1165.

87 X. Bin, I. Zawisza, J. D. Goddard and J. Lipkowski, Langmuir 21 (2005) 330.

88 X. Bin, S. L. Horswell and J. Lipkowski, Biophys. J. 89 (2005) 592.

89 J. Umemura, T. Kamata, T. Kawai and T. Takenaka, J. Chem. Phys. 94 (1990) 62.

90 I. Burgess, G. Szymanski, M. Li, S. Horswell, J. Lipkowski, J. Majewski and S. Satija, Biophys. J. 86 (2004) 1763.

91 N. Li, "Ionic and molecular adsorption at the single crystal surfaces of platinum and gold". PhD thesis, University of Guelph, 2000.

92 *http://www.hindspem.com/LIT/PDF/ HNDPEM90.PDF.*

10

Tip-enhanced Raman Spectroscopy – Recent Developments and Future Prospects

Bruno Pettinger

10.1
General Introduction

The power of vibrational spectroscopies such as infrared spectroscopy, Raman spectroscopy, and sum frequency generation is based on the fact that when used as *in-situ* tools they can deliver very important information about the electrode/electrolyte interface, such as the identity of adsorbates and intermediates, the nature of chemisorbate-surface interaction and bonding, and, last but not least, the strength of intermolecular interactions. A variant of Raman spectroscopy, surface-enhanced Raman spectroscopy (SERS), now known about for nearly 40 years, is a particularly interesting technique in this respect, as the surface enhancement process as a whole operates mainly on the adsorbed species, thereby rendering "invisible" the adjacent electrolyte regions [1]. This unique property of SERS, its widely accessible frequency range, and the local nature of the surface-enhancement process make SERS not merely an ideal *in-situ* spectroscopy. Recent developments employing these unique properties indicate that SERS is developing into a widely applicable *in-situ* vibrational spectroscopy and, in some research areas, is becoming a spectroscopy of hitherto unprecedented sensitivity and spatial resolution [2–11]. A few key phrases, highlighting also the directions of these developments, illustrate this: "extension of SERS beyond the coinage metals" [11, 12], "single-molecule Raman spectroscopy" [13–15], "tip-enhanced Raman spectroscopy" [16–19], "sub-wavelength microscopy" [20]. It has taken nearly four decades of intense experimental and theoretical research to reach this level.

Using standard Raman spectroscopy for surface studies leads to rather tedious experiments because of its generally very low sensitivity. In contrast, SERS often provides high sensitivity due to a giant enhancement. The first paper on SERS published by the Van Duyne group reported a million-fold enhanced Raman intensity for pyridine molecules adsorbed on a silver electrode compared with the Raman intensity for unbound pyridine molecules [21]. Hence, SERS makes it possible to detect submonolayer quantities of adsorbates.

Advances in Electrochemical Science and Engineering Vol. 9.
Edited by Richard C. Alkire, Dieter M. Kolb, Jacek Lipkowski and Philip N. Ross
Copyright © 2006 WILEY-VCH Verlag GmbH & Co. KGaA, Weinheim
ISBN: 3-527-31317-6

Within a few years the main mechanisms of the surface enhancement process were revealed, answering two important questions: (a) why does SERS not work equally for all adsorbates? (b) why are intense SERS bands only observed for roughened gold, silver, and copper surfaces (or for colloids made of these metals)? Experimental evidence was accumulated for two cooperative enhancement mechanisms, denoted as "chemical" enhancement and "electromagnetic" enhancement [1]. Chemical enhancement is essentially a (pre-)resonance Raman effect arising from a (pre-)resonant excitation of electrons into orbitals newly formed between the substrate and the adsorbate or into (modified) orbitals of the adsorbate itself. Electromagnetic enhancement is based on the excitation of (localized) surface plasmons at the interface, which creates an enhanced electromagnetic field in the vicinity of the adsorbate. The chemical and the electromagnetic enhancements are the driving forces of SERS. The two enhancement processes imply rather distinct presuppositions regarding molecules and substrates: briefly, the chemical enhancement mechanism (CEM) requires a special chemical affinity between the adsorbate and substrate, while the electromagnetic enhancement mechanism (EEM) assumes free-electron metals and non-planar surface structures which facilitate the excitation of surface plasmons. For the latter enhancement mechanism, the coinage metals (copper, silver and gold) are the most suitable metals. The non-planar surface structures can be achieved, for example, by roughening the surface or using suspensions of suitable metal colloids.

Apart from the quickly reached general understanding, which is briefly sketched above, the evaluation of the details of the enhancement processes required a substantial amount of experimental and theoretical work. Even today a number of details are still unclear, in particular in the field of the chemical enhancement (which is more difficult to assess given the complexity of the interfacial region and interactions to be taken into account). A counterpart of such a complexity is also present in the area of the electromagnetic enhancement, as specific arrangements of non-planar structures are required (within a fractal surface structure, or within aggregated colloids, there must be specific configurations which are suitable to mediate the excitation of localized surface plasmons) [13, 22–24]. We shall meet some implications of this a number of times in this chapter.

Given the intricacy of enhancement processes and interfacial configurations, it is not a surprise that conflicting observations have been made, resulting in long-lasting disputes on enhancement processes and their individual contributions to SERS. Some of this is lingering still today. In fact, it is not of general importance whether the CEM or the EEM is more important; this is more a matter of the actual experimental arrangement and may vary from case to case. Important is the deepening of the understanding of the SERS process as a whole and its application to the wide area of interfacial science and technology.

The aim of this book chapter is to review developments leading to the new and very promising approach of tip-enhanced Raman spectroscopy, which have taken place in recent years. The chapter is organized as follows. After the general introduction (10.1), Section 10.2 sketches the attempts at Raman spectroscopy on single-crystalline surfaces and electrodes. Section 10.3 describes important steps in the direction of single-molecule SERS. Section 10.4 describes the state of the art in the area of tip-enhanced Raman spectroscopy, a variant of SERS that combines a scanning probe tip with Raman spectroscopy and Raman microscopy. It is promising for a number of applications, including electrochemical systems.

An Outlook is given in Section 10.5, sketching first of all the most recent TERS experiments and then possible developments in a number of directions.

10.2
SERS at Well-defined Surfaces

In general, SERS stems from non-planar surfaces. The most common are rough surfaces or aggregated colloids, which are, to some extent, both of fractal character and also provide rather exotic adsorption sites in addition to the "usual" sites. Among the many configurations in the roughness/aggregation space, only a few sites support intense SERS scattering, but these few are (often) superior to all other configurations. The specific arrangement of local nanostructures and the presence of more or fewer exotic adsorption sites are important, but neither is well characterized. In contrast, surface science, including modern electrochemistry, often requires the use of single crystals or well-defined sample surfaces. Therefore, from the very beginning onwards, attempts were made to apply SERS also to single-crystalline surfaces, but very few over the last 30 years have been published [25–31]. These reports become interesting again in the view of the new possibilities provided by TERS, which – in contrast to SERS – works also for smooth and even single-crystalline surfaces.

In most SERS studies, the samples were substantially roughened by activation procedures to make them "SERS active". In fact, Pettinger et al. noted a need to activate the electrode (to make it "SERS active"), otherwise no Raman signal could be observed. In order to obtain SERS for crystalline surfaces, the authors exposed the sample to carefully designed oxidation-reduction cycles to achieve some SERS activity, but avoiding roughness [32–34]. Interestingly enough, for electrodes prepared in this way, the SERS spectra for pyridine at silver electrodes depend on the crystallographic orientation of the substrate: the band frequencies, the half-widths, and the relative intensities vary between Ag(111), Ag(110), and Ag(100) electrodes [32–34]. Mullins and Campion reported on normal, i.e. "unenhanced" Raman scattering for pyridine chemisorbed at stepped single-crystal surfaces [35]. Irish and coworkers observed surface-enhanced Raman spectra for pyridine adsorbed at smooth Au(210) electrodes (see Figs. 10.1 and 10.2), which have a structure that is composed of a high density of regular steps. However, these authors could not observe Raman signals for pyridine on smooth single-crystal Au(111) electrodes [36, 37].

All these observations support the importance of steps, i.e. of roughness, for the enhancement. Recently, the Tian group published a series of papers describ-

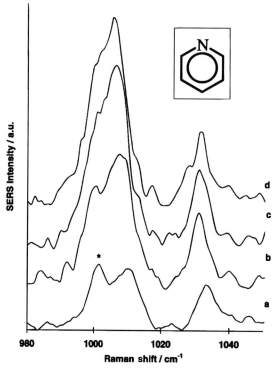

Fig. 10.1 SERS spectra of pyridine adsorbed on an Au(210) single crystal at different potentials: (a) + 200 mV; (b) –100 mV; (c) –400 mV; (d) –600 mV. (*) Band from aqueous pyridine. From a 0.1 M KClO₄/1 mM pyridine solution. (Reproduced with permission from Ref. [36].)

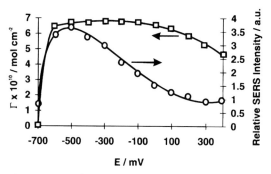

Fig. 10.2 Dependence of both the integrated SERS intensity and the surface coverage (from Ref. 18 in Ref. [36]) on the applied potential for pyridine adsorbed on an Au(210) single crystal. (Reproduced with permission from Ref. [36].)

ing the extension of SERS to transition metal surfaces [38–40]. Indeed, roughened transition metal electrodes also show some enhancement, probably a chemical enhancement of a few orders of magnitude. Again, for smooth transition metal surfaces, only extremely weak Raman scattering was found [41, 42]. Evidently, no Raman scattering is observed for adsorbates at smooth surfaces in the absence of a chemical or a resonance Raman enhancement.

Recently, Kambhampati et al. presented a study that combines SERS with electron energy loss spectroscopy on crystalline copper surfaces. The aim was to measure charge transfer excitations between an adsorbate and atomically smooth Cu(111) and Cu(100) surfaces [43]. For pyromellitic dianhydride (PMDA) as adsorbate, the authors noticed remarkable wavelength dependence of the Raman scattering: the excitation profile exhibits a sharp peak structure which matches the narrow electron energy loss peak observed at about 1.9 eV for the same system. Figure 10.3 presents Raman spectra for PMDA/Cu(111) and PMDA/Cu(100) for p- and s-polarized laser light. The comparison of the Raman spectra shows subtle differences in relative band intensities, indicating an excitation process that is sensitive to the crystal face and to the local electronic structure of the interfacial region [43].

In the field of catalysis, it is essential to achieve a better understanding of the physical and chemical properties of various well-defined crystalline orientations of catalyst particles over a wide range of temperature and pressure. Such insights permit us to interpret the catalytic behavior of small particles at the atomic and molecular level. Surprisingly, SERS is also a suitable *in-situ* spectroscopy for well-defined surfaces. This was recently demonstrated for oxygen on faceted Ag surfaces, which are not smooth but locally well ordered along the facets (over 10–1000 nm) [44].

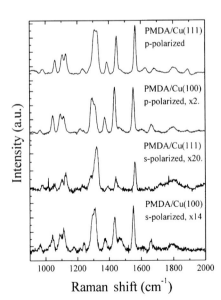

Fig. 10.3 Raman spectra of a monolayer PMDA adsorbed on Cu(111) and Cu(100) at 647 nm using both p- and s-polarized excitation. (From Ref. [43], Fig. 5; courtesy of Surface Science, Elsevier.)

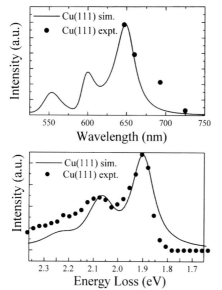

Fig. 10.4 Experimental and simulated Raman excitation profile (top) and electron energy loss spectra (bottom) of a monolayer of PMDA adsorbed on Cu(111). The Raman enhancement tracks the interfacial charge-transfer excitation for the symmetric surface carboxylate mode. (From Ref. [43], Fig. 5; courtesy of Surface Science, Elsevier.)

Silver catalysts have been used for the partial oxidation of methanol to formaldehyde; this is a very important process in the chemical industry. The role of the silver catalyst and, in particular, the influence of its atomic structure on the catalytic process have been extensively studied with various surface science tools [44–50]. In these investigations, Raman spectroscopy was employed to identify and confirm the role of the oxygen species for the catalytic process. These studies were performed under reaction conditions close to those in industrial processes using Ag(111) and Ag(110) samples. Upon extended exposure to oxygen at high temperatures, both samples restructure to (111) planes with a well-defined microstructure and with mesoscopic roughness (on a scale of 1 μm). Therefore, in the course of the oxygen pretreatment, the local nature of the surface of the two samples becomes nearly identical and, hence, their Raman spectra are quite similar [44].

Figure 10.5 presents four Raman spectra successively recorded for an Ag(111) sample in the course of an extended oxygen treatment. Exposure to 0.2 bar O_2 at 620 K had no effect; only a nearly featureless background spectrum was found. Further exposure to 0.2 bar O_2 at 780 K caused a strong band at 803 cm^{-1} to appear. This band is attributed to the O_γ species, which is strongly chemisorbed at the surface. Increasing the temperature to 830 K caused a decrease of this band and the emergence and rise of a band at around 628 cm^{-1}, denoted as O_β and assigned to subsurface oxygen [44]. At 920 K, the 803 cm^{-1} band nearly vanished and the 628 cm^{-1} band lost some intensity. These observations point to a desorption of the O_γ species from the surface and to some depletion of O_β in the subsurface region by diffusion to and desorption from the surface.

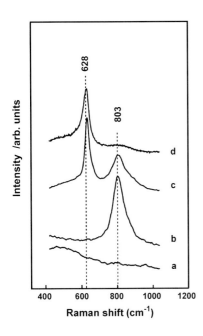

Fig. 10.5 Raman spectra at the Ag(111) surface exposed to 0.8 bar N_2 + 0.2 bar O_2 at various temperatures: (a) 620 K, (b) 780 K, (c) 830 K, and (d) 920 K. After [44].

Analogous observations were found for the Ag(110) surface exposed to a similar oxygen treatment. A series of SER spectra recorded for the Ag(110) surface is shown in Fig. 10.6. After preparation as in Fig. 10.5, the sample was kept at a constant temperature (780 K), but exposed to varying gas atmospheres. Upon prolonged exposures to O_2, the O_γ adlayer is completed and a maximum intensity of the 803 cm^{-1} peak (Fig. 10.6a) at the same frequency as for the Ag(111) sample is found. Evidently, the local adsorption sites for O_γ on both surfaces are identical. At this stage, the O_β species were present in subsurface position in addition to the O_γ species, as confirmed by X-ray photoelectron spectroscopy (XPS) [46]. Surprisingly, however, no band around 630 cm^{-1} was discernible. After exposing the Ag(110) sample to pure O_β for some time, the 803 cm^{-1} band decreased and, simultaneously, a band at around 638 cm^{-1} grew in intensity (Fig. 10.6b–h). Apparently, the desorption of O_γ caused an increase in the Raman intensity of the O_β species while the concentration of subsurface O_β species remained essentially unchanged (as verified by separate XPS measurements using analogous pretreatments and temperature cycles) [44].

After switching back to the initial atmosphere, 0.8 bar N_2 + 0.2 bar O_2, the band at 803 cm^{-1} increased, and simultaneously the 638 cm^{-1} band decreased substantially (see Fig. 10.6i–l). Obviously, the presence of the O_γ in close proximity to the O_β species changes the electronic configuration of the latter. In other words, the chemical part of SERS for the Ag-O_β species is tuned into or out of resonance depending on the O_γ coverage.

Please note that the adsorption sites for the oxygen species are exceptionally stable at these faceted Ag surfaces, which is in marked contrast to usual SERS

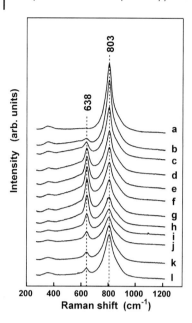

Fig. 10.6 Raman spectra recorded at an Ag(110) surface prepared as in Fig. 10.5 at 780 K, (a) in 0.8 bar N_2+0.2 bar O_2 after 30 min; (b)–(h) in 1 bar N_2 after 2, 5, 8, 12, 15, 20 and 30 min; (i)–(l) again in N_2+0.2 bar O_2 after 1, 2, 5 and 10 min. After [44].

sites. The surface is locally well defined, providing discrete sites for chemisorption and oxygen incorporation into the silver lattice. Yet, we find the usual scenario for SERS: the more or less extended faceted surfaces provide the electromagnetic enhancement, while the Ag-O_β and Ag-O_γ structures constitute the electronic configurations for the chemical enhancement.

The above results show that SERS can be found for adsorbates at crystalline surfaces having well-defined adsorption sites. Moreover, SERS can occur even at elevated temperatures up to 900 K [44].

10.3
Single-molecule Raman Spectroscopy

The investigation of single molecules in condensed phases began about 30 years ago with the pioneering work of Hirschfeld [51, 52], followed by papers by Dovichi [53], Nguyen [54], Soper [55], and Ambrose [56, 57]. Single-molecule detection in condensed phases became a rapidly growing topic with numerous applications in biology, chemistry, and physics.

Single-molecule detection (SMD) by optical means is only possible if the signal can be filtered from a large background signal. Techniques based on fluorescence are favorable because fluorescence cross sections reach levels of $\sigma \approx 10^{-16}$ cm^2 and higher and because the fluorescent species deliver photons which originate only from the fluorophore and are shifted in frequency relative to the excitation laser line. Therefore, the single fluorophore signal can be separated from other optical background signals by spatial, temporal, and spectral discrimination techniques.

Raman scattering features most of the above-sketched properties – but not the high cross section. Quite on the contrary, Raman cross sections are about 12 orders of magnitude lower than cross sections sufficient for SMD. Hence, normal Raman spectroscopy is inadequate for SMD.

Surprisingly, in the last 10 years, reports appeared on single-molecule detection by surface-enhanced Raman spectroscopy [6–8, 10, 13–15, 58–71]. Seemingly, the SERS process(es) can bridge the above-mentioned gap of about 12 orders of magnitude between the (normal) Raman and the fluorescence process. This is an astonishing result, given the usual surface enhancements of two to seven orders of magnitude. In the following, I would like to present a few examples of this rapidly growing field in more detail.

The first approach toward SMD using surface-enhanced resonance Raman scattering was published by Kneipp et al. in 1995 [58]. It involves the use of extremely low dye concentrations (down to 10^{-16} M of rhodamine 6 g) adsorbed at silver colloids which were activated by NaCl. The authors estimated that less than 100 dye molecules were enough for recording an SERS spectrum with sufficient signal-to-noise ratio.

In February 1997, Nie and Emory were the first to report on Raman spectroscopy of single molecules adsorbed at optically selected nanoparticles [15]. For this system, giant enhancement factors of 10^{14} to 10^{15} had been estimated, which were substantially larger than the ensemble averaged enhancement factors. It was also important that the single-molecule SERS spectra were both more intense and more stable than single-molecule fluorescence spectra [15].

In the same year, Kneipp et al. published single-molecule Raman spectra for crystal violet (CV) adsorbed at Ag colloids dispersed in aqueous solution. The authors used near-infrared excitation at 830 nm, which is off-resonant to the electronic excitations of both the crystal violet and the isolated silver particles.

Figure 10.7 shows a time series of SER spectra of CV adsorbed at the silver colloid. Because of the low dye concentration, the monitored scattering volume contains, on average, 0.6 dye molecule and about 100 Ag clusters. Therefore, it is unlikely that more than one dye molecule is present on a single particle. The spectra show typical Raman bands of crystal violet, for example at 915, 1174,

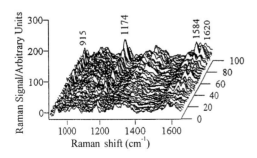

Fig. 10.7 100 SERS spectra of CV displayed in the time sequence of measurement. Scattering volume: 30 pL that contains an average of 0.6 crystal violet molecule. Each spectrum is acquired in 1 s. (Reprinted with permission from Ref. [13].)

1584 and 1620 cm^{-1}, with intensities that vary with time. This is due to the Brownian motion within and through the small scattering volume. Despite the small size of the sample (30 pL $\approx$ 12 μm diameter $\times$250 μm height), the residence time of these molecules in the probed volume has been estimated to be 10–20 s.

A statistical analysis of the Raman intensities measured in the time series is shown in Fig. 10.8. For convenient comparison, the data are normalized to the maximum intensity. The region from 0 to the maximum intensity is divided into 20 sections and the number of events found in each section was counted and presented as a frequency. Kneipp et al. compared three different experiments: (a) The analysis of the Raman data of 10^{14} methanol molecules with approximately the same (but unenhanced) intensities as the dye. The analysis yielded a Gaussian profile. (b) The analysis of 100 SERS measurements for, in

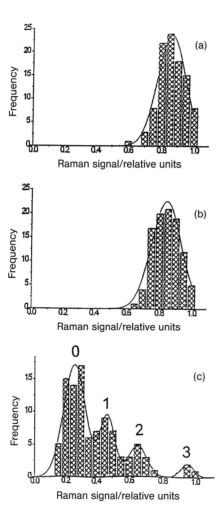

Fig. 10.8 (a) Statistical analysis of 100 "normal" Raman measurements at 1030 cm^{-1} of 10^{14} methanol molecules. (b) Statistical analysis of 100 SERS measurements (1174 cm^{-1} Raman line) of six crystal violet molecules in the probed volume. The solid lines are Gaussian fits to the data. (c) Statistical analysis of 100 SERS measurements (1174 cm^{-1} Raman line) for an average of 0.6 crystal violet molecule in the probed volume. The peaks reflect the probability of finding 0, 1, 2, or 3 molecules in the scattering volume. (Reprinted with permission from Ref. [13].)

average, 6 crystal violet molecules present in the scattering volume. Again, a Gaussian profile was found. (c) The analysis of 100 SERS measurements of an average of 0.6 crystal violet molecule present in the scattering volume. Figure 10.8c shows four maxima, and the authors associate this pattern with the presence of 0, 1, 2, or 3 dye molecules in the scattering volume and conclude that the huge enhancement varies little with the different sizes and shapes of the silver clusters [13].

Nie and Emory were also among the first to report on single-molecule surface-enhanced Raman spectroscopy using rhodamine 6g (rh6g) adsorbed at particles deposited and immobilized[1)] on a glass slide. The colloidal solution is composed of a mixture of heterogeneous particles with an average particle size of about 35 nm diameter and a typical concentration of 10^{11} particles mL^{-1}. Using wide field laser illumination, about 200 to 1000 immobilized particles were examined in a region of 100×100 µm sampling area. Among this larger ensemble, only a small fraction of immobilized nanoparticles exhibited unusually efficient enhancements. They were denoted as "hot particles", which emit bright, Stokes-shifted (toward lower energies) photons. The Raman spectra from individual particles provide evidence that the observed signals arose from adsorbed rh6g while simultaneously the fluorescence emission of rh6g was quenched by a rapid energy transfer to the metal surface. On the basis of the measured adsorption isotherms it was estimated that each particle carried an average of one dye molecule at a rh6g concentration of 2×10^{-10} M and an average of 0.1 molecule at a concentration of 2×10^{-11} M. The number of observed hot particles increased with the rh6g concentration, but in a nonlinear fashion.

To determine whether the optically hot particles were single particles or aggregates, the authors recorded high-resolution AFM images of selected hot Ag nanoparticles. Figure 10.9 indicates that the majority of them were well-separated, single particles with a narrow size range of 110 to 120 nm diameter, which points to a strong correlation between particle size and enhancement efficiency. A minor fraction consists of aggregates, each containing two to six tightly packed particles.

Nie and Emory provided evidence for the observed SER signals arising from single adsorbed molecules (or conjugated molecular aggregates that behave as single molecules) [15]. At analyte concentrations below 10^{-10} M, single colloidal particles are expected to contain mostly zero or one analyte molecule according to the Poisson distribution. Uncertainties arise, however, because of the heterogeneity of the colloidal system and the fact that most particles did not exhibit any efficient Raman enhancement. Moreover, *in-situ* AFM showed that the optically hot particles were about three times the average particle size and thus were likely to have nine times the average surface area. In addition, it is not known whether the hot particles contain adsorption sites or facets of unusually

---

1) The colloidal particles were immobilized on polylysine coated glass surfaces by the electro- static interactions between the negative charges on the particles and the positive charges on the surface.

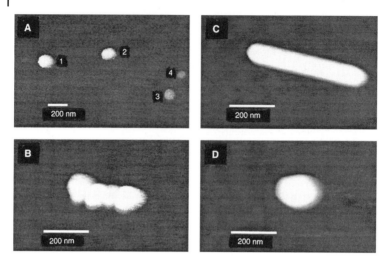

Fig. 10.9 Tapping-mode AFM images of screened Ag nanoparticles. (A) Large area survey image showing four single nanoparticles. Particles 1 and 2 were highly efficient for Raman enhancement, but particles 3 and 4 (smaller in size) were not. (B) Close-up image of a hot aggregate containing four linearly arranged particles. (C) Close-up image of a rod-shaped hot particle. (D) Close-up image of a faceted hot particle. (Reprinted with permission from Ref. [15].)

high affinities, which could preferentially accumulate rh6g molecules onto these particles [15].

A particularly interesting piece of evidence for single-molecule behavior is the report of sudden spectral changes observed in sequentially recorded spectra in the same paper [15].

Photobleaching or photochemical decomposition of the dye is significantly diminished for molecules directly adsorbed on metal nanoparticles because of the high energy transfer rate between the excited molecule and the metal. This rapidly quenches the excited electronic state and thus suppresses excited state reactions. However, after some time of continuous illumination, the enhanced Raman signals suddenly disappear or change in intensity and/or frequency. In a series of Raman spectra recorded sequentially from a single particle (Fig. 10.10), the observed changes in Raman signal frequencies amount to 10 cm^{-1}. Even when sudden spectral changes are not discernible, the Raman spectra obtained from different particles exhibited slightly different vibrational frequencies, pointing to different adsorption sites for each molecule. These single-molecule, single-particle results indicate that the total enhancement factor in SERS can reach levels of 10^{14} to 10^{15}, yielding Raman scattering cross sections of the order of 10^{-15} cm^2 per molecule. The authors conclude that these cross sections are comparable with or higher than the optical cross sections of single-chromophore fluorescent dyes. They estimate that the Raman signals are approximately four to five times higher than the integrated fluorescence of single rh6g molecules. This finding is in line with the

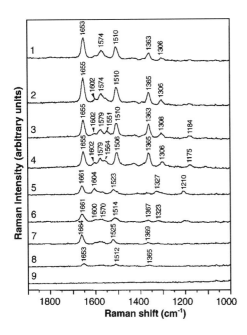

Fig. 10.10 Time-resolved surface-enhanced Raman spectra of a single R6G molecule recorded at 1-s intervals. Over 300 spectra were recorded from this particular particle before the signals disappeared. Nine spectra were selected to highlight sudden spectral changes. The Raman signals abruptly changed both in frequency and intensity three times, as shown in spectra 2, 5 and 8. The laser excitation power was approximately 10 mW. (Reprinted with permission from Ref. [15].)

observations of other groups. The authors attribute the extremely large enhancement factors to the absence of two population-averaging effects, because they deal with single molecules and single particles [15].

Sudden changes of Raman spectra have also been denoted as blinking. Blinking behavior in SERS was soon understood as a characteristic signature for single-molecule SERS spectroscopy [65, 66, 72–76] in analogy to the blinking of single fluorescent dye molecules [77–80]. The blinking of SERS spectra has been attributed to thermally activated diffusion of individual molecules into or out of the SERS active sites, to structural relaxation of these sites, and to photo-induced electron transfer processes [15].

The observation of large fluctuations of SERS due to the presence of carbonaceous species (impurities) at the interface (often in addition to the adsorbates under investigation) is related to this blinking phenomenon. Usually, SERS spectra represent features averaged over a large molecular ensemble. In recent years, evidence has accumulated that electromagnetic enhancement is not constant along the surface. On the contrary, large enhancement variations were found, which has led to the concept of hot spots [81].

For fractal surfaces, surface intensity variations by a factor of 2 to 7 have been observed [81, 83, 84]. Theoretical studies have shown variations of up to 10^4 in the surface intensity enhancement [85]. Gadenne et al. found that the intensities and spatial distribution of the hot spots were in reasonable agreement with these theoretical predictions [86].

Recently, large fluctuations of SERS signals have been reported for carbon at rough gold and silver surfaces [82, 88] and for carbon domains on Ag nanoparti-

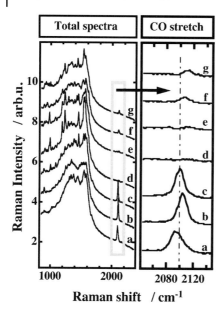

Fig. 10.11 SER spectra of carbon clusters deposited on a rough gold surface measured in air at room temperature. Each spectrum was accumulated for 1 s. The waiting time between recording consecutive spectra was 2 s. (a) Denotes the first spectrum, (g) the last. All spectra have the same intensity scaling, but are spaced upwards in the figure to enhance the clarity of the presentation. (Reproduced with permission from Ref. [82].)

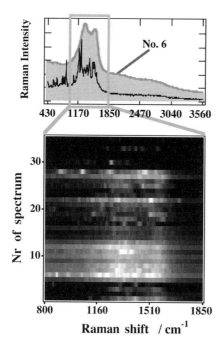

Fig. 10.12 A sequence of 36 SER spectra for the same carbon layer on rough silver in a gray color mode. Experimental conditions were the same as for Fig. 10.4a. Top panel: spectrum No. 6 (black curve) plotted together with an average of over 200 spectra showing fluctuations. (Reproduced with permission from Ref. [87].)

cles [89]. Figure 10.11 presents a particularly striking sequence of spectra that shows fluctuations not only for carbon bands but also for CO stretch vibrations. The narrow Raman lines around 1340 cm^{-1} and 1580 cm^{-1} on top of the broad background were assigned to vibrations of the carbon network. Their number, intensity, and frequency vary extensively with time. For the CO vibrations at the Au surface, a single band at around 2095 cm^{-1} with a line width of about 20 cm^{-1} (FWHM) was found. It varied substantially in intensity and, to a minor extent, in frequency, which can be clearly seen in the frequency-expanded section of Fig. 10.11. Similar observations were obtained for carbon on silver surfaces, presented in Fig. 10.12. 36 spectra are depicted in a gray color mode. Spectrum no. 6 is displayed in the upper panel together with an average spectrum taken over 200 Raman spectra showing fluctuations, to illustrate the enormous fluctuations in each spectrum. Nevertheless, the broad average spectrum is typical for amorphous carbon, displaying the D (disordered) and G (graphitic) peaks at 1340 cm^{-1} and 1580 cm^{-1}, respectively. The ratio of the integral intensities of both peaks is rather large, indicating a substantial disorder in the carbon network [87, 88]. Contrarily, the appearance of a number of narrow Raman lines in the individual spectra points to a low inhomogeneous broadening, which can only occur for small segments of the carbon network at the interface experiencing strong enhancements at a time [87, 88].

These observations were also interpreted as evidence for hot spots, proposing a generalized concept: hot spots represent those small surface locations where both the electromagnetic and the resonance Raman part of surface enhancement are large.

10.4
Tip-enhanced Raman Spectroscopy (TERS)

Tip-enhanced Raman spectroscopy (TERS) has been developed from the field of scanning near-field microscopy (SNOM), a field that is not directly related to SERS. However, one particular important concept of SNOM is also essential for the SERS process: the near-field effect. It is common to both SNOM and the electromagnetic enhancement process of SERS and it is the heart of TERS.

10.4.1
Near-field Raman Spectroscopy with or without Apertures

The essential concept of scanning near-field microscopy, abbreviated as SNOM or NSOM, emerged in the mid 1980s [90–98]. It is based on the use of a kind of waveguide, often a fiber, which has a subwavelength aperture at its end. This aperture is scanned over the sample at a distance to the object which is much smaller than the wavelength of the light. For these geometrical configurations, the interaction between the aperture (which operates as the light source and/or recorder) and the sample is a near-field interaction. Such a device has a superior spatial resolution ($\ll \lambda/2$) for image recording and local spectroscopy to that of the resolution of conventional optics, which is restricted to the diffraction limit or Abbé's limit of $\sim \lambda/2$ [99]. A comprehensive review of near-field optics is outside the scope of this chapter; the interested reader is referred to review articles in this specific field [100–103]. However, it is relevant to describe the emergence of TERS out of other near-field activities along an historic path: near-field microscopy, near-field fluorescence, near-field Raman spectroscopy, apertureless near-field Raman spectroscopy, and tip-enhanced Raman spectroscopy.

If one includes the employment of evanescent light waves in spectroscopic studies in the near-field area, Knoll and coworkers are among the first who reported on "near-field Raman spectroscopy", though this approach was termed at that time (the early 1990s) as "surface plasmon field-enhanced Raman spectroscopy and microscopy" [104].

In 1994, Tsai and coworkers presented a pioneering paper on "Raman spectroscopy using a fiber optic probe with subwavelength aperture". A tapered fiber optical probe was used in two ways simultaneously: to deliver the pump radiation to the sample and to collect the Raman scattering. The close approach of the aperture to the sample made this approach comparatively efficient [105]. One year later, Jahnke, Paesler, and Hallen presented a Raman near-field scanning microscope approach to image Rb-doped $KTiOPO_4$ crystals. Both Raman spectra and Raman images were recorded using the near-field approach [106].

Near-field Raman spectroscopy was expected to become a very attractive approach because it offers chemical information from a subwavelength region and microscopic images of the distribution of species over a sample if it is viewed by its characteristic Raman line. Most intriguing is that the spatial resolution is only limited by the size of the aperture.

The next step was to improve the stability of the system and – even more importantly – to improve the transmission of the fiber tips. Metalization of the tip is the key and produces superior properties, such as a near-field spot of less than 20 nm [107–110]. Zeisel et al. reported that metalization led to optical transmission of a fiber higher than that of conventional fibers by several orders of magnitude [111]. The authors showed images based on the fluorescence of dye-labeled polystyrene spheres and noted that fast and irreversible photobleaching takes place in the near field with enhanced intensity. In addition, they reported on surface-enhanced near-field Raman spectra of cresol fast violet and p-aminobenzoic acid adsorbed on a (rough) silver substrate, exhibiting a signal-to-noise ratio of >10 [111].

A slightly different approach was followed by Grausem and coworkers, who employed near-field Raman spectroscopy to study liquid CCl_4. The near-field probe had an aperture of <50 nm. Because of the rapid decrease of the near field with distance from the aperture, only a small volume of liquid, containing only 0.4 attomol of CCl_4, contributes to the Raman signal [112].

Emory and Nie described a particularly interesting approach: the combination of near-field effects from tapered single-mode fiber probes with surface enhanced Raman scattering from single Ag colloidal nanoparticles [113]. By using the combined enhancement effects of molecular (rhodamine 6g) and particle resonances (resulting in surface plasmon excitations of the silver particles), the authors could operate with excitation intensities as low as 10 nW, which is about the available throughput of their fiber probes. According to the authors, the comparison of near-field and far-field Raman spectra indicated similar selection rules in each case [113]. This is not surprising, as SERRS and near-field (excited) Raman spectroscopy have two important processes in common: the resonance Raman and the surface enhancement processes monitored by standard far-field optics. A very interesting additional approach, earlier suggested by Wessel [114] and realized by the authors [113], is to attach a silver particle to the fiber end and to use the near field of the excited silver particle as the probe field. This approach was soon denoted as apertureless Raman spectroscopy and paved the way for tip-enhanced Raman spectroscopy.

Webster, Batchelder, and Smith developed a near-field optical microscope that could record Raman spectra with a resolution of approximately 150 nm. They used this technique to obtain both topological and Raman images from a region of silicon wafer which had been plastically deformed. The induced stresses were monitored by the corresponding frequency changes of the 520-cm^{-1} Raman band of silicon during the scan over this region. These results indicate that near-field Raman spectroscopy can be a promising technique for optical characterization of a semiconductor surface with high spatial resolution [115, 116].

Deckert and coworkers reported near-field surface-enhanced Raman imaging of dye-labeled DNA [117]. Similarly to the work of Emory and Nie [113], the authors resorted to the concept of increasing the Raman cross section by using the resonance Raman and the surface enhancement processes provided by the adsorbate and the substrate. These processes boost the near-field Raman processes up to a detectable level solely because they are cooperative. The high spatial resolution of some surface sites led the authors to identify these as special surface sites with particularly high SERS enhancement resembling the so-called "hot spots" often encountered in SERS.

Hamann et al. were among the first to use the apertureless near-field approach to greatly enhance optical processes in the near-field region. One important result of this investigation is that the spatial resolution of this approach is correlated with the radius of the tip apex, while the high cross section arises from an antenna enhancement provided by the tip volume [118]. This illustrates the high potential of this approach for local spectroscopy and optical imaging on the nanometer scale.

S. Takahashi, M. Futamata, and I. Kojima followed a slightly different route using a photon tunneling mode and a fiber tip. This means that the sample is illuminated employing the attenuated total reflection (ATR) configuration. The evanescent wave creates radiation by the sample, which is monitored locally by a sharpened optical fiber tip. This experimental arrangement minimizes Raman scattering from the optical fiber, while the ATR configuration permits the use of a comparatively large laser intensity at the sample location. The authors observed Raman spectra of copper phthalocyanine (CuPc) in the off-resonance condition and without contributions due to surface-enhanced Raman processes [119].

A number of groups have made substantial efforts to prepare samples which can be used as sensors for an easy application of SERS to monitor unknown substances. Among the many, a particularly noteworthy approach is that of Viets and Hill, who developed SERS sensors based on optical fibers [120, 121]. The sensor ends were covered with roughened metal films. The authors compared various metal deposition techniques with respect to surface enhancement, durability, and ease of preparation and regeneration. Another important parameter was the angle under which the tip end was polished (before metal deposition). The optimal angle was 40°, presumably because at this angle the large vertical component of the EM field at the metal/sample interface is more suitable for exciting SERS [121]. The fiber has here two functions: (a) to deliver the excitation laser light to the SERS active fiber end and (b) to return a substantial part of SERS to the spectrograph. This technique permits remote Raman measurements over comparatively large distances (ca. 20–90 m) [120, 121].

Recent theoretical studies on SERS yielded a surprising and intriguing result, which is important with respect to further development and application of SERS. For a long time it was known that intense SERS requires specific arrangements of non-planar surfaces, such as aggregated Ag or Au colloids or "just" roughened metal or electrode surfaces. The explanation for this is that the EM part of the total enhancement of SERS is a near-field Raman process, leading not only to an enhanced incident electromagnetic field but also to a nearly equally enhanced emission rate. Hence, the incident field is enhanced by a factor g_i and the emission of the scattered light is also enhanced by a factor g_{sc}. The EM enhancement factor is then $F_{EM} = g_i^2 g_{sc}^2 \approx g^4$; the last part of this equation, the 4th power law, holds if and only if $g_i \cong g_{sc}$, which is likely for small Raman shifts. Roughened surfaces and aggregated colloids can often be described as fractal systems [122]. Among the many configurations of such fractal structures, there are some local configurations, which are particularly suitable for exciting SERS. Consequently, because of the 4th power law for the EM enhancement, a few very SERS-active locations can provide most of the SERS signal, while the rest of the sample is relatively inactive. This effect has been proposed by the Moskovits group as the concept of "hot spots" and was substantiated by the same group in later theoretical and experimental work [86, 123–125]. The same idea of hot spots can also be applied to surface structures that have no fractal character but some variability in the local configurations. Consider, for example, Ag clusters of the same size arranged as an ensemble of well-separated dimers with internal distances that are small but somewhat varying. Then, depending on the internal distances, the dimers will show different resonance frequencies. Among the whole ensemble of dimers, only a few can be in close resonance with an exciting frequency, and only this fraction can support SERS, but these will produce superior intensities. A number of groups have pursued related ideas such as the deliberate tuning of the optical resonance of regular particle arrays by varying their lattice constant in the *x*- and *y*-directions (see, for example, Van Duyne [126–129] and Aussenegg [130–133]).

Other theoretical studies focus on the evaluation of selection rules for near-field Raman spectroscopy, which may be different from those of the (usually employed) far-field Raman spectroscopy. A possible cause of this may be the presence of strong field gradients, an inevitable property of near fields. This means that the usual assumption of optical spectroscopy – that the spatial field variations are small over the dimension of the probe – breaks down: in fact, the near field provided by nanometer-sized structures is not constant over the dimensions of the probed molecules [134, 135]. In addition, the authors noticed that strong field gradients can produce Raman-like lines and named this effect the "gradient-field Raman" process. Contrary to the above, Emory and Nie observed no significantly different selection rules for near-field and far-field excitations [113].

10.4.2
First TERS Experiments

In recent years, a new approach, named tip-enhanced Raman spectroscopy (TERS) has evolved, with very promising advantages with respect to sensitivity, spatial resolution, and general applicability [16–19, 87, 136–146]. The first instrument for TERS, sometimes also referred to as apertureless near-field optical spectroscopy (a-SNOM), was designed by the Zenobi group. Stöckle et al. reported a significant enhancement of the Raman signal if a metalized AFM tip was brought within a few nanometers of a dye film deposited on a glass substrate [16]. Figure 10.13 reproduces these spectroscopic results (upper part of Fig. 1 in Ref. [16]). The two spectra were recorded with (a) the tip retracted from the sample and (b) the tip in contact mode. The authors report a more than 30-

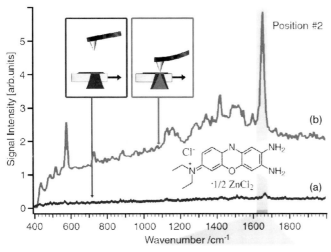

Fig. 10.13 Tip-enhanced Raman spectra of brilliant cresyl blue (BCB) dispersed on a glass support and measured with a silver-coated AFM probe. The two Raman spectra were measured with the tip retracted from the sample (a) and with the tip in contact with the sample (b). (Reprinted with permission from Ref. [16].)

fold net increase [2] of the Raman signal by the tip. Taking into account the small area of the enhanced field underneath the tip of less than 50 nm in diameter, a 2000-fold enhancement of Raman scattering in the vicinity of the tip was deduced. Similar effects were reported for C_{60} molecules and an electrochemically etched Au tip with an apex 20 nm in diameter (this tip was mounted in a shear-force set-up) [16]. In addition, the authors also illustrated that the spatial resolution of this approach is about 55 nm.

Only a few months later in the same year, Anderson described a very similar experiment. An AFM tip covered with a layer of gold grains of mean size 45 nm was employed to enhance the Raman scattering from a sulfur layer on a quartz substrate (Fig. 10.14) [136]. If the tip was moved 15 µm away from the substrate, no Raman signal of sulfur was detectable, but with the tip in contact with the layer, a spectrum exhibiting high signal-to-noise ratio was observed. An enhancement factor > 10 000 was estimated [136]. In another experiment, the AFM device was used to remove part of the C_{60} layer from the surface by scanning the AFM tip at high tip force in contact mode. In this way, C_{60} molecules were accumulated on the gold-coated cantilever tip. This tip was used as a SERS substrate. Its spectrum is shown in Fig. 10.15. The author estimates an amount of 10 fg of C_{60} collected on the tip, highlighting the enormous sensitivity of this approach [136].

In September 2000, Hayazawa and coworkers reported the next study on apertureless SNOM using a silver-coated cantilever and a dye-coated silver film on a glass slide [137]. The dye was rhodamine 6g (Rh-6g). A 40-fold enhancement of Raman scattering was observed with a 488-nm laser excitation; an enhancement of fluorescence was also noticed (see Fig. 10.16) [137]. The authors observed bleaching behavior for rhodamine 6g, but did not mention whether or not the bleaching rate was tip-enhanced. Instead, they pretreated this system with 20-min illumination until a stationary state was reached.

In a further experiment, the authors scanned the metalized cantilever within the laser focus, employing AFM contact mode. The recorded spectra are shown in Fig. 10.18 as a line image. This image was obtained with 0.5-mW laser illumination for 5 s per line scan. A lateral resolution of about 50 nm was reached in this experiment.

At the end of the same year, Pettinger et al. presented a TERS study using for the first time an STM device. The system investigated was BCB adsorbed at a smooth thin gold film (12 nm thick) evaporated on a glass slide. The STM tip was made from an electrochemically etched silver wire [87]. The set-up was otherwise similar to that of Stöckle et al. [16]. The adsorption procedure resulted in a (sub)monolayer coverage of the dye. Under these conditions, the fluorescence of the dye was effectively quenched by the metal film, and, as shown in Fig. 10.19, a weak resonance Raman signal was detectable in the absence of the STM tip. When the tip was brought into the tunneling position (approx. 1 nm

---

2) By the "net increase" we mean the n-fold increase of the overall Raman signal in the presence of the tip compared with the Raman intensity in the absence of the tip.

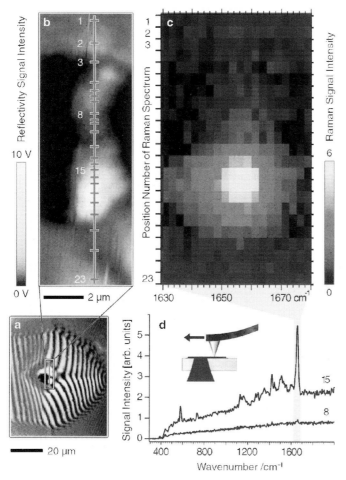

Fig. 10.14 Ultraflat, homogeneous BCB sample measured in tip-scan mode: the same sample spot remains in the laser focus while the tip is scanned [see schematic in inset of (d)]. (a) Reflectivity image. The fringes are due to an optical interference between cantilever and sample surface. The cantilever is not parallel to the sample surface. (b) Zoomed reflectivity image of tip region. Raman spectra were measured along the line in this figure at the marked positions; the intensity of the main Raman band between 1630 and 1680 cm^{-1} is displayed color-coded in (c). Note that the tip locations in (b) are not equidistant. (d) Two selected overview Raman spectra measured with tip inside position #15 and outside position #8 of the laser focus. Acquisition time: 60 s per spectrum. (Reprinted with permission from Ref. [16].)

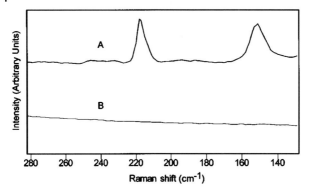

Fig. 10.15 Raman spectrum demonstrating gold-coated AFM tip that causes a local surface-enhanced Raman effect on a sulfur film (A). When the beam is focused away from the tip on the film, the Raman signal is undetectable with the same microprobe parameters (B). (Reprinted with permission from Ref. [136].)

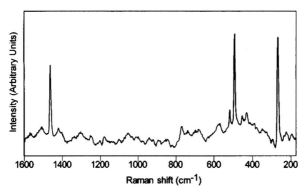

Fig. 10.16 An Au-coated AFM tip was used to remove C_{60} layers from the film surface. The spectrum results from approximately 10 fg of C_{60} on the AFM tip. The tip functions as a SERS substrate for ultratrace detection of the removed material. (Reprinted with permission from Ref. [136].)

above the surface), a 16-fold net increase could be observed [141]. Since in these experiments the tip radius was large (100–500 nm), only an average enhancement of 4–100 could be estimated, which is weak in comparison with the enhancement factors reported by the authors mentioned above. Pettinger et al. addressed the point that the enhancement must have a sharp radial profile. In the highest field zone at the center of the tip apex, it may reach values between 10^3 to 10^4 [87]. In addition, the TERS spectra shown in Fig. 10.12 and Fig. 10.19 are quite different in their relative intensities. Most likely, this has to be ascribed to the different excitation conditions used in the experiments: the two excitation lines at 488 nm (Fig. 10.12) and 633 nm (Fig. 10.19) are located on the left and

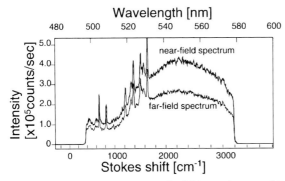

Fig. 10.17 Near-field black line and far-field gray line Raman spectra of Rh-6G with a silver-coated AFM cantilever. Near-field and far-field spectra correspond to the spectra with and without a cantilever at a tip–sample distance of 0 nm. The laser power at the sample is 0.5 mW, and the exposure time is 5 s. (Reprinted with permission from Ref. [137].)

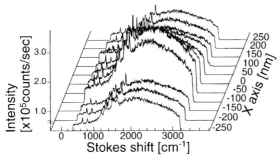

Fig. 10.18 Spectral mapping of the near-field Raman spectra. The laser power is 0.5 mW at the sample, and the exposure time is 5 s. (Reprinted with permission from Ref. [137].)

right side, respectively, of the major absorption band of this dye. Hence, for the experiment shown in Fig. 10.19, the Raman lines around 500 cm^{-1} are located closer to the absorption center and are more resonantly enhanced; an analogous argument holds for the case of Fig. 10.12, where the lines near 1640 cm^{-1} are more in resonance.

Characteristic of many of the experiments described above and other TERS studies reported in the literature are (a) the use of an inverted microscope or illumination that is closely related to such a configuration, (b) the use of thick probe films, and (c) tips covered with metal grains. Each of these points indicates a non-optimal experimental condition for TERS. In fact, a configuration optimal for TERS requires (a) illumination with a strong polarization component parallel to the tip axis, (b) the possibility to use opaque, massive substrates, (c) an easy control of the adsorbate coverage, and (d) sharp, smooth tips with a

Experimental TERS setup

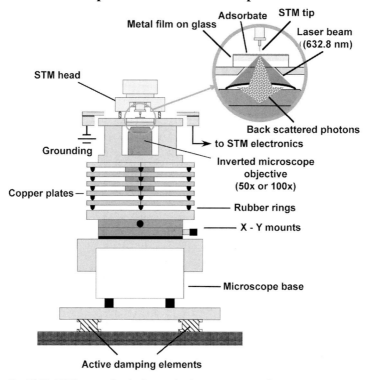

Fig. 10.19 TERS set-up for the inverted microscope approach.

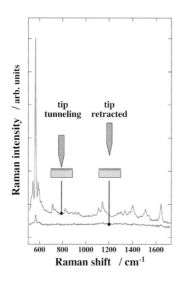

Fig. 10.20 Comparison of an RRS spectrum with the tip-enhanced Raman spectrum for brilliant cresyl blue on a smooth gold film. Coverage: 0.5 M. Tunneling distance 1 nm, retracted distance 1000 nm. Laser power at the sample ca. 5 mW at $\lambda = 633$ nm. (Reprinted with permission from Ref. [87].)

small apex and a suitable narrow cone. The production of suitable tips is a crucial part of any TERS application. In the following, we would like to describe our recent TERS development along these lines. [3]

10.4.3
TERS on Single-crystalline Surfaces

The use of Raman spectroscopy to study adsorbates on single-crystalline samples has been a dream for decades. Since the first presentation of the TERS approach by Stöckle et al. [16], it has been obvious that this new development has the potential for local Raman spectroscopy and microscopy with hitherto unprecedented sensitivity and spatial resolution. However, TERS studies have been reported so far for molecules having rather large Raman cross sections, such as dyes, C_{60} molecules, carbon nanotubes, and sulfur [16–19, 87, 136–139, 140–146]. In these cases, Raman signals of the adsorbates are detectable, even in the absence of the tip. If the tip is brought close to the surface, n-fold higher Raman intensities have been observed. Obviously, for strong Raman scatterers, local enhancement factors of 100–10000 are sufficient for getting a net amplification factor of $n > 1$. In order to be able to apply TERS in general, significantly larger enhancement factors of one or more orders of magnitude larger than 10^4 are imperative. According to theory, for an optimal excitation of localized surface plasmons, a tip-metal substrate configuration cavity seems to be quite suitable. This holds, for example, for an Au-tip/Au-substrate. For studies solely based on TERS, it is important to avoid complications associated with SERS. [4] This means that one needs tips free of adsorbates and contamination, and very smooth surfaces, which cannot support SERS. Well-annealed single-crystalline metal surfaces are suitable substrates in this respect. An easy way to prepare such crystals is to use the flame-annealing technique developed by Clavilier [147]. This simple procedure not only removes contamination, but also produces very smooth surfaces. These smooth surfaces are not SERS active in the usual sense.

In a recent paper, Pettinger et al. reported on TERS at smooth, flame-annealed Au(111) and Pt(110) surfaces [19]. The STM tips were prepared in most cases from a gold wire of 0.25 mm in diameter by electrochemical etching in a 1:1 mixture of ethanol and fuming HCl. The typical tip radii achieved by this method were about 40–60 nm [148]. STM images were recorded with the same Au tip as used for the TERS experiment. An example of these images, revealing rather smooth surfaces with mostly monatomic steps as a result of the flame-annealing pretreatment, is shown in Fig. 10.22.

Please note that in the TERS experiments, the tip adsorbate-metal configuration is operated in air (not under UHV conditions) after the adsorption of the

---

3) To our knowledge, up to today no other group has reported TERS from single crystalline surfaces.

4) However, if sensitivity is the major point, a combination of TERS and SERS may be of great advantage [143].

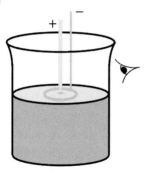

Fig. 10.21 Scheme of the set-up for the electrochemical etching of STM tips. (Reprinted with permission from Ref. [148].)

desired species from solution. Therefore, aside from the Raman spectroscopic data, there is no further information about the chemical state of the sample surface. By moving the STM tip into the tunneling position at a distance of about 1 nm above the surface (avoiding any accidental contact with the surface), an optical cavity is formed. If it is placed into the laser focus, its optical modes can be excited. They are denoted as localized surface plasmons (LSPs) and represent electron density oscillations under the action of the external electromagnetic field. The LSPs create in turn their own electromagnetic field, which is intense only in the close vicinity of the tip apex. Only the molecules exposed to this field contribute to the TERS effect. TERS is, therefore, a locally restricted phenomenon with a spatial resolution of the order of the tip curvature.

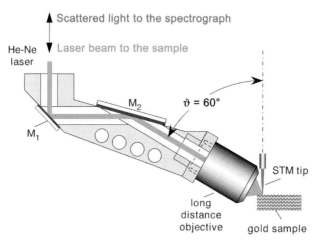

Fig. 10.22 TERS set-up for side illumination of tip and sample using the 60 arrangement. Olympus long distance microscope: 50× magnification, NA=0.5. He-Ne laser: 5 mW at the sample, λ_{ex}=632.8 nm. (Reprinted with permission from Ref. [143].)

To determine the TERS enhancement quantitatively, one needs – as mentioned above – excellent Raman scatterers in the adsorbate layer. Dye molecules are such strong Raman scatterers, provided they are resonantly excited and their fluorescence is absent or quenched. Malachite green isothiocyanate (MGITC) was chosen as an adsorbate for two reasons: (a) it binds strongly to gold via the thiol group and has, consequently, a vertical or slightly tilted orientation relative to the surface which favors the excitation of the Raman processes; (b) MGITC exhibits an intrinsic enhancement of the signal by resonance Raman scattering (RRS) if a suitable excitation energy is used, such as the 632.8-nm He-Ne laser line. In addition, the fluorescence of the adsorbed dye is effectively quenched by the metal surface. As shown in Fig. 10.23, the He-Ne laser line is well centered at the strongest excitation band of MGITC and also overlaps well with the fluorescence band. If the laser focus is placed on a dye-covered gold(111) surface, no fluorescence except a weak resonance Raman signal with a counting rate of only 1–2 cps for the strongest bands (tip in retracted position) is observed. The illumination with full laser power (5 mW) causes some photobleaching and the 1618-cm^{-1} band decreases with a time constant τ of about 800 s [19]. Since the photobleaching rate is a direct measure of the (local) intensity that acts on the dye molecules, one can compare the bleaching time constants in the presence and absence of the tip and quantify the enhancement for MGITC on Au(111). In fact, when the Au tip is brought into the tunneling position, two spectacular effects are observed: a huge enhancement for Raman scattering and a very rapid photobleaching. The strongest Raman peak intensities rise from 1–2 cps to

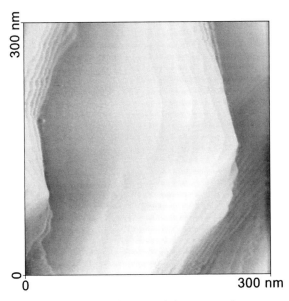

Fig. 10.23 STM image of an annealed Au(111) surface recorded with an Au tip. Tip bias 150 mV.

about 5600–16500 cps. Averaged over the six strongest bands, the net increase in the Raman intensity ($Q = I_{TERS}/I_{RRS}$) is ~8000. Q varies between 4000 and 14000 for the individual bands, which indicates an energy-dependent field enhancement or a different relative enhancement due to selection rules modified in the enhanced field. Most interestingly, a very rapid photobleaching with a time constant of 0.7 s was observed (for the same band at 1618 cm^{-1}). In order to be able to record the data and to avoid very rapid bleaching, the laser power had to be reduced to 1/10 of the full power. For convenience, all intensities in Fig. 10.24 are normalized to full laser power and 1 s accumulation time. The inset of Fig. 10.24 shows the decay of the integral intensity of the 1618 cm^{-1} band, which is assigned to the in-phase stretching of the phenyl rings [149, 150]. A bleaching time constant of $\tau_{TERS} \approx 7.3$ s was evaluated for 1/10 of the full laser power. The time constant for the bleaching of MGITC is inversely proportional to the local intensity. In the absence of the tip, only the focused laser beam acts on the dye layer with a time constant $\tau_{RRS} = 1/(I_L \gamma)$, where I_L is the laser power in the focus and γ is the bleaching constant. The approach of the tip causes a local field enhancement (but only near the tip apex) by a factor denoted as g. If one assumes a step function for the field distribution, g means an average over the probed area. For TERS, the bleaching time constant is $\tau_{TERS} = 1/[g^2 (I_L/10)\gamma]$, where the reduced laser intensity is taken into account by the divisor 10 and the enhancement of the local intensity underneath the tip by g^2. Hence, the ratio of the time constants yields directly $g^2 = 10 \times \tau_{RRS}/\tau_{TERS}$. Inserting the values of the observed bleaching time constants results in $g \approx 33$. Since the Raman enhancement F_{TERS} scales approximately with the fourth power of the field enhancement [151], we obtain $F_{TERS} \approx g^4 \approx 10^6$. Ref. [19] is the first report on the

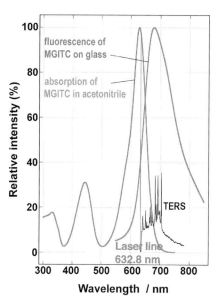

Fig. 10.24 Absorption and fluorescence spectra for malachite green isothiocyanate plotted in relative intensities. The laser line at 632.8 nm is indicated as a vertical line.

direct measurement of a near-field enhancement of TERS, indicating that six or more orders of enhancement are possible. The above data also allows us to estimate the radius a of the enhanced field along the sample surface. Its size is only a small fraction of the area of the laser focus with radius R_f. Therefore, from the intensity ratio at $t=0$ we have $Q = I_{TERS}/I_{RRS} \approx g^4 a^2/R_f^2$. With $R_f \approx 1000$ nm, the radius of the area of the sample probed by TERS results in $a \approx 90$ nm. This estimate does not depend much on the local field distribution; a Gaussian or a Lorentzian profile would also yield the above equation for a apart from a somewhat altered prefactor. The actual tip radius of 40 nm, determined by SEM for the tip used in this experiment, comes close to the above estimate.

It should be emphasized that the tip shape (roughly characterized by its cone angle, sharpness, and smoothness) has a strong influence on the net gain of the Raman signal. This varied between 100 and 15 000 for a number of different tips.

Additionally, the tip and substrate materials are of crucial importance for the undamped excitations of LSP. Fig. 10.25 compares Raman spectra for MGITC adsorbed on an Au(111) surface in the presence of an Ir tip as well as adsorbed on a Pt(110) surface in combination with an Au tip. In the former case with the tip retracted, a weak Raman spectrum is observed due to RRS with integral intensities of about 2–5 cps for the strongest bands (top panel, trace "retracted"). When the Ir tip is brought into the tunneling position, the Raman intensity rises by a factor of 2 (top panel, trace "tunneling"). It is still an open question as to what causes the two-fold signal increase. Either the Ir-tip/Au(111) cavity supports an excitation of LSP, although with strong damping, or it is a far-field

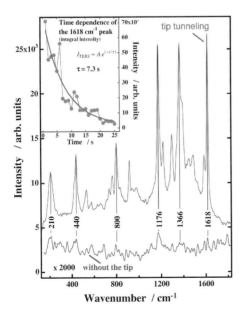

Fig. 10.25 Comparison of RRS and TERS spectra for malachite green isothiocyanate adsorbed at an Au(111) surface. The laser power in the TERS case is reduced to 0.5 mW. The spectral intensities are normalized to full laser power (5 mW) and to an acquisition time of 1 s. The actual acquisition times were 1 s for TERS and 60 s for RRS. The MGITC dye is adsorbed from a 10^{-7} M ethanol solution over 30 min. Tunneling current: 1 nA; voltage: 150 mV. Inset: Time dependence of the integral intensity of the 1618 cm^{-1} band for reduced laser power. (Reproduced with permission from Ref. [19].)

effect, for example a local increase in the intensity due to reflection of light by the illuminated tip. MGITC on a Pt(110) surface with the tip retracted shows no RRS signal (Fig. 10.25, bottom panel, trace "retracted"). However, in the presence of an Au tip, a comparatively intense spectrum (trace "tunneling") is observed, indicating the enhancement of the Raman processes by several orders of magnitude. Obviously, the choice of the tip material (here a gold tip) is more critical for the generation of localized surface plasmons than the choice of the substrate material (Au or Pt).

If an enhancement of five to six orders of magnitude can be achieved routinely, TERS for small molecules, which are not in resonance with the laser line, is within reach. These molecules have cross sections of the order of $(d\sigma/d\Omega) \sim 10^{-29}$ cm^2 sr^{-1} and are barely "seen" as adlayers on smooth interfaces by normal Raman spectroscopy. To test this, benzenethiol was chosen, which assumes an essentially vertical orientation to the surface due to its thiol group. The adlayer preparation is quick and easy: the adsorption occurs from an ethanolic solution, and a self-assembled monolayer is formed on previously flame-annealed gold or platinum surfaces. Fig. 10.26 shows TERS spectra for benzenethiol at Au(110) and Pt(110) surfaces. The comparison of the spectra reveals the characteristic benzenethiol bands but with slightly different band positions and relative intensities for the two samples and a nearly 20-fold lower intensity level for benzenethiol at Pt(110). The comparison of these data with SER spectra for

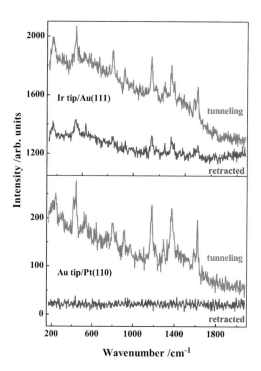

Fig. 10.26 TERS and RRS for other tip–metal configurations. Top panel: Ir tip/Au(111); acquisition time 30 s. Bottom panel: Au tip/Pt(110); acquisition time 2 s. For both configurations, the MGITC dye is adsorbed from a 10^{-6} M ethanol solution over 30 min. Laser power: 5 mW. Tunneling current: 1 nA; voltage: 150 mV. (Reproduced with permission from Ref. [19].)

benzenethiol at roughened Au and Pt surfaces, shown in Fig. 10.27, is also interesting. It is noteworthy that the spectroscopic patterns of the TERS and SERS spectra are quite similar, including the characteristic relative intensities. The frequency differences are at most about 8 cm^{-1}. In Fig. 10.28, expanded TERS spectra for benzenethiol are presented for a sequence of three experiments performed with the same Au tip. After the first experiment at Pt(110), the sample was exchanged for Au(110), and afterwards again exchanged for Pt(110).

It is well known that adsorbates interact differently with varying substrates, which results in (usually) slight variations of vibrational frequencies and relative intensities. The power of vibrational spectroscopies as analytical tools is based on this fact. Let us carefully examine the two spectra in Fig. 10.26, in particular the benzenethiol band at around 1070 cm^{-1} with an a_1-type symmetry [152]. The Raman frequency shifts from 1068 cm^{-1} on Pt to 1075 cm^{-1} on Au. This is a 7-cm^{-1} blue shift of the radiated light frequency. In addition, there is a clear change of the relative intensities in the frequency region from 950 to 1150 cm^{-1}. The assignment of the benzenethiol SERS bands has been given in Ref. [152]. Because of the minor changes between SERS and TERS spectra, this assignment can be applied also to the TERS case. TERS provides a diagnostic tool for the determination of the kind of substrate to which the benzenethiol is bound. In the sequence of experiments, Fig. 10.28 clearly shows that the vibrational frequency at 1068 cm^{-1} stems from benzenethiol bound, not to the Au-tip, but to a Pt(110) or Au(110) surface, respectively [146]. Additional evidence is provided by the change in the relative intensities: the 1075 cm^{-1} band [benzenethiol at Au(110)] is nearly twice as intense as the 1068 cm^{-1} band [benzenethiol at Pt(110)]. The other bands do not differ as much. Since all three bands in Fig. 10.28 are assigned to the vibrations possessing a_1 symmetry, the differences in the peak intensity ratios cannot be explained by different orienta-

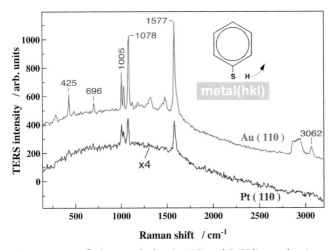

Fig. 10.27 TERS for benzenethiol at Au(110) and Pt(110) samples. Laser power: 0.5 mW.

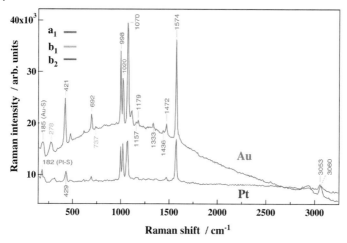

Fig. 10.28 SERS for benzenethiol at roughened Au and Pt surfaces. Laser power at the sample: 0.5 mW.

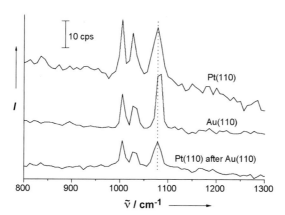

Fig. 10.29 Expanded TERS spectra for experiments performed on Pt(110), then on Au(110), and finally on Pt(110) again. All other experimental conditions were the same as those in Fig. 10.27: TERS for benzenethiol at Au(110) and Pt(110) samples. Laser power: 0.5 Mw. (Reproduced with permission from Ref. [146].)

tions of benzenethiol on the two surfaces. They are probably due to different interactions of this molecule with different metal surfaces (in analogy to the case of pyridine, see Ref. [153]).

10.5
Outlook

10.5.1
Recent Results

More and more results demonstrate the potential of TERS as a general analytical tool. TERS has been observed for dye layers on glass as well as for adsorbates on silver, gold, or platinum substrates. TERS is found not only for very strong Raman scatterers, but also for species usually exhibiting normal Raman scattering. Among the latter are CN^- and ClO_4^- ions, benzenethiol, mercaptopyridine, and the DNA bases (adenine, cytosine, guanine, and thymine), all adsorbed at smooth single-crystalline gold surfaces [19, 146, 154, 155].

TERS of the four DNA bases Preliminary results of our group show that the four DNA bases adenine, cytosine, guanine, and thymine are easily monitored when adsorbed at crystalline gold surfaces. This may permit the study of coadsorption of DNA-base pairs. Coadsorption has a significant impact on the Raman spectra of the DNA bases. For instance, we found spectral shifts due to the interaction of ions and DNA bases, e.g., perchlorate ions and adenine molecules. Another interesting route in the future may be the investigation of fragments of DNA helices. With increasing spatial resolution, TERS has the potential to permit the reading of individual DNA sequences [155].

Tip-sample distance dependence of the TER intensity Another preliminary result of our group is the distance dependence of TERS scattering. There is, in fact, a very fast decay of the TERS intensity if the tip is moved away from the surface: for a 20-nm tip radius, we observed a decrease in the TERS intensity to approximately zero within a 10-nm tip retraction. This holds not only for the individual Raman bands, but also for the inelastic background. The fast intensity decay indicates the short range of the near-field processes that govern TERS and points to a high spatial resolution of TERS. Most likely, its spatial resolution can be improved to better than 10 nm, provided that ideal tips (optimized with respect to tip curvature, surface smoothness and cone angle) become available [156].

10.5.2
New Approaches on the Horizon

Though TERS is a very promising approach, there are a few obstacles, at least in our view, which have to be overcome. In most studies, TERS was applied in an ambient environment, which may cause contamination of the tip and surface. In addition, the control of the surface coverage of the adsorbate is limited and the state of the surface is unknown.

TERS in combination with electrochemistry Modern electrochemistry is based on insights arising from studies that often combine typical electrochemical and UHV investigations. Another route is the combination of electrochemical and infrared spectroscopic studies, which can also be applied under *in-situ* conditions in solutions. The application of TERS as an *in-situ* spectroscopic tool is – at present – still a dream. To realize it, a number of technical problems (whose discussion here is beyond the scope of this section) have to be solved. Hence, EC-TERS, with the perspectives of excellent sensitivity *and* spatial resolution, remains a great challenge for scientists.

Employing TERS in UHV systems There are a number of surface science tools available for samples in UHV which allow us to characterize the state of a surface. Surface and adlayer structures can be determined by LEED (low electron energy diffraction) as well as by SPM (scanning probe microscopy) techniques. While the kind of chemical interactions can be studied, for example, with AES (Auger electron spectroscopy), EELS (energy electron loss spectroscopy) permits the identification of the chemical nature of the adsorbed species. TERS, on the other hand, may provide similar but also complementary information on the chemical identity under UHV conditions. As an additional advantage, TERS and SPM permit the identification and characterization of the spatial region from which this information is accumulated.

TERS for the field of heterogeneous catalysis A further interesting approach is to apply TERS to single particles which are catalytically active. If successful, this will open new and advanced avenues for spectroscopic investigations in the field of heterogeneous catalysis. Clearly, a study of individual particles under conditions that are close to catalytic reaction conditions will provide more conclusive information than analogous studies on large particle ensembles.

TERS and theory TERS also provides excellent stimulation and a playground for theoretical studies on enhanced optical processes. The tip–substrate configuration can be considered as a unique photonic system with tunable optical resonances that provides huge near-field enhancements for species located in the tip–substrate cavity. A suitable theoretical description of such a photonic system is, even today, a great challenge, although the general features of such a system are well understood.

Acknowledgment

The author gratefully acknowledges the careful reading of the manuscript by Katrin F. Domke and Prof. Sarah Horswell.

References

1 *Surface Enhanced Raman Scattering*, T. E. Furtak and R. K. Chang (eds) (Plenum Publ. Corp., 1982).

2 M. Moskovits, *Surface-Enhanced Spectroscopy*, Rev. Mod. Phys. **57**, 783 (1985)

3 T. M. Cotton, J. H. Kim, and G. D. Chumanov, *Application of Surface-Enhanced Raman-Spectroscopy to Biological-Systems*, J. Raman Spectrosc. **22**, 729 (1991)

4 A. Otto, I. Mrozek, H. Grabhorn, and W. Akemann, *Surface-Enhanced Raman-Scattering*, J. Phys.-Condens. Matter **4**, 1143 (1992).

5 K. C. Grabar, K. J. Allison, B. E. Baker, R. M. Bright, K. R. Brown, R. G. Freeman, A. P. Fox, C. D. Keating, M. D. Musick, and M. J. Natan, *Two-Dimensional Arrays of Colloidal Gold Particles: a Flexible Approach to Macroscopic Metal Surfaces*, Langmuir **12**, 2353 (1996)

6 A. Campion and P. Kambhampati, *Surface-Enhanced Raman Scattering*, Chem. Soc. Rev. **27**, 241 (1998)

7 K. Kneipp, H. Kneipp, I. Itzkan, R. R. Dasari, and M. S. Feld, *Ultrasensitive Chemical Analysis by Raman Spectroscopy*, Chem. Rev. **99**, 2957 (1999)

8 C. J. L. Constantino, T. Lemma, P. A. Antunes, and R. Aroca, *Single-Molecule Detection Using Surface-Enhanced Resonance Raman Scattering and Langmuir-Blodgett Monolayers*, Anal. Chem. **73**, 3674 (2001)

9 C. L. Haynes and R. P. Van Duyne, *Nanosphere Lithography: a Versatile Nanofabrication Tool for Studies of Size-Dependent Nanoparticle Optics*, J. Phys. Chem. B **105**, 5599 (2001)

10 K. Kneipp, H. Kneipp, I. Itzkan, R. R. Dasari, and M. S. Feld, *Surface-Enhanced Raman Scattering and Biophysics*, J. Phys.-Condens. Matter **14**, R597–R624 (2002)

11 Z. Q. Tian, B. Ren, and D. Y. Wu, *Surface-Enhanced Raman Scattering: From Noble to Transition Metals and From Rough Surfaces to Ordered Nanostructures*, J. Phys. Chem. B **106**, 9463 (2002)

12 Z. Q. Tian, B. Ren, and B. W. Mao, *Extending Surface Raman Spectroscopy to Transition Metal Surfaces for Practical Applications. 1. Vibrational Properties of Thiocyanate and Carbon Monoxide Adsorbed on Electrochemically Activated Platinum Surfaces*, J. Phys. Chem. B **101**, 1338 (1997)

13 K. Kneipp, Y. Wang, H. Kneipp, L. T. Perelman, I. Itzkan, R. Dasari, and M. S. Feld, *Single-Molecule Detection Using Surface-Enhanced Raman Scattering (SERS)*, Phys. Rev. Lett. **78**, 1667 (1997)

14 M. Moskovits, L. L. Tay, J. Yang, and T. Haslett, *SERS and the Single Molecule, Optical Properties of Nanostructured Random Media.* Top. Appl. Phys. **82**, 215 (2002)

15 S. M. Nie and S. R. Emery, *Probing Single Molecules and Single Nanoparticles by Surface-Enhanced Raman Scattering*, Science **275**, 1102 (1997)

16 R. M. Stöckle, Y. D. Suh, V. Deckert, and R. Zenobi, *Nanoscale Chemical Analysis by Tip-Enhanced Raman Spectroscopy*, Chem. Phys. Lett. **318**, 131 (2000)

17 N. Hayazawa, Y. Inouye, Z. Sekkat, and S. Kawata, *Near-Field Raman Scattering Enhanced by a Metallized Tip*, Chem. Phys. Lett. **335**, 369 (2001)

18 A. Hartschuh, M. R. Beversluis, A. Bouhelier, and L. Novotny, *Tip-Enhanced Optical Spectroscopy*, Philos. Trans. R. Soc. Lond. Ser. A-Math. Phys. Eng. Sci. **362**, 807 (2004)

19 B. Pettinger, B. Ren, G. Picardi, R. Schuster, and G. Ertl, *Nanoscale Probing of Adsorbed Species by Tip-Enhanced Raman Spectroscopy*, Phys. Rev. Lett. **92**, 096101 (2004)

20 D. Zeisel, V. Deckert, R. Zenobi, and T. Vo-Dinh, *Near-Field Surface-Enhanced Raman Spectroscopy of Dye Molecules Adsorbed on Silver Island Films*, Chem. Phys. Lett. **283**, 381 (1998)

21 D. L. Jeanmaire and R. P. Vanduyne, *Surface Raman Spectroelectrochemistry. 1. Heterocyclic, Aromatic, and Aliphatic-Amines Adsorbed on Anodized Silver Electrode*, J. Electroanal. Chem. **84**, 1 (1977)

22 K. Kneipp, H. Kneipp, R. Manoharan, E. B. Hanlon, I. Itzkan, R. R. Dasari, and M. S. Feld, *Extremely Large Enhancement Factors in Surface-Enhanced Raman Scattering for Molecules on Colloidal Gold Clusters*, Appl. Spectrosc. **52**, 1493 (1998)

23 V. M. Shalaev, *Optical Nonlinearities of Fractal Composites, Optical Properties of Nanostructured Random Media.* Top. Appl. Phys. **82**, 93 (2002)

24 M. Moskovits and D. H. Jeong, *Engineering Nanostructures for Giant Optical Fields*, Chem. Phys. Lett. **397**, 91 (2004)

25 A. Campion, J. K. Brown, and V. M. Grizzle, *Surface Raman-Spectroscopy Without Surface Enhancement*, J. Vac. Sci. Technol. **20**, 893 (1982)

26 A. G. Brolo, Z. Jiang, and D. E. Irish, *The Orientation of 2,2'-Bipyridine Adsorbed at a SERS-Active Au(111) Electrode Surface*, J. Electroanal. Chem. **547**, 163 (2003)

27 B. Pettinger, X. Bao, I. C. Wilcock, M. Muhler, and G. Ertl, *Surface-Enhanced Raman-Scattering From Surface and Subsurface Oxygen Species at Microscopically Well-Defined Ag Surfaces*, Phys. Rev. Lett. **72**, 1561 (1994)

28 A. Bruckbauer and A. Otto, *Raman Spectroscopy of Pyridine Adsorbed on Single Crystal Copper Electrodes*, J. Raman Spectrosc. **29**, 665 (1998)

29 P. Kambhampati, C. M. Child, M. C. Foster, and A. Campion, *On the Chemical Mechanism of Surface Enhanced Raman Scattering: Experiment and Theory*, J. Chem. Phys. **108**, 5013 (1998)

30 P. Kambhampati, O. K. Song, and A. Campion, *Probing Photoinduced Charge Transfer at Atomically Smooth Metal Surfaces Using Surface Enhanced Raman Scattering*, Phys. Status Solidi A-Appl. Res. **175**, 233 (1999)

31 E. R. Savinova, D. Zemlyanov, B. Pettinger, A. Scheybal, R. Schlogl, and K. Doblhofer, *On the Mechanism of Ag(111) Sub-Monolayer Oxidation: a Combined Electrochemical, in-situ SERS and Ex Situ Xps Study*, Electrochim. Acta **46**, 175 (2000)

32 B. Pettinger and U. Wenning, *Raman-Spectra of Pyridine Adsorbed on Silver (100) and (111) Electrode Surfaces*, Chem. Phys. Lett. **56**, 253 (1978)

33 B. Pettinger, U. Wenning, and D. M. Kolb, *Raman and Reflectance Spectroscopy of Pyridine Adsorbed on Single-Crystalline Silver Electrodes*, Ber. Bunsen-Ges. Phys. Chem. Chem. Phys. **82**, 1326 (1978)

34 B. Pettinger, A. Tadjeddine, and D. M. Kolb, *Enhancement in Raman Intensity by Use of Surface-Plasmons*, Chem. Phys. Lett. **66**, 544 (1979)

35 D. R. Mullins and A. Campion, *Unenhanced Raman-Scattering From Pyridine Chemisorbed on a Stepped Silver Surface – Implications for Proposed SERS Mechanisms*, Chem. Phys. Lett. **110**, 565 (1984)

36 A. G. Brolo, D. E. Irish, and J. Lipkowski, *Surface-Enhanced Raman Spectra of Pyridine and Pyrazine Adsorbed on a Au(210) Single-Crystal Electrode*, J. Phys. Chem. B **101**, 3906 (1997)

37 A. Iannelli, A. G. Brolo, D. E. Irish, and J. Lipkowski, *Electrochemical and Raman Spectroscopic Studies of Pyrazine Adsorption at the Au(210) Electrode Surface*, Can. J. Chem.-Rev. Can. Chim. **75**, 1694 (1997)

38 P. G. Cao, J. L. Yao, B. Ren, B. W. Mao, R. A. Gu, and Z. Q. Tian, *Surface-Enhanced Raman Scattering from Bare Fe Electrode Surfaces*, Chem. Phys. Lett. **316**, 1 (2000)

39 W. B. Cai, B. Ren, X. Q. Li, C. X. She, F. M. Liu, X. W. Cai, and Z. Q. Tian, *Investigation of Surface-Enhanced Raman Scattering from Platinum Electrodes Using a Confocal Raman Microscope: Dependence of Surface Roughening Pretreatment*, Surf. Sci. **406**, 9 (1998)

40 B. Ren, Q. J. Huang, W. B. Cai, B. W. Mao, F. M. Liu, and Z. Q. Tian, *Surface Raman Spectra of Pyridine and Hydrogen on Bare Platinum and Nickel Electrodes*, J. Electroanal. Chem. **415**, 175 (1996)

41 Z. Q. Tian and B. Ren, *Molecular-Level Investigation on Electrochemical Interfaces by Raman Spectroscopy*, Chin. J. Chem. **18**, 135 (2000)

42 T. M. Cotton and R. P. Vanduyne, *Electrochemical Investigation of the Redox Properties of Bacteriochlorophyll and Bacteriopheophytin in Aprotic Solvents*, J. Am. Chem. Soc. **101**, 7605 (1979)

43 P. Kambhampati and A. Campion, *Surface Enhanced Raman Scattering as a Probe of Adsorbate-Substrate Charge-Transfer Excitations*, Surf. Sci. **428**, 115 (1999)

44 B. Pettinger, X. Bao, I. C. Wilcock, M. Muhler, and G. Ertl, *Surface-Enhanced Raman Scattering From Surface and Subsurface Oxygen Species at Microscopically*

Well-Defined Ag Surfaces. Phys. Rev. Lett. 1561 (1994)

45 B. Pettinger, X. H. Bao, I. Wilcock, M. Muhler, R. Schlögl, and G. Ertl, *Thermal-Decomposition of Silver-Oxide Monitored by Raman- Spectroscopy – From AgO Units to Oxygen-Atoms Chemisorbed on the Silver Surface,* Angew. Chem. Int. Ed. Engl. **33**, 85 (1994)

46 X. Bao, M. Muhler, B. Pettinger, R. Schlögl, and G. Ertl, *On the Nature of the Active State of Silver During Catalytic Oxidation of Methanol,* Catal. Lett. **22**, 215 (1993)

47 X. Bao, B. Pettinger, G. Ertl, and R. Schlögl, *In-situ Raman Studies of Ethylene Oxidation at Ag(111) and Ag(110) Under Catalytic Reaction Conditions,* Ber. Bunsen-Ges. Phys. Chem. Chem. Phys. **97**, 322 (1993)

48 X. Bao, J. V. Barth, G. Lehmpfuhl, R. Schuster, Y. Uchida, R. Schlögl, and G. Ertl, *Oxygen-Induced Restructuring of Ag(111),* Surf. Sci. **284**, 14 (1993)

49 X. Bao, M. Muhler, B. Pettinger, Y. Uchida, G. Lehmpfuhl, R. Schlögl, and G. Ertl, *The Effect of Water on the Formation of Strongly Bound Oxygen on Silver Surfaces,* Catal. Lett. **32**, 171 (1995)

50 X. Bao, U. Wild, M. Muhler, B. Pettinger, R. Schlögl, and G. Ertl, *Coadsorption of Nitric Oxide and Oxygen on the Ag(110) Surface,* Surf. Sci. **425**, 224 (1999)

51 T. Hirschfeld, *Quantum efficiency independence of the time integrated emission from a fluorescent molecule,* Appl. Opt. **15**, 3135 (1976)

52 T. Hirschfeld, *Optical Microscopic Observation of Single Small Molecules,* Appl. Opt. **15**, 2965 (1976)

53 N. J. Dovichi, J. C. Martin, J. H. Jett, M. Trkula, and R. A. Keller, *Laser-Induced Fluorescence of Flowing Samples as an Approach to Single-Molecule Detection in Liquids,* Anal. Chem. **56**, 348 (1984)

54 D. C. Nguyen, R. A. Keller, J. H. Jett, and J. C. Martin, *Detection of Single Molecules of Phycoerythrin in Hydrodynamically Focused Flows by Laser-Induced Fluorescence,* Anal. Chem. **59**, 2158 (1987)

55 S. A. Soper, E. B. Shera, J. C. Martin, J. H. Jett, J. H. Hahn, H. L. Nutter, and R. A. Keller, *Single-Molecule Detection of Rhodamine-6g in Ethanolic Solutions Using Continuous Wave Laser Excitation,* Anal. Chem. **63**, 432 (1991)

56 W. P. Ambrose, T. Basche, and W. E. Moerner, *Detection and Spectroscopy of Single Pentacene Molecules in a Para-Terphenyl Crystal by Means of Fluorescence Excitation,* J. Chem. Phys. **95**, 7150 (1991)

57 W. P. Ambrose and W. E. Moerner, *Fluorescence Spectroscopy and Spectral Diffusion of Single Impurity Molecules in a Crystal,* Nature **349** , 225 (1991)

58 K. Kneipp, Y. Wang, R. R. Dasari, and M. S. Feld, *Approach to Single-Molecule Detection Using Surface-Enhanced Resonance Raman-Scattering (SERRS) – a Study Using Rhodamine 6g on Colloidal Silver,* Appl. Spectrosc. **49**, 780 (1995)

59 A. R. Bizzarri and S. Cannistraro, *Evidence of Electron-Transfer in the SERS Spectra of a Single Iron-Protoporphyrin Ix Molecule,* Chem. Phys. Lett. **395**, 222 (2004)

60 A. R. Bizzarri and S. Cannistraro, *Temporal Fluctuations in the SERRS Spectra of Single Iron-Protoporphyrin Ix Molecule,* Chem. Phys. **290**, 297 (2003)

61 E. J. Bjerneld, Z. Foldes-Papp, M. Kall, and R. Rigler, *Single-Molecule Surface-Enhanced Raman and Fluorescence Correlation Spectroscopy of Horseradish Peroxidase,* J. Phys. Chem. B **106**, 1213 (2002)

62 I. Delfino, A. R. Bizzarri, and S. Cannistraro, *Single-Molecule Detection of Yeast Cytochrome C by Surface-Enhanced Raman Spectroscopy,* Biophys. Chem. **113**, 41 (2005)

63 W. E. Doering and S. M. Nie, *Single-Molecule and Single-Nanoparticle SERS: Examining the Roles of Surface Active Sites and Chemical Enhancement,* J. Phys. Chem. B **106**, 311 (2002)

64 K. Kneipp, H. Kneipp, V. B. Kartha, R. Manoharan, G. Deinum, I. Itzkan, R. R. Dasari, and M. S. Feld, *Detection and Identification of a Single DNA Base Molecule Using Surface-Enhanced Raman Scattering (SERS),* Phys. Rev. E **57**, R6281–R6284 (1998)

65 A. J. Meixner, T. Vosgrone, and M. Sackrow, _Nanoscale Surface-Enhanced Resonance Raman Scattering Spectroscopy of Single Molecules on Isolated Silver Clusters,_ J. Lumines. **94**, 147 (2001)

66 A. Otto, _Theory of First Layer and Single Molecule Surface Enhanced Raman Scattering (SERS),_ Phys. Status Solidi A-Appl. Res. **188**, 1455 (2001)

67 A. Otto, A. Bruckbauer, and Y. X. Chen, _On the Chloride Activation in SERS and Single Molecule SERS,_ J. Mol. Struct. **661–662**, 501 (2003)

68 B. Tolaieb, C. J. L. Constantino, and R. F. Aroca, _Surface-Enhanced Resonance Raman Scattering as an Analytical Tool for Single Molecule Detection,_ Analyst **129**, 337 (2004)

69 H. X. Xu, J. Aizpurua, M. Kall, and P. Apell, _Electromagnetic Contributions to Single-Molecule Sensitivity in Surface-Enhanced Raman Scattering,_ Phys. Rev. E **62**, 4318 (2000)

70 S. L. Zou and G. C. Schatz, _Silver Nanoparticle Array Structures that Produce Giant Enhancements in Electromagnetic Fields,_ Chem. Phys. Lett. **403**, 62 (2005)

71 M. Futamata, Y. Maruyama, and M. Ishikawa, _Critical Importance of the Junction in Touching Ag Particles for Single Molecule Sensitivity in SERs,_ J. Mol. Struct. **735/736**, 75 (2005)

72 C. Blum, F. Stracke, S. Becker, K. Mullen, and A. J. Meixner, _Discrimination and Interpretation of Spectral Phenomena by Room-Temperature Single-Molecule Spectroscopy,_ J. Phys. Chem. A **105**, 6983 (2001)

73 A. Otto, _What Is Observed in Single Molecule SERS, and Why?,_ J. Raman Spectrosc. **33**, 593 (2002)

74 M. Futamata, Y. Maruyama, and M. Ishikawa, _Microscopic Morphology and SERS Activity of Ag Colloidal Particles,_ Vib. Spectrosc. **30**, 17 (2002)

75 M. Ishikawa, Y. Maruyama, J. Y. Ye, and M. Futamata, _Single-Molecule Imaging and Spectroscopy of Adenine and an Analog of Adenine Using Surface-Enhanced Raman Scattering and Fluorescence,_ J. Lumines. **98**, 81 (2002)

76 M. Futamata, Y. Maruyama, and M. Ishikawa, _Metal Nanostructures with Single Molecule Sensitivity in Surface Enhanced Raman Scattering,_ Vib. Spectrosc. **35**, 121 (2004)

77 R. M. Dickson, A. B. Cubitt, R. Y. Tsien, and W. E. Moerner, _On/Off Blinking and Switching Behaviour of Single Molecules of Green Fluorescent Protein,_ Nature **388**, 355 (1997)

78 M. F. Garcia-Parajo, G. M. J. Segers-Nolten, J. A. Veerman, J. Greve, and N. F. Van Hulst, _Real-Time Light-Driven Dynamics of the Fluorescence Emission in Single Green Fluorescent Protein Molecules,_ Proc. Nat. Acad. Sci. USA **97**, 7237 (2000)

79 L. A. Peyser, A. E. Vinson, A. P. Bartko, and R. M. Dickson, _Photoactivated Fluorescence From Individual Silver Nanoclusters,_ Science **291**, 103 (2001)

80 R. Verberk and M. Orrit, _Photon Statistics in the Fluorescence of Single Molecules and Nanocrystals: Correlation Functions Versus Distributions of On- and Off-Times,_ J. Chem. Phys. **119**, 2214 (2003)

81 P. Zhang, T. L. Haslett, C. Douketis, and M. Moskovits, _Mode Localization in Self-Affine Fractal Interfaces Observed by Near-Field Microscopy,_ Phys. Rev. B **57**, 15513 (1998)

82 A. Kudelski and B. Pettinger, _SERS on Carbon Chain Segments: Monitoring Locally Surface Chemistry,_ Chem. Phys. Lett. **321**, 356 (2000)

83 S. I. Bozhevolnyi, B. Vohnsen, I. I. Smolyaninov, and A. V. Zayats, _Direct Observation of Surface Polariton Localization Caused by Surface-Roughness,_ Opt. Commun. **117**, 417 (1995)

84 R. Fikri, T. Grosges, and D. Barchiesi, _Apertureless Scanning Near-Field Optical Microscopy: Numerical Modeling of the Lock-in Detection,_ Opt. Commun. **232**, 15 (2004)

85 V. M. Shalaev and A. K. Sarychev, _Nonlinear Optics of Random Metal-Dielectric Films,_ Phys. Rev. B **57**, 13265 (1998)

86 P. Gadenne, X. Quelin, S. Ducourtieux, S. Gresillon, L. Aigouy, J. C. Rivoal, V. Shalaev, and A. Sarychev, _Direct Observation of Locally Enhanced Electromagnetic Field,_ Physica B **279**, 52 (2000)

87 B. Pettinger, G. Picardi, R. Schuster, and G. Ertl, _Surface Enhanced Raman Spectroscopy: Towards Single Molecular Spectroscopy,_ Electrochemistry **68**, 942 (2000)

88 A. Kudelski and B. Pettinger, *Fluctuations of Surface-Enhanced Raman Spectra of Co Adsorbed on Gold Substrates*, Chem. Phys. Lett. **383**, 76 (2004)

89 P. J. Moyer, J. Schmidt, L. M. Eng, and A. J. Meixner, *Surface-Enhanced Raman Scattering Spectroscopy of Single Carbon Domains on Individual Ag Nanoparticles on a 25 Ms Time Scale*, J. Am. Chem. Soc. **122**, 5409 (2000)

90 D. W. Pohl, W. Denk, and M. Lanz, *Optical Stethoscopy – Image Recording With Resolution Lambda/20*, Appl. Phys. Lett. **44**, 651 (1984)

91 U. Durig, D. W. Pohl, and F. Rohner, *Near-Field Optical-Scanning Microscopy*, J. Appl. Phys. **59**, 3318 (1986)

92 E. Betzig, A. Harootunian, A. Lewis, and M. Isaacson, *Near-Field Diffraction by a Slit – Implications for Superresolution Microscopy*, Appl. Opt. **25**, 1890 (1986)

93 E. Betzig, A. Lewis, A. Harootunian, M. Isaacson, and E. Kratschmer, *Near-Field Scanning Optical Microscopy (NSOM) – Development and Biophysical Applications*, Biophys. J. **49**, 269 (1986)

94 E. Betzig, A. Harootunian, A. Lewis, and M. Isaacson, *Near-Field Scanning Optical Microscopy (NSOM) – Investigation of Radiation Transmitted Through Sub-Wavelength Apertures*, Biophys. J. **47**, A407 (1985)

95 U. C. Fischer, U. Durig, and D. W. Pohl, *Near-Field Optical-Scanning Microscopy and Enhanced Spectroscopy with Submicron Apertures*, Scanning Microsc. 47 (1987)

96 A. Harootunian, E. Betzig, M. Isaacson, and A. Lewis, *Superresolution Fluorescence Near-Field Scanning Optical Microscopy*, Appl. Phys. Lett. **49**, 674 (1986)

97 A. Harootunian, E. Betzig, A. Muray, A. Lewis, and M. Isaacson, *Near-Field Investigation of Submicrometer Apertures at Optical Wavelengths*, J. Opt. Soc. Am. A-Opt. Image Sci. Vis. **1**, 1293 (1984)

98 M. Isaacson, E. Betzig, A. Harootunian, and A. Lewis, *Scanning Optical Microscopy at Lambda/10 Resolution Using Near-Field Imaging Methods*, Ann. N.Y. Acad. Sci. **483**, 448 (1986)

99 E. Abbe, Arch. Mikrosk. Anat. **9**, 413 (1873)

100 D. Courjon and C. Bainier, *Near-Field Microscopy and Near-Field Optics*, Rep. Progr. Phys. **57**, 989 (1994)

101 R. C. Dunn, *Near-Field Scanning Optical Microscopy*, Chem. Rev. **99**, 2891 (1999)

102 B. Hecht, B. Sick, U. P. Wild, V. Deckert, R. Zenobi, O. J. F. Martin, and D. W. Pohl, *Scanning Near-Field Optical Microscopy With Aperture Probes: Fundamentals and Applications*, J. Chem. Phys. **112**, 7761 (2000)

103 C. Girard and A. Dereux, *Near-Field Optics Theories*, Rep. Progr. Phys. **59**, 657 (1996)

104 W. Knoll, W. Hickel, M. Sawodny, and J. Stumpe, *Polymer Interfaces and Ultra-thin Films Characterized by Optical Evanescent Wave Techniques*, Makromolekulare Chemie-Macromolecular Symposia **48/49**, 363 (1991)

105 D. P. Tsai, A. Othonos, M. Moskovits, and D. Uttamchandani, *Raman-Spectroscopy Using a Fiber Optic Probe With Subwavelength Aperture*, Appl. Phys. Lett. **64**, 1768 (1994)

106 C. L. Jahncke, M. A. Paesler, and H. D. Hallen, *Raman Imaging With Near-Field Scanning Optical Microscopy*, Appl. Phys. Lett. **67**, 2483 (1995)

107 L. Novotny, D. W. Pohl, and B. Hecht, *Scanning Near-Field Optical Probe With Ultrasmall Spot Size*, Opt. Lett. **20**, 970 (1995)

108 G. Krausch, S. Wegscheider, A. Kirsch, H. Bielefeldt, J. C. Meiners, and J. Mlynek, *Near-Field Microscopy and Lithography With Uncoated Fiber Tips – a Comparison*, Opt. Commun. **119**, 283 (1995)

109 C. Obermuller and K. Karrai, *Far-Field Characterization of Diffracting Circular Apertures*, Appl. Phys. Lett. **67**, 3408 (1995)

110 A. G. T. Ruiter, M. H. P. Moers, N. F. Vanhulst, and M. Deboer, *Microfabrication of Near-Field Optical Probes*, J. Vac. Sci. Technol. B **14**, 597 (1996)

111 D. Zeisel, B. Dutoit, V. Deckert, T. Roth, and R. Zenobi, *Optical Spectroscopy and Laser Desorption on a Nanometer Scale*, Anal. Chem. **69**, 749 (1997)

112 J. Grausem, B. Humbert, A. Burneau, and J. Oswalt, *Subwavelength Raman*

Spectroscopy, Appl. Phys. Lett. **70**, 1671 (1997)

113 S. R. Emory and S. M. Nie, *Near-Field Surface-Enhanced Raman Spectroscopy on Single Silver Nanoparticles*, Anal. Chem. **69**, 2631 (1997)

114 J. Wessel, J. Opt. Soc. Am. B **2**, 1538 (1985)

115 S. Webster, D. A. Smith, and D. N. Batchelder, *Raman Microscopy Using a Scanning Near-Field Optical Probe*, Vib. Spectrosc. **18**, 51 (1998)

116 S. Webster, D. N. Batchelder, and D. A. Smith, *Submicron Resolution Measurement of Stress in Silicon by Near-Field Raman Spectroscopy*, Appl. Phys. Lett. **72**, 1478 (1998)

117 V. Deckert, D. Zeisel, R. Zenobi, and T. Vo-Dinh, *Near-Field Surface Enhanced Raman Imaging of Dye-Labeled DNA With 100-nm Resolution*, Anal. Chem. **70**, 2646 (1998)

118 H. F. Hamann, A. Gallagher, and D. J. Nesbitt, *Enhanced Sensitivity Near-Field Scanning Optical Microscopy at High Spatial Resolution*, Appl. Phys. Lett. **73**, 1469 (1998)

119 S. Takahashi, M. Futamata, and I. Kojima, *Spectroscopy with Scanning Near-Field Optical Microscopy Using Photon Tunnelling Mode*, J. Microsc. Oxford **194**, 519 (1999)

120 C. Viets and W. Hill, *Single-Fibre Surface-Enhanced Raman Sensors with Angled Tips*, J. Raman Spectrosc. **31**, 625 (2000)

121 C. Viets and W. Hill, *Laser Power Effects in SERS Spectroscopy at Thin Metal Films*, J. Phys. Chem. B **105**, 6330 (2001)

122 C. Douketis, T. L. Haslett, V. M. Shalaev, Z. H. Wang, and M. Moskovits, *Fractal Character and Direct and Indirect Transitions in Photoemission from Silver Films*, Physica A **207**, 352 (1994)

123 S. Gresillon, J. C. Rivoal, P. Gadenne, X. Quelin, V. Shalaev, and A. Sarychev, *Nanoscale Observation of Enhanced Electromagnetic Field*, Phys. Status Solidi A; Appl. Res. **175**, 337 (1999)

124 S. Gresillon, L. Aigouy, A. C. Boccara, J. C. Rivoal, X. Quelin, C. Desmarest, P. Gadenne, V. A. Shubin, A. K. Sarychev, and V. M. Shalaev, *Experimental Observation of Localized Optical Excitations in Random Metal-Dielectric Films*, Phys. Rev. Lett. **82**, 4520 (1999)

125 V. A. Markel, V. M. Shalaev, P. Zhang, W. Huynh, L. Tay, T. L. Haslett, and M. Moskovits, *Near-Field Optical Spectroscopy of Individual Surface-Plasmon Modes in Colloid Clusters*, Phys. Rev. B **59**, 10903 (1999)

126 A. D. Ormonde, E. C. M. Hicks, J. Castillo, and R. P. Van Duyne, *Nanosphere Lithography: Fabrication of Large-Area Ag Nanoparticle Arrays by Convective Self-Assembly and their Characterization by Scanning UV-Visible Extinction Spectroscopy*, Langmuir **20**, 6927 (2004)

127 C. L. Haynes and R. P. Van Duyne, *Plasmon-Sampled Surface-Enhanced Raman Excitation Spectroscopy*, J. Phys. Chem. B **107**, 7426 (2003)

128 C. L. Haynes, A. D. Mcfarland, L. L. Zhao, R. P. Van Duyne, G. C. Schatz, L. Gunnarsson, J. Prikulis, B. Kasemo, and M. Kall, *Nanoparticle Optics: the Importance of Radiative Dipole Coupling in Two-Dimensional Nanoparticle Arrays*, J. Phys. Chem. B **107**, 7337 (2003)

129 A. D. Mcfarland and R. P. Van Duyne, *Single Silver Nanoparticles as Real-Time Optical Sensors with Zeptomole Sensitivity*, Nano Lett. **3**, 1057 (2003)

130 G. Laurent, N. Felidj, S. L. Truong, J. Aubard, G. Levi, J. R. Krenn, A. Hohenau, A. Leitner, and F. R. Aussenegg, *Imaging Surface Plasmon of Gold Nanoparticle Arrays by Far-Field Raman Scattering*, Nano Lett. **5**, 253 (2005)

131 N. Felidj, S. L. Truong, J. Aubard, G. Levi, J. R. Krenn, A. Hohenau, A. Leitner, and F. R. Aussenegg, *Gold Particle Interaction in Regular Arrays Probed by Surface Enhanced Raman Scattering*, J. Chem. Phys. **120**, 7141 (2004)

132 G. Laurent, N. Felidj, J. Aubard, G. Levi, J. R. Krenn, A. Hohenau, G. Schider, A. Leitner, and F. R. Aussenegg, *Evidence of Multipolar Excitations in Surface Enhanced Raman Scattering*, Phys. Rev. B **71**, 045430 (2005)

133 G. Laurent, N. Felidj, J. Aubard, G. Levi, J. R. Krenn, A. Hohenau, G. Schider, A. Leitner, and F. R. Aussenegg, *Surface Enhanced Raman Scattering*

Arising From Multipolar Plasmon Excitation, J. Chem. Phys. **122**, 011102 (2005)

134 E. J. Ayars, H. D. Hallen, and C. L. Jahncke, *Electric Field Gradient Effects in Raman Spectroscopy*, Phys. Rev. Lett. **85**, 4180 (2000)

135 E. J. Ayars and H. D. Hallen, *Surface Enhancement in Near-Field Raman Spectroscopy*, Appl. Phys. Lett. **76**, 3911 (2000)

136 M. S. Anderson, *Locally Enhanced Raman Spectroscopy with an Atomic Force Microscope*, Appl. Phys. Lett. **76**, 3130 (2000)

137 N. Hayazawa, Y. Inouye, Z. Sekkat, and S. Kawata, *Metallized Tip Amplification of Near-Field Raman Scattering*, Opt. Commun. **183**, 333 (2000)

138 M. S. Anderson and W. T. Pike, *Chemical Imaging with a Raman Atomic-Force Microscope*, Electron Microscopy and Analysis 2001 (Institute of Physics Conference Series, 2001) Chap. 168.

139 L. T. Nieman, G. M. Krampert, and R. E. Martinez, *An Apertureless Near-Field Scanning Optical Microscope and its Application to Surface-Enhanced Raman Spectroscopy and Multiphoton Fluorescence Imaging*, Rev. Sci. Instrum. **72**, 1691 (2001)

140 N. Hayazawa, Y. Inouye, Z. Sekkat, and S. Kawata, *Near-Field Raman Imaging of Organic Molecules by an Apertureless Metallic Probe Scanning Optical Microscope*, J. Chem. Phys. **117**, 1296 (2002)

141 B. Pettinger, G. Picardi, R. Schuster, and G. Ertl, *Surface-Enhanced and STM-Tip-Enhanced Raman Spectroscopy at Metal Surfaces*, Single Molecules **3**, 285 (2002)

142 A. Hartschuh, N. Anderson, and L. Novotny, *Near-Field Raman Spectroscopy Using a Sharp Metal Tip*, J. Microsc. Oxford **210**, 234 (2003)

143 B. Pettinger, G. Picardi, R. Schuster, and G. Ertl, *Surface-Enhanced and STM Tip-Enhanced Raman Spectroscopy of CN^- Ions at Gold Surfaces*, J. Electroanal. Chem. **554**, 293 (2003)

144 A. Hartschuh, A. J. Meixner, L. Novotny, and T. D. Krauss, *Near-Field Raman and Fluorescence Spectroscopy of Single-Walled Carbon Nanotubes*, Abstracts of Papers of the American Chemical Society **227**, U335 (2004)

145 H. Watanabe, N. Hayazawa, Y. Inouye, and S. Kawata, *Dft Vibrational Calculations of Rhodamine 6g Adsorbed on Silver: Analysis of Tip-Enhanced Raman Spectroscopy*, J. Phys. Chem. B **109**, 5012 (2005)

146 B. Ren, G. Picardi, B. Pettinger, R. Schuster, and G. Ertl, *Tip-Enhanced Raman Spectroscopy of Benzenethiol Adsorbed on Au and Pt Single-Crystal Surfaces*, Angew. Chem. Int. Ed. **44**, 139 (2005)

147 J. Clavilier, R. Faure, G. Guinet, and R. Durand, *Preparation of Mono-Crystalline Pt Microelectrodes and Electrochemical Study of the Plane Surfaces Cut in the Direction of the (111) and (110) Planes*, J. Electroanal. Chem. **107**, 205 (1980)

148 B. Ren, G. Picardi, and B. Pettinger, *Preparation of Gold Tips Suitable for Tip-Enhanced Raman Spectroscopy and Light Emission by Electrochemical Etching*, Rev. Sci. Instrum. **75**, 837 (2004)

149 W. E. Doering and S. M. Nie, *Spectroscopic Tags Using Dye-Embedded Nanoparticles and Surface-Enhanced Raman Scattering*, Anal. Chem. **75**, 6171 (2003)

150 S. Schneider, G. Brehm, and P. Freunscht, *Comparison of Surface-Enhanced Raman and Hyper-Raman Spectra of the Triphenylmethane Dyes Crystal Violet and Malachite Green*, Physica Status Solidi B-Basic Research **189**, 37 (1995)

151 M. Kerker, *Electromagnetic Model for Surface-Enhanced Raman-Scattering (SERS) on Metal Colloids*, Acc. Chem. Res. **17**, 271 (1984)

152 K. T. Carron and L. G. Hurley, *Axial and Azimuthal Angle Determination with Surface-Enhanced Raman-Spectroscopy – Thiophenol on Copper, Silver, and Gold Metal-Surfaces*, J. Phys. Chem. **95**, 9979 (1991)

153 D. Y. Wu, M. Hayashi, S. H. Lin, and Z. Q. Tian, *Theoretical Differential Raman Scattering Cross-Sections of Totally-Symmetric Vibrational Modes of Free Pyridine and Pyridine-Metal Cluster Complexes*, Spectrochimica Acta Part A

– Molec. Biomolec. Spectrosc. **60**, 137 (2004)

154 B. Pettinger, B. Ren, G. Picardi, R. Schuster, and G. Ertl, *Tip-enhanced Raman spectroscopy (TERS) of malachite green isothiocyanate at Au(111): bleaching behavior under the influence of high electromagnetic fields*, J. Raman Spectrosc. **36**, 541–550 (2005)

155 D. Zhang, K. F. Domke, and B. Pettinger, *Investigation of DNA Bases Adsorbed at Au(111) Surfaces by Tip-Enhanced Raman Spectroscopy*, to be published in Chem. Phys. Lett. (2005)

156 B. Pettinger, K. F. Domke, D. Zhang, J. Steidtner, R. Schuster, and G. Ertl, *On the Tip-Sample Distance Dependence of Tip-Enhanced Raman Spectroscopy*, to be published in Phys. Rev. Lett. (2005)

Subject Index

Advances in Electrochemical Science and Engineering Vol. 9.
Edited by Richard C. Alkire, Dieter M. Kolb, Jacek Lipkowski and Philip N. Ross
Copyright © 2006 WILEY-VCH Verlag GmbH & Co. KGaA, Weinheim
ISBN: 3-527-31317-6